ADVANCED MATHS
AS CORE
FOR EDEXCEL

Rosemary Emanuel
Contributor: John Wood

PEARSON
Longman

Pearson Education Limited
Edinburgh Gate
Harlow
Essex
CM20 2JE
England
www.longman.co.uk

First published 2004
ISBN 0 582 84237 9

Design by Ken Vail Graphic Design

Cover design by Raven Design

Typeset by Techset, Gateshead

Printed in the U.K. by Scotprint, Haddington

The publisher's policy is to use paper manufactured from sustainable forests.

Live Learning, Live Authoring and Live Player are all trademarks of Live Learning LTD.

The Publisher wishes to draw attention to the Single-User Licence Agreement situated at the back of the book. Please read this agreement carefully before installing and using the CD-ROM.

We are grateful for permission from London Qualifications Ltd. to reproduce past exam questions. All such questions have a reference in the margin. London Qualifications Ltd. can accept no responsibility whatsoever for accuracy of any solutions or answers to these questions.

Every effort has been made to ensure that the structure and level of sample question papers matches the current specification requirements and that solutions are accurate. However, the publisher can accept no responsibility whatsoever for accuracy of any solutions or answers to these questions. Any such solutions or answers may not necessarily constitute all possible solutions.

Contents

Introduction

AS Core for Edexcel is written to support the Edexcel AS Mathematics specification (teaching from September 2004). Chapters 1–11 cover material from Unit C1 and Chapters 12–20 cover Unit C2. The accompanying text *A2 Core for Edexcel* covers units C3 and C4. The authors hope that, in the course of preparing students for their AS examinations, this book will allow students to experience the intellectual excitement of advanced mathematics and to appreciate some of its powerful uses.

The book is designed to support all students, including those who need extra guidance and those who want to be challenged. It can be used by students and teachers in class, but the book and CD-ROM also provide a range of features for students working independently.

The book

Each chapter starts with a list of topics, from the Edexcel specification, to be covered in that chapter. The text has a full explanation of each topic, numerous worked examples and core exercises. The worked examples, covering all types of questions, are particularly useful for students who have to work on their own and for revision. They are presented in a user-friendly format, with solutions on the left of the page, and helpful comments – as might be made by a teacher – on the right.

The exercises in the book are carefully graded to give comprehensive coverage of the Edexcel specification. For those students who need to revise or consolidate GCSE work there are *Access exercises* on the student CD-ROM. Students who need work of a more challenging nature, or are preparing for the Advanced Extension Awards, will find both extra material and *Extension exercises* on the student CD-ROM.

At the end of each module is a *sample examination paper*.

Key Points at the end of each chapter summarise the important results of the chapter.

Review exercises at the end of each chapter test synoptic understanding.

Extension material or *extension questions* in the book are marked ■ or with a blue band in the margin.

Past Edexcel examination questions are included in four *exam practice exercises*, each covering the topics of a number of chapters.

The *glossary* of mathematical terms and a *list of notation* are available for reference.

Two approaches are offered on the topic of Proof: in the book, a short chapter 11 covers the idea of proof by logical deduction and the use of the implication signs. The student CD-ROM contains a fuller treatment of proof, including a section on necessary and sufficient conditions. The *A2 Core for Edexcel* book covers further methods of proof.

Early chapters on algebra include much that may have been covered at GCSE. Since fluency with algebra is one of the keys to success in mathematics, an abundance of worked examples and of exercises on algebra is provided in the book and on the student CD-ROM.

The student CD-ROM

The accompanying student CD-ROM is found in the back of this book. When additional support for a topic is available, a reference appears in the margin next to a CD-ROM logo .

The student CD-ROM contains:

- *'Live-authored' solutions* for the C1 and C2 sample examination paper. These talk students through model solutions to the exam questions as well as giving hints on exam technique. Details on how to install the *Live Player*® and access these solutions are included on the next page.

For most chapters, the CD-ROM also contains:

- *Access exercises* to revise and consolidate GCSE work and ease students into AS topics
- *Extension material* to extend topics beyond AS level
- *Extension exercises* to offer a challenge and help prepare students for Advanced Extension Awards papers
- *Test Yourself exercises* consisting of multiple choice questions for revision.

Also included is a *Spot the error* exercise. This gives the opportunity to identify, and hence avoid, common mistakes.

Answers are given either exactly or approximately. The usual approximation is to three significant figures or to one decimal place for angles measured in degrees. To discourage the rounding of numbers part way through a numerical calculation, numbers read from a calculator are written in the form $56.789\ldots$ The dots indicate that the first few digits of the unrounded display on the calculator have been written down.

Acknowledgements

The authors are grateful to all those who have read the manuscript and made helpful suggestions and contributions. We are particularly indebted to Barbara Morris and also to John Backhouse, Philip Cooper, Andrew Davis, Jane Dyer, Frankie Elston, John Emanuel, Tony Fisher, John Gillard, Sheila Hill, David Hodgson, Peter Horril, Peter Houldsworth, Lyn Imeson, Lawrence Pateman, Ian Potts, George Ross, Michael Ross, Rosemary Smith, John Spencer, Gareth West, Rhona West and Ben Yudkin.

We thank the excellent team at Longman, and in particular James Rees, for turning the manuscript into a book. The commitment to excellence by Longman has been a constant encouragement.

Using the CD-ROM

To use the CD-Rom, start the computer and insert the *AS Core for Edexcel* CD-Rom. The program will automatically run and a menu will appear. You can choose the file you wish to view from the menu screen by clicking on the link for that file.

When the file appears, you can use Acrobat's back and forward buttons (or previous page and next page buttons) to move between pages.

 If you wish to see and hear a worked solution to a question in one of the exam papers, click on the 'LA' icon that appears in the margin of each question paper. The first time you click on an 'LA' icon you will be asked if you want to install *Live Player*. Click 'Yes' to install and follow the instructions. After installation, the Live Player window will start.

The first time you click on an 'LA' icon a warning message will also appear. Tick the option 'do not show this message again' and click 'open' to load the *Live Player* and play the worked solution.

The solution may be paused, replayed or forwarded using the control buttons at the bottom of the Live Player window.

Some questions have long solutions, so the live authored solution is divided into parts. These questions will have more than one 'LA' icon alongside. When one part of the solution ends simply click on the next icon to continue the solution.

If the software does not automatically run on your PC:

1. Select the My Computer icon on the Windows desktop
2. Select the CD-Rom drive icon
3. Select Open
4. Select AS_Core_Edexcel.exe

Hardware requirements

Operating system: Windows 95(OS R2), 98, ME, 2000, NT or XP.

Pentium 100 (IBM Compatible PC) or equivalent PC

32 MB RAM or Higher

16 bit graphic card

4 speed CD-ROM drive (minimum 16× recommended)

SVGA colour monitor and 800 × 600 resolution

Sound card

At least 100 MB free hard disk space

Notation

Miscellaneous symbols

$=$	equals *or* is equal to
$\equiv$	is identically equal to
$>$	is greater than
$\geqslant$	is greater than or equal to
$<$	is less than
$\leqslant$	is less than or equal to
$\therefore$	therefore
$p \Rightarrow q$	p implies q; if p then q
$p \Leftarrow q$	p is implied by q
$p \Leftrightarrow q$	p implies and is implied by q; p if, and only if, q
∞	infinity

Set notation

$\in$	is an element of
$\notin$	is not an element of
$\mathbb{N}$	the set of natural numbers
$\mathbb{Z}$	the set of integers
$\mathbb{Z}^+$	the set of positive integers
$\mathbb{Q}$	the set of rational numbers
$\mathbb{R}$	the set of real numbers

Ranges on a number line

Ranges can be illustrated graphically

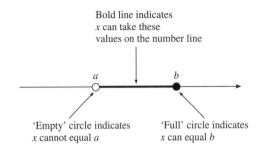

Bold line indicates x can take these values on the number line

'Empty' circle indicates x cannot equal a

'Full' circle indicates x can equal b

Functions

$f(x)$	the value of the function f at x
δx	an increment of x
$\displaystyle\lim_{x \to a} f(x)$	the limit of $f(x)$ as x tends to a
$\dfrac{\mathrm{d}y}{\mathrm{d}x}$	the derivative of y with respect to x

Operations

$\displaystyle\sum_{i=1}^{n} a_i$	$a_1 + a_2 + \cdots + a_n$
$\sqrt{a}$ or $\surd a$	the positive square root of a
$\lvert a \rvert$	the modulus of a
$n!$	n factorial
$\dbinom{n}{r}$	the binomial coefficient $\dfrac{n!}{r!(n-r)!}$ for $n \in \mathbb{Z}^+$

1 Surds and indices

*The fundamental building blocks in mathematics are **numbers** and the **operations**, such as adding and multiplying, which can be performed on them.*

There are different types of numbers and different ways of expressing them. This chapter looks at roots of numbers and numbers expressed in index form.

After working through this chapter you should be

■ *familiar with different sets of numbers*

■ *confident when using and manipulating surds*

■ *able to rationalise denominators of fractions involving surds*

■ *familiar with the laws of indices.*

1.1 Sets of numbers

Before looking at surds it is helpful to think about the different types of numbers and how they arose.

> At the dawn of civilisation, people needed no more than the numbers 1, 2, 3, ... (the **natural numbers** or **counting numbers**) to be able to count their animals and their children. Later, to be able to divide one hundred loaves equally between twelve people, for example, the need for fractions (the **rational numbers**) arose.
>
> The ancient Greeks thought that all numbers could be expressed as fractions until they tried to find a fraction which would be equal to 2 when multiplied by itself. Unable to find it, they eventually realised such a fraction did not exist. Thus **irrational numbers** were discovered (or should it be 'invented'?).

Consider these sets of numbers:

The set of **natural** (or **counting**) **numbers** $\mathbb{N}$ is the set of numbers $\{1, 2, 3, 4, 5, \ldots\}$.

The set of **integers** $\mathbb{Z}$ is the set of whole numbers $\{\ldots, -3, -2, -1, 0, +1, +2, +3, \ldots\}$.

The set of positive (+ve) integers $\mathbb{Z}^+$ is the set $\{+1, +2, +3, \ldots\}$. The sets $\mathbb{N}$ and $\mathbb{Z}^+$ are identical.

The set of negative (−ve) integers $\mathbb{Z}^-$ is the set $\{-1, -2, -3, \ldots\}$.

The set of **real numbers** $\mathbb{R}$ is the set of numbers that correspond to all points on the number line.

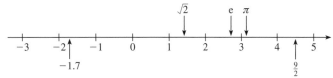

The set of real numbers can be separated into two distinct subsets: the **rational numbers** $\mathbb{Q}$ and the **irrational numbers** (no letter is given to this set).

The **rational numbers** are those that can be expressed as $\frac{p}{q}$, where p and $q \in \mathbb{Z}$ (p and q are members of the set of integers) and $q \neq 0$.

Examples of rational numbers are

> A fraction $\frac{p}{q}$ is normally written in its **lowest terms**, i.e. numerator p and denominator q share no common factor other than 1.

$$\frac{4}{7}, \frac{9}{5}, 6 = \frac{6}{1}, 3.7 = \frac{37}{10}, 0.\dot{3} = \frac{1}{3}.$$

The set of rational numbers includes

■ all decimals which terminate or recur (these can be written as $\frac{p}{q}$)

■ the integers, since an integer n can be written as $\frac{n}{1}$.

More about rational numbers can be found on the CD-ROM (A1.1).

The set of **irrational numbers** is the set of real numbers which are not rational. Examples of these are $\sqrt{2}, \sqrt[3]{7}, \pi$ and e.

> In the sixteenth century, mathematicians were frustrated by not being able to solve quadratic equations such as $x^2 + 1 = 0$. By inventing (or discovering!) the imaginary number $\sqrt{-1}$, called i, the problem was solved and all quadratic equations had solutions. So the set of **complex numbers** (numbers with an imaginary part) was added to previously defined sets.
>
> For example, the number $3 + 2i$ is a complex number. The real part of the number is 3; 2 is the imaginary part. (Complex numbers are mentioned here for completeness.)

Some equations have solutions in some of the sets of numbers, but not in others.

Example 1   Solve these equations.

a $x + 3 = 7$ 4 is a natural number
$\Rightarrow x = 4$ $4 \in \mathbb{N}$

b $x + 7 = 3$ -4 is *not* a natural number.
$\Rightarrow x = -4$ It does, however, belong to the set of integers $\mathbb{Z}$,
          $-4 \notin \mathbb{N}, -4 \in \mathbb{Z}$

c $3x = 7$ $\frac{7}{3}$ is *not* an integer.
$\Rightarrow x = \frac{7}{3}$ It does, however, belong to the set of rational numbers $\mathbb{Q}$,
          $\frac{7}{3} \notin \mathbb{Z}, \frac{7}{3} \in \mathbb{Q}$

d $x^2 = 2$ $\pm\sqrt{2}$ are *not* rational numbers.
$\Rightarrow x = \pm\sqrt{2}$ They do, however, belong to the set of irrational numbers and, therefore, to the set of real numbers.
          $\pm\sqrt{2} \notin \mathbb{Q}, \pm\sqrt{2} \in \mathbb{R}$

e $x^2 = -1$ $\pm\sqrt{-1}$ are *not* real numbers.
$\Rightarrow x = \pm\sqrt{-1}$ They do, however, belong to the set $\mathbb{C}$ of complex numbers.
          $= \pm i$ $\pm\sqrt{-1} \notin \mathbb{R}, \pm\sqrt{-1} \in \mathbb{C}$

Division by zero and the concept of infinity

One interpretation of $\frac{12}{2}$ is how many times can 2, the denominator, be subtracted from 12, the numerator, until nothing is left. The answer is 6.

Apply this to $\frac{12}{0}$. How many times can 0 be subtracted from 12 until nothing is left? The process can continue indefinitely; 0 can be subtracted an infinite number of times and 12 still remains.

The idea of infinity (symbol ∞) is not easy to grasp. It should *not* be thought of as a number, or as having a position on the number line. It can be thought of as the process of counting without end ($+\infty$ in the positive direction on the number line, $-\infty$ in the negative direction).

Since there is no numerical answer to a division by zero, such a division is not allowed. For example, $\frac{12}{0}$ is not defined and, when $x = 1$, neither is

$$\frac{3}{x-1}.$$

1.2 Surds

Consider these three roots. All three can be expressed as rational numbers.

$$\sqrt{4} = 2 \qquad \sqrt{\tfrac{9}{25}} = \tfrac{3}{5} \qquad \sqrt[3]{27} = 3$$

Roots that cannot be expressed as rational numbers are called **surds**. So surds are irrational; they cannot be expressed either as fractions or as terminating or recurring decimals. $\sqrt{2}$, $\sqrt[3]{7}$ and $\sqrt[4]{11}$ are examples of surds.

The proof that $\sqrt{2}$ is irrational is given on the CD-ROM (E1.2).

Surds are not in the least unusual. Consider, for example, $\sqrt{n}$ where $1 \leqslant n \leqslant 1000$. Only 31 out of the 1000 possible square roots are rational; 969 (or 96.9%) are surds.

Note $\sqrt{a}$ means the *positive* square root of a.

Precision

Pure mathematics often poses questions for which *exact* answers are sought. On a calculator, an *approximate* value of any root can be found. For example,

$$\sqrt{2} \approx 1.414\,213\,562$$

This gives $\sqrt{2}$ correct to 9 decimal places (9 d.p.).

The calculator provides approximations which are valuable in certain situations. For example, in measuring a distance, 3.73 m is a more useful answer than $(2 + \sqrt{3})$ m. The number 3.73 is correct to 3 significant figures (3 s.f.).

Unless a problem asks for an answer to a certain degree of accuracy, always give an **exact answer**, such as those given in the table.

Exact value	Approximate value	Degree of accuracy
$\frac{1}{6}$	0.1667	4 d.p.
$\frac{1}{3}$	0.3	1 d.p.
$0.\dot{6}$	0.67	2 d.p.
$\sqrt{2}$	1.414	4 s.f.
π	3.141 593	7 s.f.

For exact answers in pure mathematics, use fractions in preference to decimals.

Manipulation of surds

It is often necessary to manipulate expressions with surds.

Two useful results can be illustrated by substituting values for a and b:

$\sqrt{a} \times \sqrt{b} = \sqrt{ab}$ $\sqrt{4} \times \sqrt{9} = 2 \times 3 = 6$ $\sqrt{4 \times 9} = \sqrt{36} = 6$

$\dfrac{\sqrt{a}}{\sqrt{b}} = \sqrt{\dfrac{a}{b}}$ $\dfrac{\sqrt{64}}{\sqrt{16}} = \dfrac{8}{4} = 2$ $\sqrt{\dfrac{64}{16}} = \sqrt{4} = 2$

> $\sqrt{a} \times \sqrt{b} = \sqrt{ab}$
>
> $\dfrac{\sqrt{a}}{\sqrt{b}} = \sqrt{\dfrac{a}{b}}$
>
> **Note the special case:** $\sqrt{a} \times \sqrt{a} = \sqrt{a^2} = a$

Example 2 Simplify these surds.

a $\sqrt{2} \times \sqrt{18} = \sqrt{2 \times 18} = \sqrt{36} = 6$

b $\dfrac{\sqrt{50}}{\sqrt{2}} = \sqrt{\dfrac{50}{2}} = \sqrt{25} = 5$

Example 3 Square these surds.

a $\left(\sqrt{7}\right)^2 = \sqrt{7} \times \sqrt{7} = 7$

b $\left(4\sqrt{3}\right)^2 = 4\sqrt{3} \times 4\sqrt{3} = 4 \times 4 \times \sqrt{3} \times \sqrt{3} = 16 \times 3 = 48$

c $\left(\dfrac{\sqrt{2x}}{3\sqrt{5y}}\right)^2 = \dfrac{\sqrt{2x}}{3\sqrt{5y}} \times \dfrac{\sqrt{2x}}{3\sqrt{5y}} = \dfrac{2x}{9 \times 5y^2} = \dfrac{2x}{45y^2}$

1 Surds and indices

The result $\sqrt{a} \times \sqrt{b} = \sqrt{ab}$ can be used to simplify expressions containing square roots of composite numbers.

Note A composite number is a positive integer with more than two factors.

It is sometimes possible to factorise a number so that one factor is a perfect square. The root of this factor can then be taken outside the surd.

To express in terms of the *simplest* possible surd, no perfect square factor should remain under the root sign.

Example 4 Express these surds in simplest possible terms.

a $\sqrt{18}$

9 is the largest perfect square factor of 18.

$\sqrt{18} = \sqrt{9} \times \sqrt{2} = 3\sqrt{2}$

Using $\sqrt{ab} = \sqrt{a} \times \sqrt{b}$

b $\sqrt{600}$

100 is the largest perfect square factor of 600.

$\sqrt{600} = \sqrt{100} \times \sqrt{6} = 10\sqrt{6}$

c $\sqrt{968}$

It is hard to spot the largest perfect square factor of 968. There are two ways round this problem: find any perfect square factor of 968 to start with (e.g. 4) or express 968 as a product of prime factors.

Using any perfect square factor

$\sqrt{968} = \sqrt{4} \times \sqrt{242}$

$\qquad = 2\sqrt{242}$

To factorise 242, notice that 121 is a factor and is a perfect square.

$\qquad = 2\sqrt{121} \times \sqrt{2}$

$\qquad = 2 \times 11\sqrt{2}$

$\qquad = 22\sqrt{2}$

Expressing as a product of prime factors

$\sqrt{968} = \sqrt{(2^3 \times 11^2)}$

Once 968 has been completely factorised, it is possible to see that the largest square factor must be $2^2 \times 11^2$.

$\qquad = \sqrt{(2^2 \times 11^2)} \times \sqrt{2}$

$\qquad = 2 \times 11\sqrt{2}$

$\qquad = 22\sqrt{2}$

Example 5 *The method illustrated in Example 4 may be used in reverse, as in this example.*

Express these surds as the square root of a rational number.

a $2\sqrt{6} = \sqrt{(2^2 \times 6)} = \sqrt{24}$

To put 2 under the square root sign, it must be squared.

b $\dfrac{4}{3\sqrt{5}} = \sqrt{\dfrac{4^2}{3^2 \times 5}} = \sqrt{\dfrac{16}{45}}$

Note: 4 and 3 are squared under the root sign.

Exercise 1A   *This exercise may be done orally. Do not use a calculator.*

1   Square

   a $\sqrt{5}$ **b** $\sqrt{\frac{1}{2}}$ **c** $4\sqrt{3}$ **d** $\frac{1}{2}\sqrt{2}$ **e** $\sqrt{\frac{a}{b}}$

   f $\sqrt{3}\times\sqrt{5}$ **g** $\sqrt{3}\times\sqrt{7}$ **h** $\frac{\sqrt{p}}{\sqrt{q}}$ **i** $\frac{1}{2\sqrt{p}}$ **j** $\frac{3\sqrt{a}}{\sqrt{2b}}$

2   Express these in terms of the simplest possible surds.

   a $\sqrt{8}$ **b** $\sqrt{12}$ **c** $\sqrt{27}$ **d** $\sqrt{50}$ **e** $\sqrt{45}$

   f $\sqrt{1210}$ **g** $\sqrt{75}$ **h** $\sqrt{32}$ **i** $\sqrt{72}$ **j** $\sqrt{98}$

3   Express these as square roots of integers.

   a $3\sqrt{2}$ **b** $2\sqrt{3}$ **c** $4\sqrt{5}$ **d** $2\sqrt{6}$ **e** $3\sqrt{8}$

   f $6\sqrt{6}$ **g** $8\sqrt{2}$ **h** $10\sqrt{10}$ **i** $5\sqrt{7}$ **j** $14\sqrt{2}$

Rationalising the denominator

The process of removing surds from the denominator of an expression is called rationalising the denominator. An answer involving surds should usually be given with the denominator rationalised.

Differing reasons are given for rationalising the denominator. One belongs to the days before calculators, when, for example, $\sqrt{3}/3$ could be easily evaluated by looking up $\sqrt{3}$ in tables, whereas $1/\sqrt{3}$ was difficult to calculate. Rationalising the denominator is now the accepted convention and is considered more elegant.

The denominator of an expression is rationalised by multiplying both numerator and denominator by the same number. The choice of number depends on the type of expression in the denominator.

In Example 6, simply multiplying the numerator and denominator by a surd rationalises the denominator.

Example 6   Rationalise the denominators of these expressions.

   a $\dfrac{1}{\sqrt{3}}=\dfrac{1}{\sqrt{3}}\times\dfrac{\sqrt{3}}{\sqrt{3}}=\dfrac{\sqrt{3}}{3}$ Multiplying numerator and denominator by $\sqrt{3}$ makes the denominator rational.

   b $\dfrac{1}{2\sqrt{5}}=\dfrac{1}{2\sqrt{5}}\times\dfrac{\sqrt{5}}{\sqrt{5}}=\dfrac{\sqrt{5}}{10}$ Multiplying numerator and denominator by $\sqrt{5}$ makes the denominator rational. There is no need to multiply by the 2 in the denominator, which is rational already.

In more complicated examples, like Example 7, a number has to be chosen that makes use of the 'difference of squares' to give a rational denominator.

> **Difference of squares**
> $$(x - y)(x + y) = x^2 - y^2$$
> **Hence**
> $$(\sqrt{a} - \sqrt{b})(\sqrt{a} + \sqrt{b}) = a - b$$

Example 7   Use the difference of squares to rationalise the denominators of these expressions.

a $\dfrac{1}{\sqrt{5} + \sqrt{2}} = \dfrac{1}{\sqrt{5} + \sqrt{2}} \times \dfrac{\sqrt{5} - \sqrt{2}}{\sqrt{5} - \sqrt{2}}$ Multiplying the numerator and denominator by $\sqrt{5} - \sqrt{2}$ makes use of the 'difference of squares'.

$\qquad\qquad = \dfrac{\sqrt{5} - \sqrt{2}}{5 - 2}$

$\qquad\qquad = \dfrac{\sqrt{5} - \sqrt{2}}{3}$ Alternatively, write $\frac{1}{3}(\sqrt{5} - \sqrt{2})$.

b $\dfrac{\sqrt{2}}{3\sqrt{2} - 1} = \dfrac{\sqrt{2}}{3\sqrt{2} - 1} \times \dfrac{3\sqrt{2} + 1}{3\sqrt{2} + 1}$ $3\sqrt{2} + 1$ is chosen to make use of the difference of squares.

$\qquad\qquad = \dfrac{6 + \sqrt{2}}{9 \times 2 - 1}$ Note the simplification of the numerator.

$\qquad\qquad = \dfrac{6 + \sqrt{2}}{17}$

All these alternative versions are correct.

$\qquad \dfrac{6}{17} + \dfrac{\sqrt{2}}{17} \qquad \dfrac{1}{17}(6 + \sqrt{2}) \qquad \dfrac{6 + \sqrt{2}}{17}$ Any of these versions is acceptable as an answer to an examination question unless the examiner specifies a particular version by stating 'show that …'.

Note   A calculator can be used to check answers. Evaluating both

$\qquad \dfrac{\sqrt{2}}{3\sqrt{2} - 1} \quad$ and $\quad \dfrac{6 + \sqrt{2}}{17}$

will give 0.4361 correct to 4 d.p. This is no proof that the expressions are equivalent but it provides evidence and reassurance.

c $\dfrac{1 + \sqrt{2}}{3 - \sqrt{2}} = \dfrac{1 + \sqrt{2}}{3 - \sqrt{2}} \times \dfrac{3 + \sqrt{2}}{3 + \sqrt{2}}$ $3 + \sqrt{2}$ is chosen to make use of the difference of squares.

$\qquad\qquad = \dfrac{3 + \sqrt{2} + 3\sqrt{2} + 2}{9 - 2}$ Simplify the numerator.

$\qquad\qquad = \dfrac{5 + 4\sqrt{2}}{7}$

Example 8   Simplify

a $\sqrt{24} + 7\sqrt{54} = \sqrt{4} \times \sqrt{6} + 7\sqrt{9} \times \sqrt{6}$

$= 2\sqrt{6} + 7 \times 3 \times \sqrt{6}$

$= 2\sqrt{6} + 21\sqrt{6}$

$= 23\sqrt{6}$

> Expressions like this addition can be simplified only if the terms can be expressed as multiples of the same surd. In this case, both are multiples of $\sqrt{6}$.

b $\sqrt{12} \times \sqrt{15} = \sqrt{4} \times \sqrt{3} \times \sqrt{3} \times \sqrt{5}$

$= 2 \times \sqrt{3} \times \sqrt{3} \times \sqrt{5}$

$= 2 \times 3\sqrt{5}$

$= 6\sqrt{5}$

> Noticing that the numbers under the root signs have a common factor of 3, makes it possible to simplify the expression without multiplying 12 by 15.

Example 9   Find the area of a square with sides of length $(\sqrt{5}+1)$ cm in the form $(A + B\sqrt{C})$ cm^2 where A, B and C are rational numbers.

$(\sqrt{5}+1)^2 = (\sqrt{5}+1)(\sqrt{5}+1)$

$= 5 + 2\sqrt{5} + 1$

$= 6 + 2\sqrt{5}$

Area is $(6 + 2\sqrt{5})$ cm^2.

Example 10   Find A, B and C such that $\dfrac{3}{\sqrt{5}-1} = A + B\sqrt{C}$

$\dfrac{3}{\sqrt{5}-1} = \dfrac{3}{\sqrt{5}-1} \times \dfrac{\sqrt{5}+1}{\sqrt{5}+1}$

> Choose $\sqrt{5}+1$ to rationalise the denominator.

$= \dfrac{3\sqrt{5}+3}{5-1}$

$= \dfrac{3\sqrt{5}+3}{4} = \dfrac{3}{4} + \dfrac{3}{4}\sqrt{5}$

> Compare this expression with $A + B\sqrt{C}$.

So $A = \frac{3}{4}$, $B = \frac{3}{4}$, $C = 5$.

Exercise 1B   *Calculators should not be used for this exercise.*

1   Simplify

a $\sqrt{8} + \sqrt{18} - 2\sqrt{2}$ **b** $\sqrt{75} + 2\sqrt{12} - \sqrt{27}$

c $\sqrt{28} + \sqrt{175} - \sqrt{63}$ **d** $\sqrt{1000} - \sqrt{40} - \sqrt{90}$

e $\sqrt{512} + \sqrt{128} + \sqrt{32}$ **f** $\sqrt{24} - 3\sqrt{6} - \sqrt{216} + \sqrt{294}$

2   Express these in the form $A + B\sqrt{C}$, where A, B and C are rational numbers.

a $\dfrac{2}{3-\sqrt{2}}$ **b** $(\sqrt{5}+2)^2$ **c** $(1+\sqrt{2})(3-2\sqrt{2})$

d $(\sqrt{3}-1)^2$ **e** $(1-\sqrt{2})(3+2\sqrt{2})$ **f** $\sqrt{\frac{1}{2}} + \sqrt{\frac{1}{4}} + \sqrt{\frac{1}{8}}$

3 Rationalise the denominators of these fractions.

a $\dfrac{1}{\sqrt{5}}$ b $\dfrac{1}{\sqrt{7}}$ c $-\dfrac{1}{\sqrt{2}}$ d $\dfrac{2}{\sqrt{3}}$ e $\dfrac{3}{\sqrt{6}}$

f $\dfrac{1}{2\sqrt{2}}$ g $-\dfrac{3}{2\sqrt{3}}$ h $\dfrac{9}{4\sqrt{6}}$ i $\dfrac{1}{\sqrt{2}+1}$ j $\dfrac{1}{2-\sqrt{3}}$

k $\dfrac{\sqrt{3}+\sqrt{2}}{\sqrt{3}-\sqrt{2}}$ l $\dfrac{\sqrt{5}+1}{\sqrt{5}-\sqrt{3}}$ m $\dfrac{2\sqrt{2}-\sqrt{3}}{\sqrt{2}+\sqrt{3}}$

n $\dfrac{\sqrt{2}+2\sqrt{5}}{\sqrt{5}-\sqrt{2}}$ o $\dfrac{\sqrt{6}+\sqrt{3}}{\sqrt{6}-\sqrt{3}}$ p $\dfrac{\sqrt{10}+2\sqrt{5}}{\sqrt{10}+\sqrt{5}}$

4 Simplify these as far as possible, given that $x=\sqrt{3}$, $y=\sqrt{12}$.

a y^2 b x^3 c xy d $\dfrac{y}{x}$ e $x+y$ f $y-x$

g $\dfrac{12}{x}$ h $\dfrac{1}{y}$ i $\sqrt{75}y$ j $\dfrac{\sqrt{48}}{x}$ k $\dfrac{x^5}{3}$ l $5x+3y$

5 Simplify these expressions.

a $\sqrt{15}\times\sqrt{24}$ b $\sqrt{18}\times\sqrt{32}$ c $\sqrt{3}\times\sqrt{6}\times\sqrt{12}$

d $\sqrt{5}-\dfrac{1}{\sqrt{5}}-\dfrac{1}{\sqrt{125}}$ e $\dfrac{4}{\sqrt{8}}+\dfrac{6}{\sqrt{2}}$ f $\dfrac{\sqrt{8}\times\sqrt{10}\times\sqrt{12}}{\sqrt{2}\times\sqrt{20}\times\sqrt{24}}$

6 A rectangle has sides of length $2\sqrt{3}$ cm and $3\sqrt{3}$ cm. Find, in surd form, the perimeter and the area of the rectangle and the length of its diagonal.

7 Repeat Question 6 for a rectangle with sides of length $3\sqrt{7}$ cm and $\sqrt{21}$ cm.

8 A rectangle has area $(7-\sqrt{2})$ cm^2 and the length of one of its sides is $(\sqrt{2}-1)$ cm. Find, in the form $A\sqrt{B}+C$ where A, B and C are rational numbers, the length of the other side.

9 The two shorter sides of a right-angled triangle are of length $2\sqrt{5}$ m and $4\sqrt{3}$ m. Find the length of the hypotenuse in surd form.

10 The hypotenuse of a right-angled triangle is of length $3\sqrt{11}$ units and the length of one of the shorter sides is $2\sqrt{6}$ units. Find, in surd form, the length of the third side.

11 A triangle has sides AB = AC = 9 cm and BC = 6 cm. Find, in surd form, the length of the perpendicular from A to BC.

12 Find the height of an equilateral triangle of side $4\sqrt{2}$ cm, in surd form.

13 Solve, giving x in the form $A\sqrt{B}+C$, where A, B and C are rational numbers,

a $\sqrt{3}x=3$ b $\dfrac{x}{\sqrt{5}}=2\sqrt{5}$ c $1+\dfrac{x}{\sqrt{2}}=\sqrt{2}$ d $(\sqrt{2}-1)x=4$

e $\sqrt{3}x=x+\sqrt{3}$ f $2+\sqrt{3}x=\dfrac{x}{\sqrt{3}}$ g $\dfrac{1}{\sqrt{2}}+x=\dfrac{x}{\sqrt{2}}$

14 What percentage of $\sqrt{n}$, where $n\in\mathbb{Z}$, are rational, where

a $1\leqslant n\leqslant 2000$? b $1\leqslant n\leqslant 10\,000$?

1.3 Indices

x^n means x multiplied by itself n times.

$$x^n = \underbrace{x \times x \times x \times x \times x \times \cdots \times x}_{n \text{ of these}}$$

So $\quad 5^2 = 5 \times 5$

> Say: 5 squared.

$\quad 2^3 = 2 \times 2 \times 2 = 8$

> Say: 2 cubed.

$\quad a^4 = a \times a \times a \times a$

> Say: a to the power 4 (or: a to the fourth).

In the expression x^n, x is called the **base** and n is the **power** or **index** or **exponent**.

Rules for indices

Where numbers are expressed as powers of the *same* base the following rules can be used to multiply and divide the numbers.

➤
> **Multiplication:** $\quad a^m \times a^n = a^{m+n}$
> To *multiply* the numbers *add* the indices.

$2^3 \times 2^4 = (2 \times 2 \times 2) \times (2 \times 2 \times 2 \times 2)$
$= 2^{3+4} = 2^7$

➤
> **Division:** $\quad \dfrac{a^m}{a^n} = a^{m-n}$
> To *divide* the numbers *subtract* the indices.

$$\frac{6^5}{6^3} = \frac{\overset{1}{\cancel{6}} \times \overset{1}{\cancel{6}} \times \overset{1}{\cancel{6}} \times 6 \times 6}{\underset{1}{\cancel{6}} \times \underset{1}{\cancel{6}} \times \underset{1}{\cancel{6}}}$$
$$= 6^{5-3} = 6^2$$

Meaning can be given to *negative indices*.

$$\frac{6^3}{6^5} = \frac{\overset{1}{\cancel{6}} \times \overset{1}{\cancel{6}} \times \overset{1}{\cancel{6}}}{\underset{1}{\cancel{6}} \times \underset{1}{\cancel{6}} \times \underset{1}{\cancel{6}} \times 6 \times 6} = \frac{1}{6^2}$$

Subtracting the indices gives

$$\frac{6^3}{6^5} = 6^{3-5} = 6^{-2}$$

Comparing these two expressions, $6^{-2} = \dfrac{1}{6^2}$

➤
> **Negative indices**
> $a^{-n} = \dfrac{1}{a^n}$ or, more generally, $\left(\dfrac{a}{b}\right)^{-n} = \left(\dfrac{b}{a}\right)^{n}$
> A number raised to a *negative* power is the reciprocal of the number raised to the positive power.

Meaning can be given to a *zero power* by putting $m = n$ in

$$\frac{a^m}{a^n} = a^{m-n}$$

$$\frac{a^n}{a^n} = a^{n-n} = a^0$$

But for $a \neq 0$

$$\frac{a^n}{a^n} = 1$$

So $a^0 = 1$, e.g. $4^0 = 1$.

> **Raising an expression to the power zero: $a^0 = 1$ $(a \neq 0)$**
> **Any number (other than zero) to the power of *zero* equals one.**

Powers of powers

$$\left(5^2\right)^3 = 5^2 \times 5^2 \times 5^2$$
$$= 5^{3 \times 2}$$
$$= 5^6$$

> **Raising a power to a power: $(a^m)^n = a^{mn}$**
> **To *raise* to a power, *multiply* the indices.**

Fractional indices

$$2^{\frac{1}{2}} \times 2^{\frac{1}{2}} = 2^{\frac{1}{2}+\frac{1}{2}} = 2^1 = 2$$

and $\qquad \sqrt{2} \times \sqrt{2} = 2$

so $\qquad\qquad 2^{\frac{1}{2}} = \sqrt{2}$

Similarly $\quad 8^{\frac{1}{3}} \times 8^{\frac{1}{3}} \times 8^{\frac{1}{3}} = 8^{\frac{1}{3}+\frac{1}{3}+\frac{1}{3}} = 8^1 = 8$

and $\qquad \sqrt[3]{8} \times \sqrt[3]{8} \times \sqrt[3]{8} = 8$

so $\qquad\qquad 8^{\frac{1}{3}} = \sqrt[3]{8}$

> **Fractional indices: $a^{\frac{1}{n}} = \sqrt[n]{a}$**
> ***Fractional* indices correspond to *roots*.**

In general, taking the product of n factors $a^{\frac{1}{n}}$

$$a^{\frac{1}{n}} \times a^{\frac{1}{n}} \times a^{\frac{1}{n}} \times \cdots \times a^{\frac{1}{n}} = a^{\frac{1}{n}+\frac{1}{n}+\frac{1}{n}+\cdots+\frac{1}{n}} = a^1 = a$$

So $\qquad\qquad a^{\frac{1}{n}} = \sqrt[n]{a}$

$$a^{\frac{m}{n}} = \left(a^{\frac{1}{n}}\right)^m \qquad \text{For } a^{\frac{m}{n}} \text{ either}$$

$$= \left(\sqrt[n]{a}\right)^m \qquad \text{take the } n\text{th root and then raise to the power } m$$

or   or

$$a^{\frac{m}{n}} = \left(a^m\right)^{\frac{1}{n}} \qquad \text{raise to the power } m \text{ and then take the } n\text{th root.}$$

$$= \sqrt[n]{a^m}$$

Example 11   Find the value of these expressions involving indices.

a $4^{-2} = \dfrac{1}{4^2} = \dfrac{1}{16}$

b $\left(\dfrac{2}{3}\right)^{-3} = \left(\dfrac{3}{2}\right)^3 = \dfrac{27}{8} = 3\dfrac{3}{8}$

> Invert the fraction and reverse the sign of the power.

c $16^{\frac{1}{4}} = \sqrt[4]{16} = 2$

d $16^{\frac{3}{4}} = \left(\sqrt[4]{16}\right)^3 = 2^3 = 8$

    or

> The first method is easier, especially without a calculator.

    $16^{\frac{3}{4}} = \left(16^3\right)^{\frac{1}{4}} = \sqrt[4]{4096} = 8$

e $4^{-\frac{3}{2}} = \dfrac{1}{4^{\frac{3}{2}}} = \dfrac{1}{\left(\sqrt{4}\right)^3} = \dfrac{1}{2^3} = \dfrac{1}{8}$

    or

    $4^{-\frac{3}{2}} = \dfrac{1}{\sqrt{4^3}} = \dfrac{1}{\left(\sqrt{64}\right)} = \dfrac{1}{8}$

Note   When applying rules of indices the bases must be the same.

$$2^3 \times 2^5 = 2^8 \qquad\qquad 2^3 \times 3^5 \text{ cannot be simplified}$$

same base   different base

$2^3 + 2^5$ cannot be expressed as an integer power of 2.

> If, in a product or quotient, the powers
> are the same, but the bases are different,
> then some simplification is possible.
>
> $$a^m \times b^m = (ab)^m \qquad \text{and} \qquad \dfrac{a^m}{b^m} = \left(\dfrac{a}{b}\right)^m$$

> $$2^5 \times 3^5 = (2 \times 3)^5 \qquad \dfrac{2^5}{3^5} = \left(\dfrac{2}{3}\right)^5$$

Evaluating expressions containing indices

There are a number of ways of using a calculator to evaluate, for example, $\left(\frac{16}{9}\right)^{-\frac{3}{2}}$. Examples in this chapter, however, should be worked out *without* a calculator. The calculator can be used for checking and reassurance but will not help in understanding and working with indices.

Exercise 1C *This exercise may be done orally. Do not use a calculator.*

1 Find the values of

   a $25^{\frac{1}{2}}$ **b** $27^{\frac{1}{3}}$ **c** $64^{\frac{1}{6}}$ **d** $49^{\frac{1}{2}}$ **e** $\left(\frac{1}{4}\right)^{\frac{1}{2}}$ **f** $1^{\frac{1}{4}}$

   g $(-8)^{\frac{1}{3}}$ **h** $(-1)^{\frac{1}{5}}$ **i** $8^{\frac{4}{3}}$ **j** $27^{\frac{2}{3}}$ **k** $25^{\frac{3}{2}}$ **l** $49^{\frac{3}{2}}$

   m $\left(\frac{1}{4}\right)^{\frac{3}{2}}$ **n** $\left(\frac{4}{9}\right)^{\frac{1}{2}}$ **o** $\left(\frac{27}{8}\right)^{\frac{1}{3}}$ **p** $\left(\frac{16}{81}\right)^{\frac{1}{4}}$

2 Find the values of

   a 7^0 **b** 3^{-1} **c** 5^0 **d** 4^{-1} **e** 2^{-3} **f** $\left(\frac{1}{2}\right)^{-1}$

   g $\left(\frac{1}{3}\right)^{-2}$ **h** $\left(\frac{4}{9}\right)^{0}$ **i** 3^{-3} **j** $(-6)^{-1}$ **k** $\left(-\frac{1}{6}\right)^{0}$ **l** $\left(\frac{2}{3}\right)^{-2}$

   m $\left(-\frac{1}{2}\right)^{-2}$ **n** $\frac{1}{3^{-1}}$ **o** $\frac{2^{-1}}{3^{-2}}$ **p** $\frac{2^0 \times 3^{-2}}{5^{-1}}$

3 Find the values of

   a $8^{-\frac{1}{3}}$ **b** $8^{-\frac{2}{3}}$ **c** $4^{-\frac{1}{2}}$ **d** $4^{-\frac{3}{2}}$ **e** $27^{-\frac{2}{3}}$ **f** $\left(\frac{1}{4}\right)^{-\frac{1}{2}}$

   g $\left(\frac{1}{8}\right)^{-\frac{1}{3}}$ **h** $\left(\frac{1}{27}\right)^{-\frac{2}{3}}$ **i** $\left(\frac{4}{9}\right)^{-\frac{1}{2}}$ **j** $\left(\frac{8}{27}\right)^{-\frac{1}{3}}$ **k** $\left(\frac{16}{81}\right)^{-\frac{1}{4}}$ **l** $\left(\frac{27}{8}\right)^{-\frac{4}{3}}$

4 Find the values of

   a $0.16^{0.5}$ **b** $\left(\frac{4}{9}\right)^{1\frac{1}{2}}$ **c** $\left(2\frac{1}{4}\right)^{1\frac{1}{2}}$ **d** $(0.\dot{4})^{\frac{1}{2}}$

Algebra with indices

It is useful, and will be particularly so in Chapter 9 on differentiation, to be able to convert expressions with indices to the form kx^n.

Example 12 Convert these expressions to the form kx^n.

   a $\dfrac{1}{x^3} = x^{-3}$ $k = 1,\ n = -3$ Using $a^{-n} = \dfrac{1}{a^n}$

   b $\dfrac{3\sqrt{x}}{2} = \dfrac{3}{2}x^{\frac{1}{2}}$ $k = \frac{3}{2},\ n = \frac{1}{2}$ Using $\sqrt{a} = a^{\frac{1}{2}}$

   c $\dfrac{2}{5\sqrt[4]{x}} = \dfrac{2}{5x^{\frac{1}{4}}} = \dfrac{2}{5}x^{-\frac{1}{4}}$ $k = \frac{2}{5},\ n = -\frac{1}{4}$ Using $\sqrt[n]{a} = a^{\frac{1}{n}}$ and $a^{-n} = \dfrac{1}{a^n}$

Exercise 1D

This exercise may be done orally.

1 Convert these to the form x^n.

    a $\dfrac{1}{x}$ **b** $\dfrac{1}{x^2}$ **c** $\sqrt{x}$ **d** $\sqrt[3]{x}$ **e** $\dfrac{1}{x^4}$ **f** $\dfrac{1}{x^{-4}}$

    g $\dfrac{1}{\sqrt{x}}$ **h** $\dfrac{1}{x^{-7}}$ **i** $\dfrac{1}{\sqrt[3]{x}}$ **j** $\sqrt[3]{x^2}$ **k** $\dfrac{1}{\sqrt[4]{x^3}}$ **l** $\dfrac{1}{\sqrt{x^5}}$

2 Convert these to the form kx^n.

    a $\dfrac{3}{x}$ **b** $\dfrac{4}{3x}$ **c** $6\sqrt{x}$ **d** $\dfrac{5}{x^3}$ **e** $\dfrac{1}{4x^4}$ **f** $\dfrac{\sqrt{x}}{3}$

    g $\dfrac{1}{5\sqrt{x}}$ **h** $\dfrac{6}{x^{-7}}$ **i** $\dfrac{4}{5\sqrt[3]{x}}$ **j** $7\sqrt[3]{x^2}$ **k** $\dfrac{2}{\sqrt[4]{x^3}}$ **l** $\dfrac{8}{3\sqrt{x}}$

Example 13

In this example, to simplify algebraic products, the coefficients are multiplied and any powers of the same letter are combined using $a^m \times a^n = a^{m+n}$.

a $\begin{aligned}
5a^2 \times 7a^{-3} \times a^5 &= 5 \times 7 \times a^2 \times a^{-3} \times a^5 \\
&= 35a^{2-3+5} \\
&= 35a^4
\end{aligned}$

b $\begin{aligned}
8a^2bx^4 \times 5a^3b^2 &= 8 \times 5 \times a^{2+3} \times b^{1+2} \times x^4 \\
&= 40a^5b^3x^4
\end{aligned}$

> Put the letters in alphabetical order.

Example 14

This example uses $(a^m)^n = a^{mn}$ to simplify algebraic products.

a $(x^x)^2 = x^{2x}$

b $(x^x)^x = x^{x^2}$

Example 15

In this example, the coefficients are divided and any powers of the same letter are combined using $\dfrac{a^m}{a^n} = a^{m-n}$

a $\dfrac{27x^5}{9x^3} = \dfrac{\overset{3}{\cancel{27}} \times x^5}{\underset{1}{\cancel{9}} \times x^3} = 3x^{5-3} = 3x^2$

b $\begin{aligned}
35a^3b^2c^3 \div -7ab^4c &= \dfrac{\overset{5}{\cancel{35}} \times a^3b^2c^3}{\underset{1}{\cancel{-7}} \times ab^4c} \\
&= -\dfrac{5a^2c^2}{b^2}
\end{aligned}$

> Alternatively, write $-5a^2b^{-2}c^2$.

Example 16

This example uses $(a^m)^n = a^{mn}$.

$$(10c^{-3})^2 \div (5c^{-1})^3 = \frac{(10c^{-3})^2}{(5c^{-1})^3}$$

$(10c^{-3})^2 = 10^2(c^{-3})^2 = 100c^{-6}$

$$= \frac{\overset{4}{\cancel{100}}c^{-6}}{\underset{5}{\cancel{125}}c^{-3}}$$

$\dfrac{c^{-6}}{c^{-3}} = c^{-6-(-3)} = c^{-6+3} = c^{-3}$

$$= \frac{4}{5c^3}$$

Alternatively, write $\frac{4}{5}c^{-3}$.

Example 17

This example uses $\sqrt[n]{a^m} = a^{\frac{m}{n}}$.

a $\sqrt{4b^6} = \sqrt{4} \times \sqrt{b^6} = 2 \times b^{\frac{6}{2}} = 2b^3$

The *square* root of the coefficient 4 is found and the power of b is divided by 2.

b $\sqrt[3]{27b^{12}c^3} = \sqrt[3]{27} \times \sqrt[3]{b^{12}} \times \sqrt[3]{c^3}$

$$= 3 \times b^{\frac{12}{3}} \times c^{\frac{3}{3}}$$

$$= 3b^4c$$

The *cube* root of the coefficient 27 is found and the powers are divided by 3.

c $(16b^{12})^{\frac{1}{4}} = 16^{\frac{1}{4}} \times b^{\frac{12}{4}} = \sqrt[4]{16} \times b^3 = 2b^3$

Example 18

Solve these equations involving indices.

a $x^{\frac{1}{2}} = 7$

$x = 7^2 = 49$

Squaring both sides will give x. $x^{\frac{1}{2}} \times x^{\frac{1}{2}} = x$

b $3x^3 = -81$

$x^3 = -27$

$x = \sqrt[3]{-27} = -3$

Divide both sides by 3.

Taking the cube root of both sides gives x.

$x = \sqrt[3]{-27}$ or $x = (-27)^{\frac{1}{3}}$

c $3^{2x+1} = 9^{2x-1}$

$$= (3^2)^{2x-1}$$

$$= 3^{4x-2}$$

So $2x + 1 = 4x - 2$

$2x = 3$

$\therefore \quad x = \frac{3}{2}$

Recognise 9 as a power of 3. $9 = 3^2$

Use $(a^m)^n = a^{mn}$

d $x^{\frac{2}{5}} = 7$

$(x^{\frac{2}{5}})^{\frac{5}{2}} = 7^{\frac{5}{2}}$

So $x = 7^{\frac{5}{2}}$

$$= 130 \ (3 \text{ s.f.})$$

Raise both sides to the power $\frac{5}{2}$.

$(x^{\frac{2}{5}})^{\frac{5}{2}} = x^{\frac{2}{5} \times \frac{5}{2}} = x^1 = x$

On the calculator work out 7 to the power $\frac{5}{2}$ (or 2.5).

Exercise 1E

1 Simplify

a $a \times a \times a$ **b** $4 \times b \times b$ **c** $x^3 \times x^5$ **d** $y^2 \times y^2 \times y^7$

e $3b^3 \times 2b^2$ **f** $2a \times 3a^2 \times 4a^3$ **g** $3x^2y \times 5x^3y^2$ **h** $a^{1+n} \times a^{1-n}$

i $x^4 \div x^2$ **j** $\dfrac{y^7}{y^3}$ **k** $(4a^2)^3$ **l** $4(a^2)^3$

m $(a^{m+1})^m$ **n** $(a^{m+1})^2$ **o** $\dfrac{4y^2z}{2xyz}$ **p** $(u^{-2})^{-3}$

q $(v^x)^2$ **r** $(v^x)^3$ **s** $(v^{2x})^3$ **t** $(v^{2x})^x$

u $x^{\frac{1}{2}} \times x^{-\frac{1}{2}}$ **v** $x^{\frac{1}{2}} \times x^{-\frac{1}{3}}$ **w** $a^x \times a^{-y} \times a^z$ **x** $\dfrac{3y^{19}}{y^{11}}$

y $p^{-1} \times p^{-4}$ **z** $q^2 \times q^{-2}$

2 Simplify these, giving each answer in two forms: x^{-n} and $\dfrac{1}{x^n}$.

a $\dfrac{s^2}{s^4}$ **b** $\dfrac{y^3}{y^8}$ **c** $\dfrac{t^a}{t^{a+b}}$ **d** $\dfrac{x^p}{x^q}$

3 Solve these for x, giving the answers exactly or to 3 significant figures.

a $x^{\frac{1}{2}} = 5$ **b** $2x^3 - 128 = 0$ **c** $x^{-\frac{2}{3}} = \frac{1}{100}$ **d** $2x^{\frac{1}{4}} = \dfrac{64}{x}$

e $32x^{-3} = \dfrac{\sqrt{x}}{4}$ **f** $\sqrt[3]{x} = 8$ **g** $\sqrt[3]{2x} = 2$ **h** $27x^{-\frac{3}{2}} = 1$

i $5^{6x-1} = 25$ **j** $3^{2x+1} = 27$ **k** $3^{3x+1} = 9^{x-1}$ **l** $5 \times 5^x = 25^{x^2}$

m $2^{x^2+x} = \dfrac{1}{4^{x+1}}$ **n** $3x^{\frac{1}{3}} = 8$ **o** $x^{-\frac{5}{3}} = 4$ **p** $x^{\frac{2}{3}} = 5$

4 Square

a $4\pi r$ **b** $2a^3$ **c** b^2c^3 **d** $\dfrac{2a}{b}$

e $3y^3$ **f** $\dfrac{10^2a^3}{b^2}$ **g** $2a^{\frac{1}{2}}$ **h** $\frac{3}{4}a^{\frac{1}{4}}$

5 Simplify

a $\sqrt{x^8}$ **b** $\sqrt{4y^2}$ **c** $\sqrt{25a^2b^4}$ **d** $(36x^6)^{\frac{1}{2}}$

e $\sqrt[4]{a^{-8}b^{12}}$ **f** $\sqrt[3]{-8a^3}$ **g** $\sqrt[3]{27x^3y^9}$ **h** $\sqrt{\dfrac{100}{81x^{10}y^8}}$

6 Simplify

a $\dfrac{12x^2y}{3x^4y}$ **b** $\dfrac{6abc^2}{18a^2bc^2}$ **c** $\dfrac{(2st)^2}{3s^3}$

d $\dfrac{pq^2 \times p}{(pq)^2}$ **e** $(2a)^3 \div (4a)^2$ **f** $(4b)^{-1} \times (6b)^2$

g $3x^{\frac{1}{3}} \times 4x^{\frac{2}{3}}$ **h** $3x^2y \times 4xy^2 \times 5x$ **i** $8a^3 \times \dfrac{1}{2a} \times a^2$

$$\textbf{j} \quad \frac{ab^2 \times 2ab^3}{4a^2b} \qquad\qquad \textbf{k} \quad \frac{20x^3y^5}{5x^{-3}y^{-5}} \qquad\qquad \textbf{l} \quad \frac{9x^9}{3x \times x^2 \times 3x^6}$$

$$\textbf{m} \quad \frac{8b^3c^5}{-4b} \qquad\qquad \textbf{n} \quad \frac{-32l^5m}{8l^3} \qquad\qquad \textbf{o} \quad \frac{-4x^6y^3}{2x^5y}$$

$$\textbf{p} \quad (d^2e^3)^5 \qquad\qquad \textbf{q} \quad -3a^2 \times 4ab^3 \times -5 \qquad \textbf{r} \quad m^3n^5 \times (mn)^{-4}$$

$$\textbf{s} \quad (-m)^5 \qquad\qquad\quad \textbf{t} \quad (-n)^6 \qquad\qquad\qquad\quad \textbf{u} \quad 6h^2 \div 3h^{-3}$$

$$\textbf{v} \quad (4p^{-2})^3 \qquad\qquad\quad \textbf{w} \quad (9a^{-3})^2 \div 6a^{-1}$$

Exercise 1F (Review)

1 Simplify as far as possible, given that $x = \sqrt{5}$, $y = \sqrt{20}$

   a x^2 **b** y^2 **c** $x + y$ **d** $2x - 3y$

   e $\dfrac{y}{x}$ **f** x^3 **g** $\sqrt{xy}$

2 Express these in terms of the simplest possible surds.

   a $\sqrt{80}$ **b** $\sqrt{32}$ **c** $\sqrt{72}$ **d** $\sqrt{180} + \sqrt{125}$

3 Rationalise the denominator of these fractions.

   a $\dfrac{1}{\sqrt{3}}$ **b** $\dfrac{1}{\sqrt{2}}$ **c** $\dfrac{4}{\sqrt{7}}$ **d** $\dfrac{1}{4 - \sqrt{10}}$

   e $\dfrac{2}{\sqrt{6} + 2}$ **f** $\dfrac{3}{2\sqrt{6}}$ **g** $\dfrac{3}{\sqrt{6} - \sqrt{5}}$

4 Solve these for x, giving the answers exactly or to 3 significant figures.

   a $2^{x^2+x} = 64$ **b** $2x^{\frac{1}{2}} = 14$ **c** $2x^{-3} = 15$ **d** $x^{\frac{2}{3}} = 9$

   e $4^{2x+2} = 8^{3x-2}$ **f** $\sqrt[3]{x} = -2$ **g** $3^x \times 3^{2x+1} = 9^x$ **h** $x^{-\frac{3}{2}} = 2.17$

5 A square has area $6\,\text{cm}^2$.
   Find its perimeter.

6 A rectangle has sides $(4 - \sqrt{7})\,\text{cm}$ and $(3 + 2\sqrt{7})\,\text{cm}$.
   Find its perimeter and its area.

7 A rectangle has area $(6 - \sqrt{3})\,\text{cm}^2$ and the length of one of its sides is $(2 + \sqrt{3})\,\text{cm}$. Find the length of the other side.

8 The sides of a rectangle are in the ratio 2:3. The diagonal is of length $26\,\text{cm}$.
   Find the perimeter.

9 A cube has volume $10\,\text{cm}^3$.
   Find the sum of the lengths of the edges, and the total surface area.

10 Given that $72 = 2^x \times 3^y$, find x and y, given that $x, y \in \mathbb{Z}$.

11 Find the values of

a 4^0 b 4^3 c 4^{-1} d $4^{\frac{1}{2}}$ e $4^{\frac{3}{2}}$

f $4^{-\frac{5}{2}}$ g $\left(\frac{1}{5}\right)^{-1}$ h $\left(\frac{2}{3}\right)^3$ i $\left(\frac{2}{3}\right)^{-2}$ j $\left(\frac{27}{8}\right)^{-1}$

k $\left(\frac{27}{8}\right)^{\frac{1}{3}}$ l $\left(\frac{1}{36}\right)^0$ m $36^{\frac{1}{2}}$ n $36^{-\frac{1}{2}}$ o $64^{\frac{4}{3}}$

12 Simplify

a $3^{\frac{1}{4}} \times 3^{\frac{3}{4}}$ b $\sqrt{64} \times \sqrt[3]{64} \times \sqrt[6]{64}$ c $7^{\frac{1}{2}} \times 7^{\frac{1}{3}} \times 7^{\frac{1}{6}}$

d $(0.2)^4 \times 5^4$ e $(2.5)^3 \times 4^3$ f $\dfrac{6^{\frac{1}{4}} \times 36^{\frac{1}{8}}}{\sqrt{6}}$

13 Work out these values, giving each answer in index form where possible.

a $7^3 \times 7^2 \times 7^4$ b $3^4 \times 3^5 \times 3^2$ c $4^3 + 4^2$

d $5^2 \times 5^4 \times 5$ e $2^{10} - 2^5$ f $\dfrac{2^{10}}{2^5}$

g $\dfrac{8^3}{8}$ h $6^3 + 6$ i $\dfrac{7^6}{7^3}$

j $(7^2)^2$ k $(5^3)^3$

14 Simplify these, where possible.

a $a^4 \times a^3 \times a$ b $b^5 + b^3$ c $2c^3 + 3c^3$

d $d^3 \times d^4 \times d^2$ e $e^2 + e^3$ f $3f^2 \times 2f^3$

g $4g^3 \times 5g^2$ h $(h^4)^3$ i $(i^3)^4$

j $(3j)^2$ k $(2k^3)^2$ l $(7l^6)^2$

m $(m^2n)^3$ n $(p^2q^4)^3$ o $(3rs)^2$

p $v^2w \times vw^2$ q $\dfrac{x^4}{x^3}$ r $\dfrac{y^6}{y^2}$

s $\dfrac{z^6}{z}$

There is a 'Test yourself' exercise (Ty1) and an 'Extension exercise' (Ext1) on the CD-ROM.

➤ Key points

Surds

$\sqrt{a}$ is the positive square root of a

$\sqrt{a} \times \sqrt{b} = \sqrt{ab}$　　　　　Special case: $\sqrt{a} \times \sqrt{a} = a$

$$\frac{\sqrt{a}}{\sqrt{b}} = \sqrt{\frac{a}{b}}$$

Difference of squares

$$(x - y)(x + y) = x^2 - y^2$$

Hence

$$(\sqrt{a} - \sqrt{b})(\sqrt{a} + \sqrt{b}) = a - b$$

To **rationalise the denominator**

$$\frac{a}{b\sqrt{c}}$$　　　　　Multiply numerator and denominator by $\sqrt{c}$.

$$\frac{a}{b\sqrt{c} \pm d\sqrt{e}}$$　　　　Multiply numerator and denominator by $b\sqrt{c} \mp d\sqrt{e}$.

Indices

Multiplication

$$a^m \times a^n = a^{m+n}$$　　　　　To multiply, add the indices.

Division

$$a^m \div a^n = \frac{a^m}{a^n} = a^{m-n}$$　　　　To divide, subtract the indices.

Negative indices

$$a^{-n} = \frac{1}{a^n} \qquad \left(\frac{a}{b}\right)^{-n} = \left(\frac{b}{a}\right)^n$$

Raising to the power zero $(a \neq 0)$

$$a^0 = 1$$

Raising a power to a power

$$(a^m)^n = a^{mn}$$

Fractional indices correspond to roots

$$a^{\frac{1}{n}} = \sqrt[n]{a} \qquad a^{\frac{m}{n}} = (\sqrt[n]{a})^m = \sqrt[n]{a^m}$$

Different bases, same powers

$$a^m \times b^m = (ab)^m \text{ and } \frac{a^m}{b^m} = \left(\frac{a}{b}\right)^m$$

2 Algebraic expressions

Chapter 1 looked at sets of numbers and operations involving them. In this chapter, on algebra, we deal with variables and operations on them, such as adding and multiplying. Algebra enables us to generalise results in an abstract way.

A key to success in mathematics is to become fluent in the use of algebra and this requires plenty of practice. In this chapter we revise some of the techniques you met at GCSE and extend the topics further.

After working through this chapter you should be confident working with algebraic expressions and, in particular, be able to

- *manipulate polynomials*
- *expand brackets and collect like terms*
- *factorise expressions.*

2.1 Algebraic expressions

Algebraic expressions can be of many forms, for example

$$3x^2 - 2x - 1 \qquad 3x^2 + 2x + \frac{1}{x} \qquad -6x^2y - 2 \qquad \frac{3x^2 + 2x}{x^2 - 3}$$

A **term** consists of products of numbers and letters, so $3x^2$, $-6x^2y$ are terms.

The number multiplying the letters is the **coefficient** of the term.

$$3x^2$$
An x^2 term with coefficient 3

$$-2x$$
An x term with coefficient -2

$$-1$$
A constant term -1

2.2 Function notation

Many relationships are studied in mathematics, for example $y = 3x^2 - x$, $y = 2^x$ and $y = \sin x$. Such relationships can often be written using function notation, e.g.

$$f(x) = 3x^2 - x$$

> Read this as: f of x equals $3x^2 - x$.

Example 1   Find $f(x) = 3x^2 - x$ for given values of x.

a $x = 4$

$$f(4) = 3 \times 4^2 - 4 = 44$$

> $f(4)$ means evaluate the function with $x = 4$.

b $x = -\frac{1}{2}$

$$f\left(-\tfrac{1}{2}\right) = 3 \times \left(-\tfrac{1}{2}\right)^2 - \left(-\tfrac{1}{2}\right)$$

> Replace x by $-\frac{1}{2}$.

$$= \tfrac{3}{4} + \tfrac{1}{2}$$

$$= \tfrac{5}{4}$$

c $x = a$

$f(a) = 3a^2 - a$

Replace x by a.

d $x = a + h$

$f(a + h) = 3(a + h)^2 - (a + h)$

Replace x by $a + h$ and simplify.

$\qquad = 3(a^2 + 2ah + h^2) - a - h$

$\qquad = 3a^2 + 6ah + 3h^2 - a - h$

Exercise 2A

1 Given that $f(x) = x^2 - 2x$, find

  a $f(9)$ **b** $f(-4)$ **c** $f(\frac{1}{2})$ **d** $f(n)$ **e** $f(3k)$ **f** $f(a + 1)$

2 Given that $g(x) = x^3 + 1$, find

  a $g(0)$ **b** $g(5)$ **c** $g(\frac{3}{4})$ **d** $g(-2)$

  e $g(1)$ **f** $g(-1)$ **g** $g(a)$ **h** $g(3k)$

3 Given that $f(x) = x^2$, express as simply as possible

  a $f(5 + h)$ **b** $\dfrac{f(5 + h) - f(5)}{h}$ $(h \neq 0)$ **c** $\dfrac{f(a + h) - f(a)}{h}$ $(h \neq 0)$

2.3 Polynomials

A polynomial in x has the form

$$a_n x^n + a_{n-1} x^{n-1} + a_{n-2} x^{n-2} + \cdots + a_2 x^2 + a_1 x + a_0$$

where all the coefficients, $a_0, a_1, a_2, \ldots, a_{n-1}, a_n$, are constants and n is a positive integer (i.e. $n \in \mathbb{Z}^+$).

The **degree** (or order) of the polynomial is the highest power of x.

Note If $a_n \neq 0$ in the polynomial above, the polynomial is of degree n.

A polynomial of degree 1 is called **linear**

degree 2 is called **quadratic**

degree 3 is called **cubic**

degree 4 is called **quartic**.

For example, $4x^7 + 3x^2 - 2x + 1$ is a polynomial in x of degree 7. The polynomial has four terms. It is expressed in *descending* powers of x (highest power of x first). In *ascending* powers of x, it would be written as $1 - 2x + 3x^2 + 4x^7$. The x^2 term is $3x^2$; the coefficient of the x^2 term is 3.

Adding and subtracting algebraic expressions

Only like terms (those with identical letters and powers) can be added or subtracted. So

- $3x^2y - 7x^2y = -4x^2y$

- $3x^2 - 7x$ cannot be expressed as a single term, and

- $3a^2b - b^2 - 2ab + 2b^2 + 5a^2b + 7b = 8a^2b + b^2 - 2ab + 7b$

Multiplication of algebraic expressions

When expressions in two brackets are multiplied, *each* term in the first bracket must multiply *each* term in the second bracket.

So $(a + b + c)(d + e) = a(d + e) + b(d + e) + c(d + e)$
$$= ad + ae + bd + be + cd + ce$$

Example 2 Expand and simplify the brackets (two terms in each bracket).

a $(3x - 1)(4x + 3) = 12x^2 + 9x - 4x - 3$
$$= 12x^2 + 5x - 3$$

> Be methodical: multiply
> - **F**irst term in each bracket
> - **O**uter pair
> - **I**nner pair
> - **L**ast term in each bracket.
>
> *Remember* **FOIL**

b $(x + y)(3x - y) = 3x^2 - xy + 3xy - y^2$
$$= 3x^2 + 2xy - y^2$$

c $(x + 4)(x - 4) = x^2 - 4x + 4x - 16$
$$= x^2 - 16$$

> The first line of working can be omitted if the two middle terms can be combined mentally.

Expressions like $x^2 - 16$ are called **difference of squares**: $x^2 - y^2 = (x + y)(x - y)$.

Example 3 Expand and simplify the brackets in $(x + 1)(2x - 1)(x + 5)$.

$(x + 1)(2x - 1)(x + 5) = (x + 1)(2x^2 + 9x - 5)$
$$= x(2x^2 + 9x - 5) + (2x^2 + 9x - 5)$$
$$= 2x^3 + 9x^2 - 5x + 2x^2 + 9x - 5$$
$$= 2x^3 + 11x^2 + 4x - 5$$

> With three brackets, multiply any two first and then multiply their product by the third bracket.

> Collect like terms.

Example 4 If $f(x) = 3 - x + 2x^3$ and $g(x) = 7 - x^2 + 5x^3$, find each of the following:

a $-2x\,f(x) = -2x(3 - x + 2x^3)$
$$= -6x + 2x^2 - 4x^4$$

> Each term in the bracket is multiplied by $-2x$.

b $f(x) + g(x) = (3 - x + 2x^3) + (7 - x^2 + 5x^3)$
$$= 3 - x + 2x^3 + 7 - x^2 + 5x^3$$
$$= 10 - x - x^2 + 7x^3$$

> Brackets are not necessary here; they show, however, the two expressions being added.

c $f(x) - g(x) = (3 - x + 2x^3) - (7 - x^2 + 5x^3)$

$\qquad = 3 - x + 2x^3 - 7 + x^2 - 5x^3$

$\qquad = -4 - x + x^2 - 3x^3$

> Brackets are needed here for $g(x)$ because of the preceding $-$ sign. Remove brackets. Each term in the second bracket is multiplied by -1.

d $f(x)g(x) = (3 - x + 2x^3)(7 - x^2 + 5x^3)$

$\qquad = 3(7 - x^2 + 5x^3) - x(7 - x^2 + 5x^3) + 2x^3(7 - x^2 + 5x^3)$

$\qquad = 21 - 3x^2 + 15x^3 - 7x + x^3 - 5x^4 + 14x^3 - 2x^5 + 10x^6$

$\qquad = 21 - 7x - 3x^2 + 30x^3 - 5x^4 - 2x^5 + 10x^6$

Note Substituting a value for x such as $x = 1$ in the first and the last line of Example 4d will give some indication of whether the answer is correct.

e $4x - (x - 2)g(x) = 4x - (x - 2)(7 - x^2 + 5x^3)$

$\qquad = 4x - [x(7 - x^2 + 5x^3) - 2(7 - x^2 + 5x^3)]$

$\qquad = 4x - (7x - x^3 + 5x^4 - 14 + 2x^2 - 10x^3)$

$\qquad = 4x - (7x - 11x^3 + 5x^4 - 14 + 2x^2)$

$\qquad = 4x - 7x + 11x^3 - 5x^4 + 14 - 2x^2$

$\qquad = 14 - 3x - 2x^2 + 11x^3 - 5x^4$

> All of $(x - 2)$ $(7 - x^2 + 5x^3)$ is being subtracted so it must be evaluated within a bracket.

Example 5 Find the coefficient of x^3 in the expansion of $(x^2 + 3x + 4)(3x^2 - 2x - 7)$.

$(x^2 + 3x + 4)(3x^2 - 2x - 7)$

x^3 term $= x^2 \times (-2x) + 3x \times 3x^2$

$\qquad = -2x^3 + 9x^3$

$\qquad = 7x^3$

So the coefficient of the x^3 term is 7.

> The term in x^3 will come from two products
>
>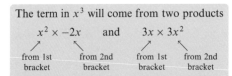
>
> $x^2 \times -2x$ $\qquad$ and $\qquad$ $3x \times 3x^2$
>
> from 1st $\quad$ from 2nd $\qquad$ from 1st $\quad$ from 2nd
> bracket $\qquad$ bracket $\qquad$ bracket $\qquad$ bracket
>
> The other terms in the multiplication can be ignored in this case.

Useful products and factorisations

$a(b + c) = ab + ac$

$(a + b)(c + d) = ac + ad + bc + bd$ $\qquad$ $(x + a)(x + b) = x^2 + (a + b)x + ab$

$(a + b)(a - b) = a^2 - b^2$

$(a + b)^2 = a^2 + 2ab + b^2$ $\qquad\qquad$ $(a - b)^2 = a^2 - 2ab + b^2$

$(a + b)^3 = a^3 + 3a^2b + 3ab^2 + b^3$ $\qquad$ $(a - b)^3 = a^3 - 3a^2b + 3ab^2 - b^3$

$a^3 - b^3 = (a - b)(a^2 + ab + b^2)$

$a^3 + b^3 = (a + b)(a^2 - ab + b^2)$

Note: $a^2 + b^2$ *cannot* be factorised.

$\qquad$ $a^2 - b^2$ is the difference of squares.

Example 6   Expand brackets for squares and the difference of squares.

a $(x+2)^2 = x^2 + 4x + 4$

> Parts **a**, **b** and **c** use $(a \pm b)^2 = a^2 \pm 2ab + b^2$.

b $(2x+3)^2 = (2x)^2 + 2 \times 2x \times 3 + 3^2$
$$= 4x^2 + 12x + 9$$

c $(5x-3)^2 = (5x)^2 + 2 \times 5x \times (-3) + (-3)^2$
$$= 25x^2 - 30x + 9$$

d $(2x-3)(2x+3) = (2x)^2 - 3^2$
$$= 4x^2 - 9$$

> Part **d** uses the difference of squares.

Exercise 2B

1 Express these polynomials in descending powers of x and state their degree.

    a $2x^3 - x + x^4 - 5$ **b** $1 + 3x^3 - x^5$ **c** $6x^4 - 2x^2 - x^3$ **d** -10

2 For the polynomials in Question 1 state the coefficient of the x^3 term and the constant term.

3 Simplify, by collecting like terms

    a $7x + 3y^3 - 2xy + 2xy - 5x - 6y^3$

    b $2x^2 - 4xy + 3y^2 - (8x^2 - 2xy + 6y^2)$

    c $(2x^3 + 5x) - (3x^2 + 2x) + (x^3 - 6x) - (x + 4x^2)$

    d $5ab + 6a^2b - 7ab^2 + 2a^2b - 3ab$

4 Expand and simplify

    a $(x+1)^2$ **b** $(x+2)^2$ **c** $(2x+1)^2$

    d $(x-5)^2$ **e** $(x+1)(x+2)$ **f** $(x+3)(x+4)$

    g $(x+3)(x-4)$ **h** $(x-2)(x+1)$ **i** $(2x+1)(x+1)$

    j $(x-4)(3x-2)$ **k** $(4x-3)(4x+3)$ **l** $(x-2)(x+2)$

5 Expand and simplify

    a $(2a+1)(a-3)$ **b** $(4d+3e)(d+2e)$ **c** $(2-x)(3-4x)$

    d $(8-7w)(8w-7)$ **e** $(3l-5)(3l+5)$ **f** $(3x+4)^2$

    g $(2h+5)(2h-7)$ **h** $(5x+7)(2x-3)$ **i** $(2x+1)(3x+4)$

    j $(2-7x)(2+7x)$ **k** $(2y-3)^2$ **l** $(b^2+4)(4-b^2)$

    m $(x-1)^3$ **n** $\left(a^{\frac{1}{2}} - b^{\frac{1}{2}}\right)\left(a^{\frac{1}{2}} + b^{\frac{1}{2}}\right)$ **o** $(x-2)(x+4)(1-x)$

6 Expand and simplify

    a $(k+7)(2k+3) + (2k+1)(k-1)$ **b** $(m+2)(3m-1) - (m+3)(m-4)$

    c $(s+4)(3s-2) + (2s-1)^2$ **d** $(v-3)^2 + (v+3)^2$

    e $4(y-1)^2 - 2(y+3)^2 - (2y+1)^2$ **f** $(2x+3)(x-1)(x+2) - x^2(1-x)$

7 Given that $f(x) = x^2 + 3x - 1$ and $g(x) = x^3 - 2x^2 + 5$, find

  a $f(x) + g(x)$ **b** $f(x) - g(x)$ **c** $f(x)g(x)$

  d $(x + 1)f(x)$ **e** $(f(x))^2$ **f** $3x\,f(x) - 2g(x)$

8 If $f(x)$ and $g(x)$ are polynomials of degree 7 and 8 respectively, state the degree of

  a $f(x) + g(x)$ **b** $(f(x))^2$ **c** $f(x)g(x)$

9 Find the coefficient of the x^2 term in these expansions. (Do not work out all the terms.)

  a $(x^2 + 2x - 1)(x^2 + 3x + 4)$ **b** $(2x^2 - x + 4)(x^2 + x + 3)$

  c $(x^3 - 2x + 2)(4x^2 + 3x - 1)$ **d** $(4 - x - x^2)(2 + x - x^2)$

2.4 Factorising

In algebra, factorising is the reverse process of multiplying out brackets, i.e. putting into brackets. If there is a factor common to all terms (the HCF of all the terms) it can be taken outside a bracket.

Note The highest common factor (HCF) of two (or more) integers or algebraic expressions is the largest factor common to both (or all).

For more help with HCFs and LCMs see the CD-ROM (A2.4).

Example 7   **a** $x^2 + 2x = x(x + 2)$

  b $3x^2 - 12x^2y + 9xz$

    $= 3x(x - 4xy + 3z)$ *$3x$ is the HCF of the terms $3x^2$, $12x^2y$ and $9xz$.*

  c $\frac{1}{4}\pi r^2 + \frac{1}{3}\pi rh$

    $= \frac{1}{12}\pi r(3r + 4h)$ *When the coefficients contain fractions the factor taken outside the bracket should be chosen so that there will be no fractions inside the bracket. (The denominator, 12, is the LCM of 4 and 3.)*

  d $(3m - 2n)^2 - 5n(3m - 2n)$ *The two parts have a common factor $(3m - 2n)$.*

    $= (3m - 2n)(3m - 2n - 5n)$

    $= (3m - 2n)(3m - 7n)$

Example 8   *This example shows how looking for common factors can make calculations easier.*

$16 \times 14 + 19 \times 14 - 15 \times 14 = 14(16 + 19 - 15)$ *14 is the HCF of the three terms.*

$= 14 \times 20 = 280$

Cancelling fractions

To reduce a fraction to its **lowest terms**, divide the numerator and denominator by their HCF.

Example 9

a $\dfrac{16}{24} = \dfrac{2}{3}$

> The HCF of 16 and 24 is 8. Divide numerator and denominator by 8.

b $\dfrac{6a^2b}{9ab^2} = \dfrac{2a}{3b}$

> The HCF of $6a^2b$ and $9ab^2$ is $3ab$. Divide numerator and denominator by $3ab$.

c $\dfrac{x^2 + xy}{xy + y^2} = \dfrac{x(x + y)}{y(x + y)} = \dfrac{x}{y}$

> Numerator and denominator must be factorised first. Divide both by their HCF, $(x + y)$.

d $\dfrac{a - b}{b - a} = \dfrac{-(b - a)}{b - a} = -1$

> Note: $a - b = -(b - a)$
> Cancel $b - a$.

Exercise 2C

1 Cancel to lowest terms (*without* using a calculator)

a $\frac{8}{24}$ **b** $\frac{17}{51}$ **c** $\frac{30}{81}$ **d** $\frac{14}{91}$ **e** $\frac{105}{147}$ **f** $\frac{99}{264}$ **g** $\frac{345}{405}$ **h** $\frac{252}{1728}$

2 Cancel to lowest terms

a $\dfrac{10x^2}{5x}$
b $\dfrac{2ab}{4a}$
c $\dfrac{6a^2b}{3a \times 2b}$
d $\dfrac{6x^2}{3xy}$

e $\dfrac{abc^2}{a^2bc}$
f $\dfrac{18p^4q^2}{27q^2}$
g $\dfrac{12a^2b^3c^2}{8ab^2c}$
h $\dfrac{x^2yz^3}{x^3y^2z}$

i $\dfrac{(3x + 1)(x - 2)}{(x - 2)(x + 1)}$
j $\dfrac{x - 4}{4 - x}$
k $\dfrac{2x^2(x + y)}{(x + y)(x - y)}$
l $\dfrac{16x(x + 1)}{24x^2(x + 1)}$

3 Factorise, where possible, by taking out the HCF.

a $b^2 - b^3$ **b** $a^2 + ab$ **c** $5a^2 - 10a$

d $y^5 - y^4$ **e** $18c^3 - 9cd^2$ **f** $4a^2 - 16a^2b$

g $81x - 54$ **h** $3x^3 - 6x^2 + 9x$ **i** $4x^3 + x^2 - x$

j $2a^3 - 4a^2 - 2a$ **k** $7a^2 - 7a^2b + 14ab^3$ **l** $3x^4y - 6x^3y^2 + 9x^2y^3$

4 Factorise

a $\frac{1}{2}g^2h + \frac{1}{2}ghj$ **b** $\frac{1}{2}ah + \frac{1}{2}bh$

c $\frac{1}{3}l^2m - \frac{2}{3}lm^2$ **d** $\frac{4}{3}\pi r^3 + \frac{1}{2}\pi r^2h$

5 Without a calculator, use factors to find

a $34 \times 48 + 34 \times 52$ **b** $61 \times 87 - 61 \times 85$

c $29 \times 31 + 29 \times 104 - 29 \times 35$ **d** $16.14 \times 19 - 16.14 \times 9$

e $3.5^2 - 3.5 \times 0.5$ **f** $158 \times 7 + 158 \times 3$

g $\frac{3}{4} \times 134 - \frac{3}{4} \times 94$ **h** $27 \times 354 + 27 \times 646$

Factorising by grouping

Some expressions can be factorised by grouping the terms in pairs.

Example 10   Factorise $2ac + ad - 2bc - bd$.

> There are two ways of grouping the terms:
> $(2ac + ad) - (2bc + bd)$
> or
> $(2ac - 2bc) + (ad - bd)$

$2ac + ad - 2bc - bd$

> Each pair has a common factor. Take the common factor out.

$= a(2c + d) - b(2c + d)$

> Take care with the signs. The same bracket, $(2c + d)$, appears.

$= (2c + d)(a - b)$

> $(2c + d)$ is a common factor so it can be taken outside a bracket.

Alternatively
$2ac - 2bc + ad - bd$

> Here the terms are grouped in different pairs.

$= 2c(a - b) + d(a - b)$

> Take the common factor out of each pair.

$= (a - b)(2c + d)$

> $(a - b)$ is a common factor and can be taken outside a bracket.

Note   Not all expressions with four terms can be factorised by grouping.
The method will work if, and only if, after grouping and taking out the common factors, the same bracket appears, as it does in Example 10.

Factorising a quadratic expression of the form $ax^2 + bx + c$

The technique of factorising a quadratic expression (i.e. putting it into brackets using integers) requires plenty of practice, strong numeracy skills and some trial and error but is a skill worth acquiring.

The methods given here are a *suggested* approach to factorising. Once the skill has been acquired, many factorisable quadratic expressions can be factorised without consciously applying a method. Aids and short cuts are available: the quadratic equation formula can be used (see page 53) as can calculators and computer software.

Note   Most quadratic expressions cannot be factorised with integers.
Check that the expression is either in ascending or descending powers of x, and rearrange if necessary.
Always check that the factorisation is correct by multiplying out the brackets.

Before considering all the possible cases of factorising quadratic expressions, here are some observations on the signs and the numbers.

Numbers

$$x^2 + (p + q)x + pq = (x + p)(x + q)$$

The **coefficient** of the x term, $(p + q)$, is the *sum* of the constant terms in the brackets.

The **constant term**, pq, is the *product* of the constant terms in the brackets.

Signs

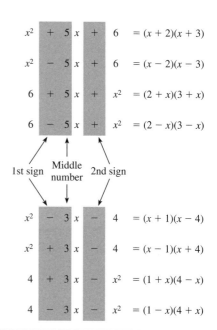

When the second sign of the quadratic is $+$, both brackets will contain the *same* sign:

■ both $+$ if the first sign is $+$

■ both $-$ if the first sign is $-$

The middle number, 5, is the *sum* of 2 and 3; and 6 is the *product* of 2 and 3.

When the second sign is $-$, the brackets will contain *different* signs, one $+$, one $-$.

The middle number, 3, is the *difference* between 1 and 4, and 4 is the *product* of 1 and 4.

In factorising $ax^2 + bx + c$ different techniques are required depending on the values of a, b and c. These are illustrated in Examples 11 to 16.

Example 11 $c = 0$

No constant term. Take the common factor outside the bracket.

$$3x^2 - 12x = 3x(x - 4)$$

$3x$ is the common factor and is put outside the bracket.

Note   This type of expression occurs frequently: be on the lookout for it.

Example 12 $b = 0$

No x term. This type can only be factorised if of the form $a^2 - b^2$ (difference of squares) or $ka^2 - kb^2$. $a^2 + b^2$ cannot be factorised.

a $x^2 - 16 = x^2 - 4^2$

Difference of squares: $a^2 - b^2 = (a - b)(a + b)$

$$= (x - 4)(x + 4)$$

b $25x^2 - 9y^2 = (5x)^2 - (3y)^2$

Difference of squares: $a^2 - b^2 = (a - b)(a + b)$

$$= (5x + 3y)(5x - 3y)$$

c $3x^2 - 12 = 3(x^2 - 4)$

Take out the common factor, 3.

$$= 3(x - 2)(x + 2)$$

Then use difference of squares.

d $x^2 + 16$

Cannot be factorised (using real numbers).

Example 13 $a = \pm 1$

| | | Factorising $x^2 + bx + c$ depends on the signs of b and c. |

a $x^2 - 7x + 6$

1st sign   2nd sign

The brackets will both start with x. $\qquad$ $(x\ \ )(x\ \ )$

Second sign is +ve, so both signs in brackets are the *same*.

First sign is −ve, so both signs in brackets are −ve. $\qquad$ $(x-\ )(x-\ )$

Second sign is +ve, so *sum* is needed.
Find two numbers whose product is 6 and whose sum is 7.
Numbers are 1 and 6. $\qquad$ $(x-1)(x-6)$

So $x^2 - 7x + 6 = (x-1)(x-6)$ $\qquad$ Check by expanding the brackets.

b $x^2 + 5x - 6$

The brackets will both start with x. $\qquad$ $(x\ \ )(x\ \ )$

Second sign is −ve, so the signs in the brackets are *different*.
Don't put the signs in yet!

Second sign is −ve, so the *difference* is needed.
Find two numbers whose product is 6 and whose difference is 5.
Numbers are 1 and 6. Place signs by trial and error. $\qquad$ $(x-1)(x+6)$

So $x^2 + 5x - 6 = (x-1)(x+6)$. $\qquad$ Check by expanding the brackets.

c $12 + 4x - x^2$

The brackets will both end with x. $\qquad$ $(\ \ x)(\ \ x)$

Second sign is −ve, so the signs in the brackets are *different*.
Don't put the signs in yet!

Second sign is −ve, so the *difference* is needed.
Find two numbers whose product is 12 and whose difference is 4.
Numbers are 2 and 6. Place signs by trial and error. $\qquad$ $(6-x)(2+x)$

So $12 + 4x - x^2 = (6-x)(2+x)$. $\qquad$ Check by expanding the brackets.

Example 14 $a \neq \pm 1$ and a is a prime number.

$3x^2 - 11x - 4$

The coefficient of x^2 is 3. Since 3 is prime its only factors
are 3 and 1. So the brackets will start with $3x$ and x. $\qquad$ $(3x\ \ )(x\ \ )$

Second sign is −ve, so the signs in the brackets are *different*.
Don't put the signs in yet!

The numbers at the ends of the brackets have a product of 4.
They could be 1 and 4, 4 and 1, or 2 and 2.
Try out the possibilities until the correct pair is found. $\qquad$ $(3x+1)(x-4)$

So $3x^2 - 11x - 4 = (3x+1)(x-4)$. $\qquad$ Check by expanding the brackets.

Example 15

$a \neq \pm 1$ and c is a prime number, e.g. $6x^2 + 13x - 5 = (2x + 5)(3x - 1)$.

The numbers at the ends of the brackets have a product of 5.
Since 5 is prime, its only factors are 5 and 1.
So the brackets must end with 5 and 1.

Since the second sign is $-$ve, the brackets will contain different signs.
The numbers at the start of the brackets will have a product of 6.
They could be 1 and 6 or 6 and 1 or 2 and 3 or 3 and 2.

Try out the possibilities until the correct pair is found.

Example 16

This example illustrates the 'worst' case: neither a nor c are prime. There are various methods for factorising expressions like this. The one suggested can also be applied to polynomials of the type in Examples 13, 14 and 15.

Neither a nor c are prime, e.g. $15x^2 + x - 6$.

Multiply the coefficient of the x^2 term by the constant term. $15 \times -6 = -90$. Find two numbers whose product is -90 and whose sum is $+1$ (the coefficient of the x term). The numbers are $+10$ and -9.

$15x^2 + x - 6$ | Rewrite the x term as $+10x - 9x$.

$= 15x^2 + 10x - 9x - 6$ | Factorise by grouping.

$= 5x(3x + 2) - 3(3x + 2)$ | $(3x + 2)$ is a common factor. Take this outside the bracket.

$= (3x + 2)(5x - 3)$

Note When factorisation is difficult, the formula for solving a quadratic equation can be used. See page 53.

Example 17

This example could be done by multiplying out the brackets. A neater method, using difference of squares, is given.

Factorise $(2x + 3)^2 - (2x - 7)^2$.

$(2x + 3)^2 - (2x - 7)^2$ | Using $a^2 - b^2 = (a + b)(a - b)$.

$= (2x + 3 + 2x - 7)(2x + 3 - (2x - 7))$ | Note the effect of the minus sign outside the bracket when the bracket is removed. $-(2x - 7) = -2x + 7$

$= (4x - 4)(2x + 3 - 2x + 7)$

$= (4x - 4) \times 10$

$= 4(x - 1) \times 10$

$= 40(x - 1)$

Example 18

Find, *without using a calculator*, $8.91^2 - 1.09^2$.

$8.91^2 - 1.09^2 = (8.91 - 1.09)(8.91 + 1.09)$ | Using $a^2 - b^2 = (a - b)(a + b)$.

$= 7.82 \times 10$

$= 78.2$

Example 19   Factorise, as fully as possible, these polynomials.

a $5x^2 + 56xy + 11y^2$

> Both brackets will contain a + sign. 5 and 11 are both prime so there are only a few possibilities to try.

$= (5x + y)(x + 11y)$

b $4x^3 - 6x + 10x^2$

$= 4x^3 + 10x^2 - 6x$

> Arrange in descending powers of x. Then, take the common factor, $2x$, outside a bracket.

$= 2x(2x^2 + 5x - 3)$

$= 2x(2x - 1)(x + 3)$

Exercise 2D

1   Factorise

   a $x^2 - 36x$ **b** $6x^2 - 36x$ **c** $2x^2 - 4x$

   d $5x^2 - 2x$ **e** $x^2 + 3x$ **f** $4x^2 + 4x$

2   Factorise, where possible

   a $x^2 + 3x + 2$ **b** $x^2 + 8x + 7$ **c** $x^2 + 6x + 8$

   d $x^2 + 9x + 8$ **e** $x^2 - 4x + 3$ **f** $x^2 - 8x + 15$

   g $x^2 - 12x + 20$ **h** $x^2 - 3x - 4$ **i** $x^2 - 6x - 8$

   j $x^2 + 7x - 8$ **k** $x^2 + 3x - 4$ **l** $x^2 - x - 90$

   m $x^2 - 3x - 180$ **n** $x^2 + 4x + 5$ **o** $x^2 - 9x - 10$

3   Factorise, where possible

   a $3x^2 + 14x + 11$ **b** $3x^2 + 22x + 7$ **c** $2x^2 + 10x + 5$

   d $2x^2 + 5x + 2$ **e** $4x^2 + 6x + 2$ **f** $3x^2 + 6x + 3$

   g $7x^2 + 47x + 7$ **h** $3x^2 + 7xy + 3y^2$ **i** $5x^2 + 11xy + 2y^2$

   j $8p^2 - 14pq + 5q^2$ **k** $6x^2 - 25x + 4$ **l** $48 + 6q^2 - 44q$

   m $9 + 10y^2 - 21y$ **n** $6a^2 - 13a + 6$ **o** $5c^2 - 15c + 10$

   p $10p^2 - 13p + 5$ **q** $9z^2 - 12z + 4$ **r** $12x^2 - 35x + 3$

   s $3x^2 - 3x - 36$ **t** $2 + 25x^2 - 15x$

4   Factorise, where possible

   a $a^2 - 5a - 14$ **b** $b^2 + 4b - 21$ **c** $c^2 + 2c - 8$

   d $d^2 + 5d - 6$ **e** $e^2 - 5e - 6$ **f** $2f^2 - f - 1$

   g $2g^2 + g - 1$ **h** $2h^2 - 2h - 1$ **i** $x^2 + 4xy - 21y^2$

   j $3j^2 - 4j + 1$ **k** $2k^2 - 5k - 3$ **l** $15 - 2l - l^2$

   m $1 - 2m - 8m^2$ **n** $n^2 + 2np - 8p^2$

5 Factorise, using difference of squares

a $a^2 - 4$ **b** $b^2 - 144$ **c** $c^2 - 9d^2$

d $25e^2 - 16f^2$ **e** $36g^2 - 1$ **f** $49h^2 - 64j^2$

g $16k^2 - 100$ **h** $25l^2 - 225$ **i** $3l^2 - 12$

j $2m^2 - 50n^2$ **k** $9z^4 - a^2$ **l** $e^2 f^4 g^6 - 121$

m $81h^4 - 16$ **n** $l^8 - 256$ **o** $169 - m^2$

6 Factorise

a $2 + 9z - 5z^2$ **b** $2 - 7y + 5y^2$ **c** $2 - 5w - 3w^2$

d $3 + 5y - 2y^2$ **e** $3 + 2t - 5t^2$ **f** $1 - j - 2j^2$

g $6 + x^2 - 7x$ **h** $2 - x - x^2$ **i** $18 + 3x - x^2$

7 Factorise, where possible

a $f^2 - 4f + 4$ **b** $x^2 + 3x + 2$ **c** $2k^2 - 32$

d $1 + v - 20v^2$ **e** $7x^2 - 7x - 14$ **f** $7 - 28c^2$

g $4h^2 - 24h + 9$ **h** $6 - 7j + 3j^2$ **i** $1 - 9u + 20u^2$

j $ax^2 - 3ax - 4a$ **k** $19s^2 - 27s + 8$ **l** $2l^2 - 12l$

m $10g^2 + 13g + 4$ **n** $6a^2 + 19a - 25$

8 Factorise by grouping

a $my + ny + mz + nz$ **b** $cx + dx - cy - dy$ **c** $5x - 5y - nx + ny$

d $a^2 - ac + ab - bc$ **e** $5a + ab + 5b + b^2$ **f** $6ac - 2cy - 3a + y$

g $x^4 + x^3 + 2x + 2$ **h** $y^3 - y^2 + y - 1$ **i** $ab + b - a - 1$

9 Find, *without* a calculator, the value of

a $36^2 - 34^2$ **b** $102^2 - 98^2$ **c** $42.5^2 - 57.5^2$

d $0.7^2 - 0.3^2$ **e** $48^2 - 52^2$ **f** $1007^2 - 1000^2$

10 Factorise, where possible

a $t^2 - t - 6$ **b** $6t^2 + 11t - 10$ **c** $7 + 40t - 12t^2$

d $6t^2 - 7rt + 2r^2$ **e** $12t^2 - 25t + 12$ **f** $t^2 + 6 - t$

g $6t^2 + tu - 15u^2$ **h** $12t^2 - 12t - 24$ **i** $24t^2 - 2t - 15$

j $24t^2 + 23t - 12$ **k** $5x - 6x^2 + 1$ **l** $6t^2 - 17t + 12$

m $8 - 15t^2 + 14t$ **n** $7 - 28c^2$ **o** $12t^2 - 52t - 8$

p $12t^2 - 23t - 9$ **q** $4t^2 - 23t - 6$ **r** $4t^2 - 12t + 9$

s $4t^2 - 17t + 15$ **t** $4t^2 - 59t - 15$ **u** $100t^2 - 4$

11 Factorise, where possible

a $n^4 - 9n^2$ **b** $\pi R^2 - \pi r^2$ **c** $ax^2 + bx^2 + 2a + 2b$

d $e^4 + 4e^2 + 3$ **e** $6 - d - d^2$ **f** $2x^4 - x^3 + 4x - 2$

g $c^2 - 4d^2$ **h** $4t^2 - 5t - 1$ **i** $2ax + ay + 2bx + by$

j $12t^2 - 18t - 9$ **k** $4t^2 - 8t + 4$ **l** $2 - s + 4s^2$

m $l - 2l^2 - 3l^3$ **n** $x^2 - x - 56$ **o** $12 - 6x^2 - 6x$

12 Factorise, as fully as possible

a $(x + 2)(x - 1) - (x + 2)^2$ **b** $x(x + 3) - (x + 1)(x + 3)$

c $(x + 1)^2 - (x - 1)^2$ **d** $x(x + 1)(2x + 7) + 6(x + 1)(x + 3)$

e $N^2(N + 1)^2 + 4(N + 1)^3$ **f** $(a + b)^3 - 3ab(a + b)$

g $x(x + 7)^2(3x - 1) - x^2(x + 7)$ **h** $(3x + 7)^2 - (3x + 5)^2$

i $(3x + 7)^2 - (2x + 5)^2$ **j** $(3x + 7)^2 - (2x + 7)^2$

2.5   Fractions

Algebraic fractions can be added, subtracted, multiplied and divided, just like numerical fractions.

Adding and subtracting fractions

Fractions like $\frac{1}{4}$ and $\frac{1}{3}$ can be added only if they are written with the same denominator. Quarters can be added to quarters, thirds to thirds, etc., but before quarters can be added to thirds, the fractions must be expressed with equal denominators.
The same principle applies to algebraic fractions.

➤

$$\frac{a}{b} + \frac{c}{d} = \frac{ad}{bd} + \frac{bc}{bd} = \frac{ad + bc}{bd}$$

Note   In the following examples you will need to use LCMs. The lowest (or least) common multiple of two (or more) integers or algebraic expressions is the lowest multiple common to both (or all).

 For more help on LCMs and HCFs see the CD-ROM (A2.4).

Example 20   Add and subtract these numerical fractions, using LCMs.

a $\dfrac{2}{15} + \dfrac{7}{10} = \dfrac{2 \times 2}{15 \times 2} + \dfrac{7 \times 3}{10 \times 3}$

> Common denominator is 30, the LCM of 15 and 10.

$\qquad\qquad = \dfrac{4}{30} + \dfrac{21}{30}$

> In all there are 25 thirtieths.

$\qquad\qquad = \dfrac{25}{30}$

> Cancel to lowest terms by dividing top and bottom by 5, the HCF of 25 and 30.

$\qquad\qquad = \dfrac{5}{6}$

b $173\tfrac{5}{12} - 108\tfrac{3}{4}$

$\qquad 173 - 108 = 65$

> Deal with whole numbers separately.

$\qquad \tfrac{5}{12} - \tfrac{3}{4} = \tfrac{5-9}{12} = \tfrac{-4}{12} = -\tfrac{1}{3}$

> Then the fractions.

$\qquad 173\tfrac{5}{12} - 108\tfrac{3}{4} = 65 - \tfrac{1}{3} = 64\tfrac{2}{3}$

Example 21   Add and subtract these algebraic fractions, using LCMs.

a $\dfrac{x}{6} + \dfrac{5x}{21} = \dfrac{7 \times x}{42} + \dfrac{2 \times 5x}{42}$

> Create equivalent fractions using 42, the LCM of 6 and 21, as the denominator.

$\qquad\qquad = \dfrac{7x + 10x}{42}$

$\qquad\qquad = \dfrac{17x}{42}$

b $\dfrac{a}{3x} + \dfrac{b}{2x} = \dfrac{2a}{6x} + \dfrac{3b}{6x}$

> $6x$ is the LCM of $3x$ and $2x$.

$\qquad\qquad = \dfrac{2a + 3b}{6x}$

c $\dfrac{3a}{bc} + \dfrac{2b}{ac} = \dfrac{3a \times a + 2b \times b}{abc}$

> abc is the LCM of bc and ac.

$\qquad\qquad = \dfrac{3a^2 + 2b^2}{abc}$

d $5 - \dfrac{x-2}{3} - \tfrac{3}{4}(2x+1) - x$

> 12 is the LCM of 3 and 4.

$\qquad = \dfrac{5 \times 12 - 4(x-2) - 9(2x+1) - 12x}{12}$

> Take care of signs when removing brackets.

$\qquad = \dfrac{60 - 4x + 8 - 18x - 9 - 12x}{12}$

> Collect like terms in the numerator.

$\qquad = \dfrac{59 - 34x}{12}$

Note $\tfrac{3}{4}(2x+1) = \dfrac{3(2x+1)}{4}$

e $\quad\dfrac{2}{x+1}-\dfrac{3}{2x+1}=\dfrac{2(2x+1)-3(x+1)}{(x+1)(2x+1)}$

$\qquad\qquad\qquad\quad=\dfrac{4x+2-3x-3}{(x+1)(2x+1)}$

$\qquad\qquad\qquad\quad=\dfrac{x-1}{(x+1)(2x+1)}$

> Common denominator is $(x+1)(2x+1)$.

> Take care of signs when removing brackets.

> Leave denominator in factors.

f $\quad 5-\dfrac{4}{x+3}+\dfrac{10}{x^2-9}$

$\qquad=\dfrac{5(x^2-9)-4(x-3)+10}{x^2-9}$

$\qquad=\dfrac{5x^2-45-4x+12+10}{x^2-9}$

$\qquad=\dfrac{5x^2-4x-23}{x^2-9}$

> $x^2-9=(x+3)(x-3)$.
> So LCM of $x+3$ and x^2-9 is x^2-9.

> Take care of signs when removing brackets.

> The numerator cannot be factorised so no further simplification is possible.

Multiplying fractions

The product of two fractions is a fraction whose numerator is the product of the numerators and whose denominator is the product of the denominators.

➤
$$\dfrac{a}{b}\times\dfrac{c}{d}=\dfrac{ac}{bd}$$

Before multiplying, cancel any factor which appears in both the numerator and denominator.

Never cancel part of a bracket, only the whole bracket.

Example 22   Multiply these fractions.

a $5\tfrac{1}{3}\times\tfrac{3}{8}=\dfrac{\overset{2}{\cancel{16}}}{\underset{1}{\cancel{3}}}\times\dfrac{\overset{1}{\cancel{3}}}{\underset{1}{\cancel{8}}}=2$

> Before multiplying, $5\tfrac{1}{3}$ must be expressed as an improper fraction. 3 and 8 can be cancelled.

b $6\times\tfrac{2}{5}=\dfrac{12}{5}=2\tfrac{2}{5}$

> Think: 6 lots of two fifths is twelve fifths. $\tfrac{6}{1}\times\tfrac{2}{5}=\tfrac{12}{5}$.
> *Note*: The integer $6=\tfrac{6}{1}$, so when multiplying an integer by a fraction only the numerator of the fraction changes.

c $60\times\tfrac{3}{4}=\overset{15}{\cancel{60}}\times\dfrac{3}{\underset{1}{\cancel{4}}}=45$

> 4 can be cancelled.

Example 23 Multiply these algebraic fractions.

a $\dfrac{3x}{2y^2} \times \dfrac{6y}{x^2}$

Numerator and denominator can be divided by x, y and 2.

$$\dfrac{3\cancel{x}}{_1\cancel{2}y^\cancel{2}} \times \dfrac{^3\cancel{6}\cancel{y}}{x^\cancel{2}} = \dfrac{9}{xy}$$

b $\dfrac{x^2 - 4}{xy^2} \times \dfrac{2xy}{x^2 - 4x + 4}$

$$= \dfrac{(x+2)(\cancel{x-2})}{\cancel{x}y^\cancel{2}} \times \dfrac{2\cancel{x}\cancel{y}}{(x-2)^{\cancel{2}}} = \dfrac{2(x+2)}{y(x-2)}$$

Factorise where possible. Cancel x, y, $(x - 2)$.

Dividing fractions

Dividing by a fraction is equivalent to multiplying by its reciprocal.

- Dividing by 2 is equivalent to multiplying by $\frac{1}{2}$. $6 \div 2 = 6 \times \frac{1}{2} = 3$
- Dividing by $\frac{1}{3}$ is equivalent to multiplying by 3. $6 \div \frac{1}{3} = 6 \times \frac{3}{1} = 18$
- Dividing by $\frac{2}{7}$ is equivalent to multiplying by $\frac{7}{2}$. $8 \div \frac{2}{7} = 8 \times \frac{7}{2} = 28$

Dividing by $\dfrac{a}{b}$ is equivalent to multiplying by $\dfrac{b}{a}$.

$$\boxed{\dfrac{a}{b} \div \dfrac{c}{d} = \dfrac{a}{b} \times \dfrac{d}{c} = \dfrac{ad}{bc}}$$

When simplifying fractions, one of two methods is used:
Either numerator and denominator are multiplied by the same number or expression
  – this does not change the value of the fraction.
Or   division by a fraction is replaced by multiplication by its reciprocal.

Example 24 *This example shows how fractions can be simplified where numbers are divided.*

a $\dfrac{15}{2\frac{1}{2}} = \dfrac{30}{5} = 6$

Numerator and denominator are both multiplied by 2.

b $4\frac{2}{3} \div 3\frac{1}{2} = \dfrac{14}{3} \div \dfrac{7}{2}$

Before dividing, express $4\frac{2}{3}$ and $3\frac{1}{2}$ as improper fractions.

$$= \dfrac{^2\cancel{14}}{3} \times \dfrac{2}{\cancel{7}_1} = \dfrac{4}{3}$$

Division by $\frac{7}{2}$ is equivalent to multiplication by $\frac{2}{7}$.

Example 25 | *This example shows how fractions can be simplified where expressions are divided.*

a $\dfrac{a}{\frac{b}{c}} = \dfrac{ac}{b}$

Numerator and denominator are both multiplied by c.

b $\dfrac{\frac{a}{b}}{\frac{c}{d}} = \dfrac{a}{b} \times \dfrac{d}{c} = \dfrac{ad}{bc}$

Division by $\dfrac{c}{d}$ is equivalent to multiplication by $\dfrac{d}{c}$.

c $\dfrac{1}{\frac{x}{y}} = 1 \times \dfrac{y}{x} = \dfrac{y}{x}$

Division by $\dfrac{x}{y}$ is equivalent to multiplication by $\dfrac{y}{x}$.

d $(3x + 2) \div \frac{1}{2} = (3x + 2) \times 2$
$\qquad\qquad\quad = 2(3x + 2)$

Division by $\frac{1}{2}$ is equivalent to multiplication by 2.

e $\dfrac{7x - 1}{\frac{2}{5}} = \dfrac{5}{2}(7x - 1)$

Division by $\frac{2}{5}$ is equivalent to multiplication by $\frac{5}{2}$.

f $\dfrac{c^2 - 16}{c^2 - 8c + 16} \div \dfrac{2c + 8}{3c - 9}$

Division by $\dfrac{2c + 8}{3c - 9}$ is equivalent to multiplication by $\dfrac{3c - 9}{2c + 8}$.

$= \dfrac{\cancel{(c - 4)}(c + 4)}{\cancel{(c - 4)}(c - 4)} \times \dfrac{3(c - 3)}{2\cancel{(c + 4)}} = \dfrac{3(c - 3)}{2(c - 4)}$

Expressions must be factorised before cancelling.

Example 26 (Extension) | *These examples could have been simplified by expressing both numerator and denominator as fractions, and then dividing the numerator by the denominator, i.e. multiplying the numerator by the reciprocal of the denominator. The method used here, though, is much more concise.*

a $\dfrac{x + \frac{a^2}{x}}{x - \frac{a^4}{x^3}} = \dfrac{x^4 + a^2 x^2}{x^4 - a^4}$

Multiply numerator and denominator by x^3, the LCM of x and x^3.

$= \dfrac{x^2 \cancel{(x^2 + a^2)}}{(x^2 - a^2)\cancel{(x^2 + a^2)}}$

Factorise the denominator using difference of squares.

$= \dfrac{x^2}{x^2 - a^2}$

Note $\dfrac{x^2}{x^2 - a^2} = \dfrac{x^2}{(x + a)(x - a)}$ so no further cancelling is possible.

b $\dfrac{\frac{4}{x} + \frac{x}{2} - 3}{\frac{x}{6} - \frac{1}{3} - \frac{4}{3x}} = \dfrac{24 + 3x^2 - 18x}{x^2 - 2x - 8}$

To remove fractions from the numerator and denominator multiply both by $6x$.

$= \dfrac{3(x^2 - 6x + 8)}{(x - 4)(x + 2)}$

Factorise the numerator and denominator before cancelling.

$= \dfrac{3\cancel{(x - 4)}(x - 2)}{\cancel{(x - 4)}(x + 2)}$

$= \dfrac{3(x - 2)}{x + 2}$

Practice with numerical fractions can be found on the CD-ROM (A2.5).

Exercise 2E   *This exercise involves algebraic fractions.*

1   Fill in the missing terms.

a $\dfrac{2c}{ab} = \dfrac{6c^2}{\square}$ b $\dfrac{2mn}{\square} = \dfrac{m}{4nt}$ c $\dfrac{3a}{2b} = \dfrac{\square}{8bc}$ d $2ac = \dfrac{6ace}{\square}$

e $\dfrac{a-b}{3ab} = \dfrac{\square}{6a^2b}$ f $\dfrac{3}{17y} = \dfrac{\square}{51y^2}$ g $\dfrac{4}{13} = \dfrac{28x}{\square}$

2   Express each of these as a single fraction.

a $\dfrac{2}{a} + \dfrac{3}{a}$ b $\dfrac{a}{4} + \dfrac{a}{3}$ c $\dfrac{6}{x} - \dfrac{5}{x}$

d $\dfrac{y}{3} + \dfrac{y}{4} - \dfrac{y}{6}$ e $\dfrac{2}{a} - \dfrac{3}{b}$ f $\dfrac{1}{3x} + \dfrac{1}{5x}$

g $\dfrac{1}{4a} - \dfrac{1}{6a} - 3$ h $\dfrac{b}{4} + \dfrac{3b}{8} + \dfrac{1}{2}$ i $\dfrac{x}{4} + \dfrac{5x}{12} - \dfrac{x}{3}$

j $\dfrac{x}{yz} + \dfrac{z}{xy}$ k $\dfrac{m}{n} + \dfrac{n}{m}$ l $\dfrac{6a^2}{9a} - \dfrac{b^2}{a^2}$

m $5 + \dfrac{31}{3y}$ n $\dfrac{2b}{a} + \dfrac{a-b}{b}$ o $\dfrac{x}{a-x} - \dfrac{y}{a-y}$

3   Simplify

a $\frac{5}{6}(b - 3c) + \frac{3}{8}(4b + 3c)$ b $\frac{3}{4}(x + 4) - \frac{2}{5}(3x - 1)$

c $x - \frac{1}{3}(x + 3) + \frac{1}{5}(x - 2) + 2$ d $\dfrac{3(2x - 3y)}{2} - \dfrac{4(3x + y)}{9} - \dfrac{3x - 17}{18}$

4   Simplify

a $\dfrac{a}{b} \times \dfrac{b^2}{a^2}$ b $\dfrac{3x}{2y} \times \dfrac{6x^2y}{5z}$ c $\dfrac{ab^2}{2} \times 6a$ d $3 \times \dfrac{2x^2}{9}$

e $3 \times \dfrac{2x^2}{y}$ f $3 \times \dfrac{x^2}{2y}$ g $\dfrac{ab^2}{4c^2d} \times \dfrac{2cd}{ab}$ h $\dfrac{18ab}{15bc} \times \dfrac{20ce}{24de}$

i $\dfrac{mn}{2} \times \dfrac{2}{mn}$ j $\dfrac{3ab^2}{5b^3c} \times \dfrac{15b^2c^2}{9a^2b}$ k $\dfrac{a-b}{c} \times \dfrac{c}{b-a}$ l $\dfrac{6}{a+b} \times \dfrac{a^2 - b^2}{2}$

5   Simplify, expressing as a single fraction

a $\dfrac{a}{2b} \div \dfrac{b}{2a}$ b $c \div \dfrac{1}{1-d}$ c $\dfrac{12xy^3}{5xy} \div \dfrac{2xz}{3xyz}$ d $\dfrac{1}{\frac{x}{y}}$

e $\dfrac{2}{\frac{a}{b}}$ f $\dfrac{1}{a + \frac{b}{c}}$ g $\dfrac{1}{\frac{x}{y} - z}$ h $1 - \dfrac{1}{a}$

i $\dfrac{x+1}{1 - \frac{1}{x^3}}$ j $\dfrac{\frac{1}{b} + \frac{1}{a}}{\frac{a}{b} - \frac{b}{a}}$

Exercise 2F (Review)

1 Given that $p(x) = 3x + 1$, $q(x) = 2x - 3$, and $r(x) = 5x^2 + x - 7$, find

  a $[p(x)]^2$

  b $[q(x)]^2$

  c $p(x)q(x)$

  d $6p(x) - q(x)$

  e $p(x)q(x) + r(x)$

  f $q(x) - r(x)$

  g $2r(x) - q(x)$

  h $r(x) - q(x) - p(x)$

2 Factorise fully.

  a $a^2 + 6a + 8$

  b $a^2 + 5a - 14$

  c $6a^2 - 11a - 7$

  d $a^2 - 9$

  e $12a^2 - 4a$

  f $4a^2 - 25$

  g $ab + a + b + 1$

  h $a^3 - 7a^2 + 12a$

3 Given $a = 3$, $b = \frac{1}{4}$, $c = 5$, $d = \frac{2}{7}$ and $e = 2$, find the value of these, giving the answer as simply as possible.

  a $\dfrac{a + c}{e}$

  b $\dfrac{a + b}{c}$

  c $\dfrac{a + c}{b}$

  d $3d$

  e $\dfrac{a + 3b}{2d}$

  f $\dfrac{b}{d}$

  g $\dfrac{a - c}{c - a}$

  h $12bcd$

  i $b \div e$

  j $7(b + d)$

  k $a + b + ab$

  l $10b^2$

  m $\dfrac{12}{d^2}$

  n $\dfrac{c - e}{e - c}$

  o $\dfrac{(2b)^2}{3d}$

  p $\sqrt{a - 3b}$

4 Given $a = x + 1$, $b = x^2$, $c = \frac{3}{x}$ and $d = 4 - \frac{1}{x}$, find, giving the answer as simply as possible,

  a bc

  b bd

  c bd^2

  d $4a - d$

  e $c + d$

  f $\dfrac{c}{d}$

  g $c(a^2 - b - 1)$

  h bc^2

  i $\sqrt{b + 2a - 1}$

  j $\dfrac{d}{c^2}$

  k $\dfrac{c + 3d}{3c + d}$

  l $\dfrac{1 - b}{a}$

There is a 'Test yourself' exercise (Ty2) and as 'Extension exercise' (Ext2) on the CD-ROM.

➤ Key points

Polynomials

A polynomial in x of degree n has the form

$$a_n x^n + a_{n-1} x^{n-1} + \cdots + a_2 x^2 + a_1 x + a_0$$

where $n \in \mathbb{Z}^+$ and $a_n \neq 0$.

Factorisation

$a(b + c) = ab + ac$

$(a + b)(c + d) = ac + ad + bc + bd$ $\qquad (x + a)(x + b) = x^2 + (a + b)x + ab$

$(a + b)(a - b) = a^2 - b^2$

$(a + b)^2 = a^2 + 2ab + b^2$ $\qquad (a + b)^3 = a^3 + 3a^2b + 3ab^2 + b^3$

$(a - b)^2 = a^2 - 2ab + b^2$ $\qquad (a - b)^3 = a^3 - 3a^2b + 3ab^2 - b^3$

$a^3 - b^3 = (a - b)(a^2 + ab + b^2)$

$a^3 + b^3 = (a + b)(a^2 - ab + b^2)$

$a^2 + b^2$ *cannot* be factorised.

$a^2 - b^2$ is the **difference of squares**.

$\dfrac{a - b}{b - a} = -1 \; (b \neq a)$

Fractions

■ To add or subtract fractions express them with equal denominators.

$$\frac{a}{b} + \frac{c}{d} = \frac{ad}{bd} + \frac{bc}{bd} = \frac{ad + bc}{bd}$$

The LCM of the two denominators is the most efficient denominator to use.

■ *Any* fractions can be multiplied or divided.

$$\frac{a}{b} \times \frac{c}{d} = \frac{ac}{bd} \qquad \frac{a}{b} \div \frac{c}{d} = \frac{a}{b} \times \frac{d}{c} = \frac{ad}{bc}$$

Cancel any factor common to the numerator and denominator before multiplying.

Never cancel part of a bracket, only the whole bracket.

Division by $\dfrac{a}{b}$ is the same as multiplication by $\dfrac{b}{a}$. So $\dfrac{x}{y} \div \dfrac{a}{b} = \dfrac{x}{y} \times \dfrac{b}{a}$.

Multiplying or dividing both the numerator and denominator of a fraction by the *same* number or expression does not change the value of the fraction.

3 Equations and quadratic functions

It is remarkable that mathematics, which is a human invention, provides such a powerful tool for solving problems in other areas such as physics, chemistry and biology. Many of these problems can be expressed as equations. In this chapter we will solve algebraic equations, revising some of the techniques you have met at GCSE and extending them to deal with more complex ideas.

After working through this chapter you should

- *be able to solve linear equations*
- *be able to solve quadratic equations by factorisation, by completing the square and by using the formula*
- *be familiar with quadratic functions and their graphs*
- *know about the discriminant of a quadratic expression.*

3.1 Relationships

In Chapter 2 algebraic expressions, such as $x(x + 1) + 2x$, were studied.

Note Expressions do not contain an equals sign.

A **single** expression can be

- simplified, or

    $$x(x + 1) + 2x = x^2 + x + 2x = x^2 + 3x$$

- factorised, or

    $$x^2 + 3x = x(x + 3)$$

- evaluated for a particular value.

    $$\text{When } x = 5, \, x^2 + 3x = 5^2 + 3 \times 5 = 40$$

Chapters 3 and 4 consider relationships between *two* expressions. These will be statements about the expressions, such as equations and inequalities. The statements *always* contain an equals sign ($=$) or an inequality sign ($<$, $>$, $\geqslant$, $\leqslant$).

Variables and unknowns

In algebra, letters take the place of numbers. There are two main situations where this occurs.

- The letter represents a **variable** and can take various values.

    For example, in a formula: $\qquad\qquad\qquad v = u + at$

    or in function notation: $\qquad\quad \mathrm{f}(x) = 2x + 1$

    or in the equation of a graph: $\qquad y = 3x^2 - 4$

- The letter represents a specific **unknown** value (or specific unknown values).

    For example, in an equation: $\qquad 3x^2 - 2x - 1 = 0$

    or in an inequality: $\qquad\qquad 2x > 5$

Equations, inequalities and identities

Consider these statements and try substituting values of x to see when they are true.

$$x^2 - x = x(x - 1) \qquad 3x - 1 = 0 \qquad x^2 = 9 \qquad x^2 = -4$$
$$x - 1 > 0 \qquad\qquad x^2 + 1 > 0 \qquad x^2 + 1 < 0$$

Example 1 $x^2 - x \equiv x(x - 1)$

> This is true for all values of x. It is an **identity**. The symbol $\equiv$ can be used, rather than $=$. Read $\equiv$ as 'is identically equal to'.

> An *identity* is an equation which is true for all values of the variable.

Example 2   **a** $3x - 1 = 0 \Rightarrow x = \frac{1}{3}$

> This is an equation. The result holds for just one value of x. The number $\frac{1}{3}$ is the **root** of the equation. The **solution** is $x = \frac{1}{3}$.

            b $x^2 = 9 \Rightarrow x = \pm 3$

> This is an equation with two real roots, i.e. two possible values of x satisfy the equation. The solution is $x = \pm 3$.

            c $x^2 = -4$

> This equation has no real roots. (It does, however, have a solution in the set of complex numbers.)

> An equation may be true for some real values of the unknown.

Example 3   **a** $x - 1 > 0 \Rightarrow x > 1$

> This inequality holds for values of x where $x > 1$.

            b $x^2 + 1 > 0$

> This inequality is true for all real values of x. x^2 is always positive or zero, therefore $x^2 + 1$ is always positive.

            c $x^2 + 1 < 0$

> This inequality is not true for any real value of x. x^2 is always positive or zero, therefore $x^2 + 1$ cannot be negative.

> An inequality may be true for a set of values of the unknown.

This chapter deals with solving equations and rearranging formulae. Chapter 4 covers solving inequalities.

3.2 Solving linear equations

A **linear equation in x** is of the form $ax + b = 0$, $a \neq 0$, or can be arranged in this form. In linear equations, the highest power of the unknown is *one*. Linear equations have *one* root.

The graph of a linear equation is a straight line which cuts the x-axis ($y = 0$) once.

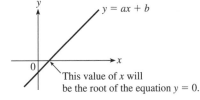

> **Technique for solving equations:**
> **Do the same to both sides.**

Since both sides of an equation are equal, any operation performed on both sides will not affect the equality. For example, if one side is multiplied by 2, the other side must also be multiplied by 2, to maintain equality.

Example 4

In this example, a lot of steps are shown to illustrate the method clearly. With practice, it is possible to combine some of the steps to produce a more streamlined solution.

Solve $7x - 3 = 3(x + 3)$.

> When x terms appear on both sides of a linear equation all x terms must be collected on one side and all other terms on the other side.

$$7x - 3 = 3(x + 3)$$

> Remove brackets.

$$7x - 3 = 3x + 9$$

> Collect all x terms on one side by subtracting $3x$ from both sides.

$$7x - 3x - 3 = 3x - 3x + 9$$

> Tidy up by collecting like terms.

$$4x - 3 = 9$$

> Collect all numbers (constant terms) on the other side by adding 3 to both sides.

$$4x - 3 + 3 = 9 + 3$$

> Tidy up.

$$4x = 12$$

> Divide both sides by 4.

$$\frac{4x}{4} = \frac{12}{4}$$

> The solution can be checked by substituting $x = 3$ in the original equation.
> LHS $= 7x - 3 = 7 \times 3 - 3 = 18$
> RHS $= 3(x + 3) = 3 \times (3 + 3) = 3 \times 6 = 18$

$$x = 3$$

Example 5

In this example, an operation in brackets on the left indicates what is to be done to both sides of an equation.

Solve $3x - 5 = 2(x - 1) - (x + 7)$.

$$3x - 5 = 2(x - 1) - (x + 7)$$

> Simplify the RHS. *Note:* $-(x + 7) = -x - 7$.

$$= 2x - 2 - x - 7$$

So $\quad 3x - 5 = x - 9$

> Collect all terms in x on one side.

$(-x) \quad 2x - 5 = -9$

> Collect all numbers (constant terms) on the other side.

$$(+5) \qquad 2x = -9 + 5$$
$$= -4$$
$$(\div 2) \qquad x = -\frac{4}{2}$$
$$= -2$$

So the solution is $x = -2$.

Example 6 Solve $\dfrac{x}{2} + \dfrac{x-1}{3} = 7$.

$$\frac{x}{2} + \frac{x-1}{3} = 7$$

Multiply every term by 6 (the LCM of the denominators) to eliminate the fractions.

$$\overset{3}{\cancel{6}} \times \frac{x}{\underset{1}{\cancel{2}}} + \overset{2}{\cancel{6}} \times \frac{x-1}{\underset{1}{\cancel{3}}} = 6 \times 7$$

$$3x + 2(x-1) = 42$$

Remove the brackets.

$$3x + 2x - 2 = 42$$

Collect like terms.

$$5x - 2 = 42$$

$$5x = 44$$

Divide both sides by 5.

$$x = \frac{44}{5}$$

$$= 8\frac{4}{5}$$

Example 7 Solve $\dfrac{3}{x} + 2 = \dfrac{6}{x}$.

$$\frac{3}{x} + 2 = \frac{6}{x}$$

Multiply by x to eliminate the fractions. Each term must be multiplied by x.

$$\cancel{x} \times \frac{3}{\cancel{x}} + 2x = \cancel{x} \times \frac{6}{\cancel{x}}$$

$$3 + 2x = 6$$

$$2x = 3$$

$$x = \frac{3}{2}$$

Cross-multiplying

When two fractions are equal, the technique called cross-multiplying can be used.

$$\frac{3}{4} = \frac{6}{8} \quad \Leftrightarrow \quad 3 \times 8 = 6 \times 4$$

$$\frac{a}{b} = \frac{c}{d} \quad \Leftrightarrow \quad ad = bc \quad (b \neq 0, d \neq 0)$$

Example 8   Solve $\dfrac{x-1}{x+4} = \dfrac{x+3}{x-7}$.

$$\dfrac{x-1}{x+4} = \dfrac{x+3}{x-7}$$ — Two equal fractions so cross-multiplying can be used.

$$(x-1)(x-7) = (x+3)(x+4)$$ — Remove the brackets.

$$x^2 - 8x + 7 = x^2 + 7x + 12$$ — Subtract x^2 from both sides.

$$-8x + 7 = 7x + 12$$ — Adding $8x$ to both sides gives a positive coefficient to the x term.

$$7 = 15x + 12$$

$$-5 = 15x$$ — The LHS and RHS can be interchanged to put the x term on the left.

$$15x = -5$$

$$x = -\tfrac{5}{15} = -\tfrac{1}{3}$$

Example 9   Solve $\tfrac{1}{5}(2x + 7) - x = 6 - (1 - x)$.

$$\tfrac{1}{5}(2x + 7) - x = 6 - (1 - x)$$ — Remove brackets from the RHS.

$$\tfrac{1}{5}(2x + 7) - x = 6 - 1 + x$$ — Add x to both sides.

$$\tfrac{1}{5}(2x + 7) = 5 + 2x$$ — Multiply both sides by 5 to eliminate the fraction.

$$2x + 7 = 5(5 + 2x)$$

$$= 25 + 10x$$ — Subtract $2x$ from both sides.

$$7 = 25 + 8x$$ — Subtract 25 from both sides.

$$-18 = 8x$$ — Divide both sides by 8.

$$x = -\tfrac{18}{8} = -\tfrac{9}{4}$$

Example 10   A cleaner is paid a normal hourly rate during the week, time and a half for evenings and double time for weekends.

One week the cleaner works 32 hours during weekdays, 5 hours in the evening and 6 hours at the weekend.

The total pay for the week is £267.80. Find the normal hourly rate.

Solution   Let the normal hourly rate be £x per hour. — Choose a letter to represent the quantity to be found.

Then the evening rate is £$\dfrac{3x}{2}$ per hour and the weekend rate is £$2x$ per hour.

So $\quad 32x + 5 \times \dfrac{3x}{2} + 6 \times 2x = 267.8$ — Form an equation from the information. No units in an equation. (Both sides are expressed in £.)

$$64x + 15x + 24x = 535.6$$

$$103x = 535.6$$

$$x = \dfrac{535.6}{103} = 5.2$$ — State the answer as a sum of money.

So the normal hourly rate is £5.20.

Exercise 3A

1 Solve these equations.

a $8x + 2 = x - 3$ **b** $2x + 1 = 3 - x$

c $3x + 1 = 4x - 2$ **d** $5x + 2 = 4 + x$

e $5 - 2x = 3 + 4x$ **f** $x + 7 + 5x - 8 = 3 + 2x$

g $4 - 3x = 6 - 4x$ **h** $3y + 3 + 4y = 7y + 2 - 2y$

i $4(a - 2) = 3(a - 1)$ **j** $4(b + 1) + 2(b + 3) = 10$

k $2(c - 2) - (1 - c) = 18$ **l** $2(d - 1) - 3(2 - d) = 3(d + 1)$

m $7 + 2(e - 3) = 1$ **n** $5m - 3(m + 2) = 0$

o $2(2v - 1) - 3(3v + 1) = 5$ **p** $5 - 3(2x - 5) = 7 - x$

q $8k - 7 = 4(k - 3) - 2(3k - 5) - (3 - 2k)$

2 Solve these equations.

a $\dfrac{a}{4} = 12$ **b** $\frac{2}{3}x = \frac{9}{8}$

c $\dfrac{a}{4} + 7 = 13$ **d** $\dfrac{2}{x} = \dfrac{3}{x + 1}$

e $\dfrac{x}{3} + 4 = 2x + 7$ **f** $\dfrac{x}{2} + \dfrac{x}{3} + \dfrac{x}{6} = 5$

g $\dfrac{y}{2} + 5 = \dfrac{y}{5}$ **h** $\dfrac{6}{x} + \dfrac{3}{2x} = \dfrac{5}{2}$

i $\dfrac{3}{y} + 5 = 11$ **j** $\dfrac{3x}{4} - \dfrac{2}{3} = \dfrac{7}{12}$

k $\dfrac{7x + 2}{5} = \dfrac{4x - 1}{2}$ **l** $\dfrac{3x - 13}{7} + \dfrac{11 - 4x}{3} = 0$

m $2x - \frac{1}{3}(x + 27) = 16$ **n** $\frac{1}{3}(1 - 2x) - \frac{1}{6}(4 - 5x) + \frac{13}{42} = 0$

o $6 - \dfrac{x - 1}{2} = 3x - 11$ **p** $\dfrac{x}{4} + \dfrac{5x + 8}{6} = \dfrac{2x + 9}{3}$

q $0.4x = 1.3 - 0.2x - 1$ **r** $(x + 1)(x + 2) = x(x + 7) - 6$

s $\dfrac{x + 1}{2} - \dfrac{x - 7}{5} = \dfrac{x + 4}{3}$ **t** $\dfrac{2a - 1}{3} - \dfrac{a + 5}{4} = \frac{1}{2}$

u $3.2x + 0.99 = 3.13x + 0.15$ **v** $\frac{7}{6}(3x - 1) - 8\frac{1}{3} = \frac{3}{2}(2x - 5)$

w $\dfrac{x}{5} + 1\frac{1}{2}x - \dfrac{11}{20} = \dfrac{5x - 1}{4}$ **x** $\dfrac{4x - 3}{5} - \dfrac{5x - 3}{8} = 1$

y $\dfrac{2x + 1}{x - 3} = \dfrac{2x + 3}{x - 1}$ **z** $\dfrac{3}{x^2 + 4x - 1} = \dfrac{6}{2x^2 - 3x + 5}$

3 The length of a swimming pool is 2.5 m greater than its width.
The pool's perimeter is 29 m. Find the length of the pool.

4 Some of the balls in bag A are divided into two bags B and C. If five are put in bag B and a third of the remainder are put in bag C, twelve remain.
Find the number of balls originally in bag A.

5 The temperature, C, in degrees Celsius can be calculated from the temperature, F, in degrees Fahrenheit using the formula $C = \frac{5}{9}(F - x)$.
A cook has forgotten the value of x, but remembers that the boiling point of water is $100°C$, which is equivalent to $212°F$. Find the value of x.

6 When an object initially travelling at a speed u is subject to a constant acceleration a, the object's speed v after time t has elapsed is given by the equation $v = u + at$.
A motorist driving at $14\,\text{ms}^{-1}$ brakes steadily for $3.5\,\text{s}$, reducing her speed by half. Find the acceleration of the car during this time.

Practice in rearranging formulae can be found on the CD-ROM (A3.3).

3.3 Solving quadratic equations

A quadratic equation is of the form $ax^2 + bx + c = 0$, $a \neq 0$, or can be arranged in that form. In quadratic equations, the highest power of the unknown is *two*.
Quadratic equations have *two* roots: these may be

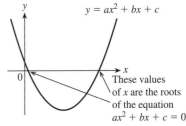

$y = ax^2 + bx + c$

- two distinct real roots
- two equal roots, or
- no real roots (two complex roots).

The graph of a quadratic equation cuts the x-axis ($y = 0$) at most twice.

These values of x are the roots of the equation $ax^2 + bx + c = 0$

Three techniques for solving quadratic equations are covered in this chapter.

- Factorisation (this page)
- Completing the square (page 50)
- Using the quadratic formula (page 53)

Solving quadratic equations using factorisation

Example 11 Solve the equation $x^2 - 3x - 4 = 0$.

$$x^2 - 3x - 4 = 0$$
$$(x - 4)(x + 1) = 0$$
$$\Rightarrow \quad x - 4 = 0 \text{ or } x + 1 = 0$$
$$\therefore \quad x = 4 \text{ or } x = -1$$

> Factorise. If the product of two numbers is zero, then one or other (or both) of the numbers must be zero.

So the solution of $x^2 - 3x - 4 = 0$ is $x = 4$ or $x = -1$.

Note If the solution of a quadratic equation is $x = 4$ and $x = -1$, then, working backwards, the equation must be $(x - 4)(x + 1) = 0$. So given the solution one can work backwards to the factorised form.

Example 12   Find the values of x which satisfy the equation $3x^2 + 12x = 0$.

$$3x^2 + 12x = 0$$

> 3 is a common factor and can be divided out.

$$x^2 + 4x = 0$$

> x is a common factor which can be taken outside a bracket.

$$x(x + 4) = 0$$

> *Remember*: Do not divide through by x; it could be zero.

$$\Rightarrow \qquad x = 0 \text{ or } x + 4 = 0$$

$$\therefore \qquad x = 0 \qquad x = -4$$

The values of x which satisfy the equation $3x^2 + 12x = 0$ are 0 and -4.

Example 13   Solve $x^2 - 25 = 0$.

$$x^2 - 25 = 0$$

> One method is to factorise using difference of squares.

$$(x - 5)(x + 5) = 0$$

$$\Rightarrow \qquad x - 5 = 0 \text{ or } x + 5 = 0$$

$$\text{So} \qquad x = \pm 5$$

Alternatively

$$x^2 - 25 = 0$$

$$x^2 = 25$$

> Take the square root of both sides.

$$x = \pm\sqrt{25}$$

> *Remember*: This can be $+5$ or -5.

$$\therefore \qquad x = \pm 5$$

So the solution to $x^2 - 25 = 0$ is $x = \pm 5$.

Example 14   Solve $(x + 1)(2x + 3) = 4x^2 - 22$.

$$(x + 1)(2x + 3) = 4x^2 - 22$$

> Multiply out brackets.

$$2x^2 + 5x + 3 = 4x^2 - 22$$

> Put all of the terms on one side.

$$2x^2 + 5x + 3 - 4x^2 + 22 = 0$$

$$-2x^2 + 5x + 25 = 0$$

> Multiply all terms by -1 so that the x^2 term is +ve.

$$2x^2 - 5x - 25 = 0$$

$$(2x + 5)(x - 5) = 0$$

$$\Rightarrow \qquad 2x + 5 = 0 \quad \text{or} \quad x - 5 = 0$$

$$\therefore \qquad x = -\tfrac{5}{2} \qquad x = 5$$

So the solution is $x = -\tfrac{5}{2}$ or $x = 5$.

Example 15   Solve $x^3 - x = 0$.

> This is a **cubic equation**, (highest power of x is 3).

$$x^3 - x = 0$$

> x is a common factor and can be taken outside a bracket.

$$x(x^2 - 1) = 0$$

$$x(x + 1)(x - 1) = 0$$

$$\Rightarrow \qquad x = 0 \text{ or } x = \pm 1$$

Note   Cubic equations have, at most, *three* real roots.

Examples 16 and 17 involve quadratic equations in some function of x.

Example 16   *This example involves a quadratic equation in $x^{\frac{1}{3}}$, most easily seen by substituting $y = x^{\frac{1}{3}}$. Alternatively, factorise the original equation: $(2x^{\frac{1}{3}} - 1)(x^{\frac{1}{3}} + 3) = 0$*

Solve $2x^{\frac{2}{3}} + 5x^{\frac{1}{3}} - 3 = 0$.

Substituting $y = x^{\frac{1}{3}}$ gives

$$2y^2 + 5y - 3 = 0$$

$$(2y - 1)(y + 3) = 0$$

$$\Rightarrow \qquad 2y - 1 = 0 \quad \text{or} \quad y + 3 = 0$$

$$y = \tfrac{1}{2} \qquad\qquad y = -3$$

$$\therefore \qquad x^{\frac{1}{3}} = \tfrac{1}{2} \quad \text{or} \qquad x^{\frac{1}{3}} = -3$$

> Having found the values of y substitute $y = x^{\frac{1}{3}}$ and then find the values of x.

$$x^{\frac{1}{3}} = \tfrac{1}{2} \Rightarrow x = \left(\tfrac{1}{2}\right)^3 = \tfrac{1}{8}$$

$$x^{\frac{1}{3}} = -3 \Rightarrow x = (-3)^3 = -27$$

So the solutions are $x = -27$ or $x = \tfrac{1}{8}$.

Example 17   *This example involves a **quartic equation** (highest power of x is 4) which is a quadratic equation in x^2.*

Solve $9x^4 + 8x^2 - 1 = 0$.

$$9x^4 + 8x^2 - 1 = 0$$

> If preferred, the substitution $y = x^2$ can be made, giving $9y^2 + 8y - 1 = 0$.

$$(9x^2 - 1)(x^2 + 1) = 0$$

$$\Rightarrow \qquad 9x^2 - 1 = 0 \quad \text{or } x^2 + 1 = 0$$

$$x^2 = \tfrac{1}{9} \qquad\qquad x^2 = -1$$

$$\therefore \qquad x = \pm \tfrac{1}{3} \qquad \text{No real roots}$$

So the solutions to $9x^4 + 8x^2 - 1 = 0$ are $x = \pm \tfrac{1}{3}$.

Note   Quartic equations have, at most, *four* real roots. The quartic in Example 17 has two real roots.

Completing the square

Writing a quadratic expression, $ax^2 + bx + c$, in the form $a(x+p)^2 + q$ is called completing the square.

Consider the form of expressions which are perfect squares:

- $(x+3)^2 = x^2 + 6x + 9$

  Notice 3 is $\frac{6}{2}$.

- $(x-5)^2 = x^2 - 10x + 25$

  Notice -5 is $\frac{-10}{2}$.

- $(x+k)^2 = x^2 + 2kx + k^2$

  Notice k is $\frac{2k}{2}$.

The constant in each bracket is half the coefficient of x.

A quadratic expression, such as $x^2 - 12x$, can be written as

$$x^2 - 12x \equiv (x-6)^2 - 36$$

(-6) is half the coefficient of x. Subtracting $(-6)^2 = 36$ from $(x-6)^2$ makes both sides of the equation equal. Check this yourself.

> To complete the square for expressions of the form $x^2 + kx$, put $\frac{k}{2}$ (half the coefficient of x) at the end of the bracket and subtract $\left(\frac{k}{2}\right)^2$ (its square).

Example 18

This example illustrates possible cases for completing the square for expressions of the form $x^2 + bx$.

a $\quad x^2 + 8x = (x+4)^2 - 16$

Half the coefficient of $x = \frac{8}{2} = 4$. Subtract $4^2 = 16$.

b $\quad x^2 - 3x = \left(x - \frac{3}{2}\right)^2 - \frac{9}{4}$

Half the coefficient of $x = \frac{-3}{2} = -\frac{3}{2}$. Subtract $\left(-\frac{3}{2}\right)^2 = \frac{9}{4}$.

c $\quad x^2 + \dfrac{b}{a}x = \left(x + \dfrac{b}{2a}\right)^2 - \dfrac{b^2}{4a^2}$

Half the coefficient of $x = \dfrac{b}{2a}$. Subtract $\left(\dfrac{b}{2a}\right)^2 = \dfrac{b^2}{4a^2}$.

Example 19

This example illustrates possible cases for completing the square for quadratic expressions $ax^2 + bx + c$.

a $\quad x^2 - 2x + 4 = (x-1)^2 - 1 + 4$

$$= (x-1)^2 + 3$$

Completing the square: $x^2 - 2x \equiv (x-1)^2 - 1$.

b $\quad 4x^2 + 12x - 8 = (2x+3)^2 - 9 - 8$

$$= (2x+3)^2 - 17$$

Alternatively write $4(x^2 + 3x - 2)$ and complete the square for $x^2 + 3x - 2$.

c $\quad x^2 + 5x - 1 = \left(x + \frac{5}{2}\right)^2 - \frac{25}{4} - 1$

$$= \left(x + \frac{5}{2}\right)^2 - \frac{29}{4}$$

d $\quad 2x^2 + 12x + 1 = 2\left(x^2 + 6x + \frac{1}{2}\right)$

$$= 2\left((x+3)^2 - 9 + \frac{1}{2}\right)$$

$$= 2\left((x+3)^2 - 8\frac{1}{2}\right)$$

$$= 2(x+3)^2 - 17$$

If the coefficient of x^2 is not 1, take the coefficient outside the bracket first.

e $7 - 6x - x^2 = -(x^2 + 6x - 7)$ Here the coefficient of x^2 is -1.

$$= -((x + 3)^2 - 9 - 7)$$

$$= -((x + 3)^2 - 16)$$

$$= 16 - (x + 3)^2$$

Example 20   *In this example another method of completing the square is shown. The two expressions $4x^2 + 12x - 5$ and $p(x + q)^2 + r$ are identically equal. By comparing (equating)*
i   *the coefficients of the x^2 terms,*
ii *the coefficients of the x terms and*
iii *the constant terms*
the values of p, q and r can be found.

Express $4x^2 + 12x - 5$ in the form $p(x + q)^2 + r$

$4x^2 + 12x - 5 \equiv p(x + q)^2 + r$ This is an identity. The two expressions are identically equal.

$\equiv px^2 + 2pqx + pq^2 + r$ Multiplying out the bracket and collecting like terms makes it easier to compare coefficients.

Equating coefficients of x^2 terms: The coefficient of x^2 on the LHS is 4 and on the RHS is p. These must be equal.

$$p = 4$$

Equating coefficients of x terms: The value of p has been found, so it can be substituted here to find q.

$$12 = 2pq$$

But $p = 4$

So $12 = 2 \times 4 \times q$

$$\therefore \quad q = \frac{12}{8}$$

$$= \frac{3}{2}$$

Equating constant terms: The value of p and q have been found, so they can be substituted here to find r.

$$-5 = pq^2 + r$$

But $p = 4$ and $q = \frac{3}{2}$

So $-5 = 4 \times \left(\frac{3}{2}\right)^2 + r$

$$= 4 \times \frac{9}{4} + r$$

$$\therefore \quad r = -5 - 9$$

$$= -14$$

So $4x^2 + 12x - 5 \equiv 4(x + \frac{3}{2})^2 - 14$ Multiply out the brackets and collect like terms to check.

Completing the square is a useful method for

- finding maximum and minimum values (see also Chapter 15)
- sketching curves
- solving quadratic equations, although other methods are usually preferable.

It is essential at a higher level, in calculus, for certain integrals.

Example 21   Use the method of completing the square to solve $x^2 - 5x - 2 = 0$.

$$x^2 - 5x - 2 = \left(x - \tfrac{5}{2}\right)^2 - \tfrac{25}{4} - 2$$

$$= \left(x - \tfrac{5}{2}\right)^2 - \tfrac{33}{4}$$

So the equation $x^2 - 5x - 2 = 0$ can be written as $\left(x - \tfrac{5}{2}\right)^2 - \tfrac{33}{4} = 0$.

$$\left(x - \tfrac{5}{2}\right)^2 - \tfrac{33}{4} = 0$$

> Add $\tfrac{33}{4}$ to both sides.

$$\left(x - \tfrac{5}{2}\right)^2 = \tfrac{33}{4}$$

> Square root both sides, remembering $\pm$ sign.

$$x - \tfrac{5}{2} = \pm\sqrt{\tfrac{33}{4}}$$

$$= \pm\frac{\sqrt{33}}{2}$$

> Add $\tfrac{5}{2}$ to both sides.

$$x = \frac{5}{2} \pm \frac{\sqrt{33}}{2}$$

> Express as a single fraction.

$$= \frac{5 \pm \sqrt{33}}{2}$$

So the solution of $x^2 - 5x - 2 = 0$ is $x = \dfrac{5 \pm \sqrt{33}}{2}$.

Example 22   Solve $ax^2 + bx + c = 0$ by the method of completing the square.

$$ax^2 + bx + c = 0$$

> Divide through by a.

$$x^2 + \frac{b}{a}x + \frac{c}{a} = 0$$

> Complete the square.

$$\left(x + \frac{b}{2a}\right)^2 - \frac{b^2}{4a^2} + \frac{c}{a} = 0$$

$$\left(x + \frac{b}{2a}\right)^2 = \frac{b^2}{4a^2} - \frac{c}{a}$$

> Express the fractions with a common denominator, $4a^2$.

$$= \frac{b^2 - 4ac}{4a^2}$$

> Square root both sides, remembering $\pm$ sign.

So $$x + \frac{b}{2a} = \pm\sqrt{\frac{b^2 - 4ac}{4a^2}}$$

> *Note*: $\sqrt{4a^2} = 2a$

$$= \pm\frac{\sqrt{b^2 - 4ac}}{2a}$$

52

$$x = -\frac{b}{2a} \pm \frac{\sqrt{b^2 - 4ac}}{2a}$$

$$= \frac{-b \pm \sqrt{b^2 - 4ac}}{2a}$$

The fractions have a common denominator so can be combined.

$x = \dfrac{-b \pm \sqrt{b^2 - 4ac}}{2a}$ is the **quadratic equation formula**.

Using the quadratic equation formula

Example 30 led to the quadratic equation formula:

$$ax^2 + bx + c = 0 \Leftrightarrow x = \frac{-b \pm \sqrt{b^2 - 4ac}}{2a}$$

The quantity under the root sign, $b^2 - 4ac$, is called the **discriminant** of the quadratic equation $ax^2 + bx + c = 0$. The discriminant gives useful information about the roots of the quadratic equation; see the table on page 60.

To solve a quadratic equation using the quadratic equation formula:

■ **Arrange the equation in the form $ax^2 + bx + c = 0$.**

■ **Compare the equation to be solved with $ax^2 + bx + c = 0$ to find a, b and c.**

■ **Substitute the values of a, b and c in the formula. Take great care with the signs.**

Example 23 Solve the equation $x^2 + 8x + 5 = 0$ giving the answer both exactly (i.e. in surd form) and correct to 3 significant figures.

$x^2 + 8x + 5 = 0$

Compare $x^2 + 8x + 5 = 0$ with $ax^2 + bx + c = 0$.

$a = 1,\ b = 8,\ c = 5$

With practice this line can be omitted.

$$x = \frac{-8 \pm \sqrt{8^2 - 4 \times 1 \times 5}}{2 \times 1}$$

$$= \frac{-8 \pm \sqrt{64 - 20}}{2}$$

$$= \frac{-8 \pm \sqrt{44}}{2}$$

$\sqrt{44} = \sqrt{4 \times 11} = 2\sqrt{11}$.

$$= \frac{-8 \pm 2\sqrt{11}}{2}$$

Both terms of the numerator can be divided by 2.

$$= -4 \pm \sqrt{11}$$

The **exact answer** includes surds.

Exact solution is $x = -4 \pm \sqrt{11}$.

Use the calculator to find the values of x correct to the required accuracy.

Correct to 3 significant figures, the solution is $x = -0.683$ or $x = -7.32$.

Example 24 Solve the equation $2x^2 - 5x - 6 = 0$, giving the answer both exactly (i.e. in surd form) and to 2 decimal places.

$2x^2 - 5x - 6 = 0$

> Compare $2x^2 - 5x - 6 = 0$ with $ax^2 + bx + c = 0$.

$a = 2, b = -5, c = -6$

> With practice this line can be omitted.

$$x = \frac{-(-5) \pm \sqrt{(-5)^2 - 4 \times 2 \times (-6)}}{2 \times 2}$$

> Take care with signs under the root sign.

$$= \frac{5 \pm \sqrt{25 + 48}}{4}$$

$$= \frac{5 \pm \sqrt{73}}{4}$$

> The **exact** answer includes surds.

Exact solution is $x = \dfrac{5 \pm \sqrt{73}}{4}$.

> Use a calculator to find the values of x correct to the required accuracy.

To 2 decimal places, the solution is $x = -0.89$ or $x = 3.39$.

Example 25 An open box with volume $48\,\text{cm}^3$ is to be made by cutting $4\,\text{cm}$ squares from each corner of a square piece of metal and folding up the sides. Find the length, to the nearest mm, of the side of the square piece of metal.

Let $x\,\text{cm}$ be the length of the side of the square.

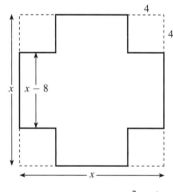

All measurements in cm

Then   volume $= 4(x - 8)^2\,\text{cm}^3$

But   volume $= 48\,\text{cm}^3$

So $4(x - 8)^2 = 48$

> No units are written in an equation (both sides are in cm^3).

$(x - 8)^2 = 12$

> Take square root. Remember $\pm$ signs.

$x - 8 = \pm\sqrt{12}$

$x = 8 \pm \sqrt{12}$

The length of the side of the square must allow two $4\,\text{cm}$ squares to be cut out.
So $x > 8$ and $x = 8 - \sqrt{12}$ is not a possible solution.
So $x = 8 + \sqrt{12}$ is the solution.

The length of the side of the square piece of metal is $11.5\,\text{cm}$ to the nearest millimetre.

- Before solving a quadratic equation: arrange the equation in the form $ax^2 + bx + c = 0$, $a > 0$.

- Always look for a common factor.
  If there is a numerical one, divide through by it before proceeding.
  Do not divide through by an unknown quantity which could be zero.

- Try to factorise, remembering particularly that the form $ax^2 + bx = 0$ gives $x(ax + b) = 0$ and difference of squares $x^2 - a^2 = 0$ gives $(x - a)(x + a) = 0$.

- For cases where factorisation is difficult, or not possible, use the formula on page 53.

 Another method of factorising quadratic functions can be found on the CD-ROM (E3.3).

Exercise 3B

1 Solve these equations by factorising.

a $(x - 2)(2x + 1) = 0$ **b** $x^2 - 7x + 12 = 0$ **c** $y^2 - 5y + 6 = 0$

d $y^2 - 16 = 0$ **e** $3x^2 - 27x = 0$ **f** $6x^2 = 10x$

g $x^2 - 4x - 21 = 0$ **h** $x^2 + x - 12 = 0$ **i** $x^2 - 6x + 9 = 0$

j $x^2 + 5x + 6 = 0$ **k** $2x^2 - 7x + 6 = 0$ **l** $3e^2 + 4e - 4 = 0$

m $3d^2 - 5d - 8 = 0$ **n** $3e^2 - 4e - 4 = 0$ **o** $12f^2 - f - 6 = 0$

p $5g^2 + 29g - 6 = 0$ **q** $49 - y^2 = 0$ **r** $25x^2 = 9$

s $x^3 - 6x^2 = 0$ **t** $90x^2 - 40 = 0$ **u** $6y^2 = 150$

2 Solve these equations by factorising.

a $3a^2 + 2 = 5a$ **b** $b(2b - 5) + 3 = 0$

c $2c^2 = 7c - 3$ **d** $2(d^2 - 3d + 3) = d + 1$

e $3(e + 1)^2 = 1 - e$ **f** $3f = 2(f^2 - 1)$

g $(g + 3)(2 - g) = g^2$ **h** $h^2 - h = 0$

i $6k^2 = 35 + k$ **j** $2l^2 + 4l = 0$

k $4m^2 - 16m = 0$ **l** $100t - t^3 = 0$

m $u^3 - 7u^2 = 0$ **n** $251w^2 = 251$

o $17x^2 - 51x + 34 = 0$ **p** $24y^2 - 48y + 24 = 0$

q $3(1 - 6k^2) = 25k$ **r** $(3x + 1)(x + 3) = 19 - 2(x + 2)^2$

3 Solve these by completing the square, giving an exact answer.

a $x^2 + 2x = 5$ **b** $x^2 - 4x = 7$ **c** $t^2 - 10t - 2 = 0$

d $s^2 + 6s - 3 = 0$ **e** $r^2 + 5r - 1 = 0$ **f** $t^2 - 3t - 7 = 0$

g $a^2 - a - 1 = 0$ **h** $b^2 + 7b + 3 = 0$ **i** $2x^2 - x - 1 = 0$

j $3x^2 = 6x + 4$

4 Express, in the form $(px + q)^2 + r$, $p > 0$

    a $16x^2 - 8x + 11$ **b** $9x^2 + 3x + 1$

5 Find a, b and c such that

    a $3x^2 + 6x - 4 \equiv a(x + b)^2 + c$ **b** $5 - 6x - x^2 \equiv a(x + b)^2 + c$

    c $4x - 5 - 2x^2 \equiv a(x + b)^2 + c$ **d** $2x^2 - 2x + 1 \equiv a(x + b)^2 + c$

6 Solve these by using the formula, giving the answer in an exact form and, if appropriate, correct to 2 decimal places.

    a $x^2 - x - 1 = 0$ **b** $2x^2 + 7x + 3 = 0$ **c** $2x^2 - 7x + 3 = 0$

    d $2x^2 + 7x - 3 = 0$ **e** $2x^2 - 7x - 3 = 0$ **f** $3 + 7x - 2x^2 = 0$

    g $x^2 - 5 = 0$ **h** $6x^2 - x - 4 = 0$ **i** $(5x + 2)(x - 1) = x(2x + 1)$

    j $(3x - 1)(2x + 7) = 1$

7 Solve

    a $x^2 - x = 0$ **b** $x^3 - x^2 = 0$

    c $(x - 1)(2x - 1)(3x - 1) = 0$ **d** $x^3 - x = 0$

    e $x^4 - 16 = 0$ **f** $(x + k)(x^2 - l^2) = 0$

8 Solve these equations.

    a $x^4 - 2x^2 - 3 = 0$ **b** $y^{\frac{2}{3}} - 2y^{\frac{1}{3}} - 3 = 0$ **c** $2z + 3z^{\frac{1}{2}} - 2 = 0$

    d $x^6 = 7x^3 - 12$ **e** $\dfrac{20}{x^2} - \dfrac{1}{x} = 1$ **f** $12 = a^4 - 11a^2$

9 The square of Adjua's age six years ago is equal to her age in six years' time. Find her present age.

10 The diagram shows the floor plan of a room. All lengths are given in metres. An architect uses this plan to design a room with a floor area of $22.5\,\text{m}^2$. Find the value of l correct to 3 significant figures.

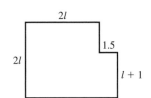

11 A factory produces metal discs of radius $(r - 1)\,\text{mm}$ and negligible thickness for making washers, and metal spheres of radius $r\,\text{mm}$ for ball bearings. The total surface area of a disc (both sides) and a ball bearing is $132\pi\,\text{mm}^2$. Find r. (Surface area of a sphere is $4\pi r^2$.)

12 A ball is thrown in the air. After t seconds, its height, s, in metres above ground is given by the equation $2s = -10t^2 + 16t + 3$.

    a   Find t when the ball is 4.5 metres above the ground.

    b   Comment on the physical interpretation of the solution to part **a**.

3.4 Sketching quadratic functions

The graph of any equation of the form $y = ax^2 + bx + c$ $(a \neq 0)$ is a parabola.

- If $a > 0$ the parabola is $\cup$-shaped.
- If $a < 0$ the parabola is $\cap$-shaped.

The turning point on a parabola is called its **vertex**.

To sketch a parabola, either complete the square to find the coordinates of the vertex and the equation of the axis of symmetry, or, if the points of intersection with the x-axis are required, solve $ax^2 + bx + c = 0$.

Example 26   Sketch $y = 3x^2 + 5x - 2$.

Consider $3x^2 + 5x - 2 = 0$

$$(3x - 1)(x + 2) = 0$$

$\Rightarrow$ $x = \tfrac{1}{3}$ or $x = -2$

> Discriminant $= 5^2 - 4 \times 3 \times (-2) = 25 + 24 > 0$
> $\therefore$ curve cuts x-axis in two distinct points.

> Curve cuts x-axis at $(-2, 0)$ and $(\tfrac{1}{3}, 0)$.

Coefficient of x^2 term $= 3 > 0$ $\therefore$ curve is $\cup$-shaped.

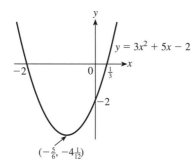

> The graph has an axis of symmetry midway between the roots so the vertex should be to the **left** of the y-axis in this case.

> When $x = 0$, $y = -2$
> $\therefore$ curve cuts y-axis at $(0, -2)$.

> Vertex at $x = \dfrac{-2 + \tfrac{1}{3}}{2} = -\dfrac{5}{6}$
> When $x = -\tfrac{5}{6}$, $y = -4\tfrac{1}{12}$

Example 27   Use the method of completing the square to

a   solve the quadratic equation $x^2 + 8x - 1 = 0$

b   find the minimum value of $x^2 + 8x - 1$ and the value of x for which this occurs

c   sketch the curve $y = x^2 + 8x - 1$, showing where the curve cuts the axes.

Solution   **a** $x^2 + 8x - 1 \equiv (x + 4)^2 - 16 - 1$

> Complete the square.

$$\equiv (x + 4)^2 - 17$$

So the equation $x^2 + 8x - 1 = 0$ can be written as $(x + 4)^2 - 17 = 0$.

Solving this:

$(x + 4)^2 - 17 = 0$

> Add 17 to both sides.

$(x + 4)^2 = 17$

> Take square root of both sides.

$x + 4 = \pm\sqrt{17}$

> *Remember*: This can be $+\sqrt{17}$ or $-\sqrt{17}$.

$x = -4 \pm \sqrt{17}$

b $x^2 + 8x - 1 \equiv (x + 4)^2 - 17$

$(x + 4)^2 \geqslant 0$ for all values of x so the minimum value for $(x + 4)^2$ is zero and this occurs when $x = -4$.

When $x = -4$, $y = (x + 4)^2 - 17 = -17$.

So the minimum value of $x^2 + 8x - 1$ is -17 and this occurs when $x = -4$.

Note   The minimum value and the value of x where it occurs give the coordinates of the vertex of the parabola.

c $y = x^2 + 8x - 1$

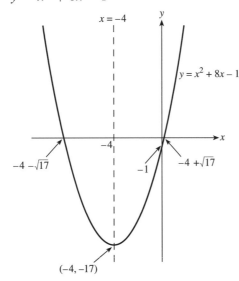

$a > 0$ ∴ parabola is ∪-shaped.

From part **b**, the graph has its vertex, a minimum point, at $(-4, -17)$.

When $x = 0$, $y = -1$.
∴   curve cuts y-axis at $y = -1$.

When $y = 0$, $x = -4 \pm \sqrt{17}$.
∴   curve cuts x-axis at
$x = -4 \pm \sqrt{17}$.

The axis of symmetry (shown with a dotted line), is midway between the roots $-4 - \sqrt{17}$ and $-4 + \sqrt{17}$, i.e. $x = -4$ or $x + 4 = 0$.

➤   **The information given by the completed square form**:

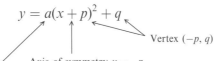

$y = a(x + p)^2 + q$

Vertex $(-p, q)$

Axis of symmetry $x = -p$

$a > 0 \Rightarrow$ ∪-shaped, vertex is a minimum point.
$a < 0 \Rightarrow$ ∩-shaped, vertex is a maximum point.

Example 28   Sketch $y = (x - 1)(x - 3)$, showing where the curve cuts the axes and the coordinates of the vertex.

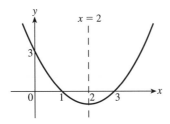

Coefficient of x^2 is +ve.
∴ ∪-shaped.

$y = 0 \Rightarrow x = 1$ or $x = 3$
∴ intercepts on x-axis are $(1, 0)$ and $(3, 0)$.

$x = 0 \Rightarrow y = 3$
∴ intercept on y-axis is $(0, 3)$.

Axis of symmetry (shown dotted) is mid-way between the roots, i.e. at $x = 2$.

When $x = 2$, $y = (2 - 1)(2 - 3) = -1$.

∴   by symmetry, the vertex is at $(2, -1)$.

Example 29 Use the method of completing the square to sketch the curve $y = 7 - 6x - x^2$.

$7 - 6x - x^2 \equiv 16 - (x + 3)^2$

> Complete the square. See Example 19e on page 51.

So $\qquad y = -(x + 3)^2 + 16$

The coefficient of x^2 is $-$ve so the parabola has a maximum point.

Maximum point is $(-3, 16)$. Axis of symmetry is $x = -3$.

> Parabola is $\cap$-shaped.

When $x = 0$, $y = 7$. So, the intercept on y-axis is 7.

The curve cuts the x-axis where $y = 0$.

$$16 - (x + 3)^2 = 0$$
$$(x + 3)^2 = 16$$
$$x + 3 = \pm 4$$
$$x = -7 \text{ or } x = 1$$

So curve cuts x-axis at $(-7, 0)$ and $(1, 0)$.

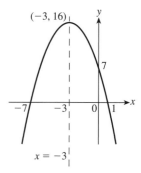

Exercise 3C

1 Sketch the graphs of these parabolas, showing where the curves cut the x- and y-axes, the equation of the line of symmetry and the coordinates of each vertex.

    a $\quad y = (x - 2)(x - 4)$ **b** $\quad y = (1 - x)(3 + x)$ **c** $\quad y = x(x - 2)$

    d $\quad y = 3x(1 - x)$ **e** $\quad y = 2(x - 4)(x + 6)$ **f** $\quad y = (2x + 1)(1 - x)$

    g $\quad y = 5(x + 1)(3x - 1)$ **h** $\quad y = (2 - 3x)(2 + 3x)$

2 By factorising these quadratic functions, or otherwise, sketch these graphs, marking the points of intersection with the x- and y-axes.

    a $\quad y = x^2 + 3x$ **b** $\quad y = x^2 - 4x + 3$ **c** $\quad y = 4 - x^2$

    d $\quad y = x^2 - 4x - 12$ **e** $\quad y = 2x^2 + 11x - 6$ **f** $\quad y = 4x^2 - 1$

    g $\quad y = 3 - 11x - 4x^2$ **h** $\quad y = 8x^2 - 26x + 15$

3 Express in the form $a(x + b)^2 + c$ and hence find the vertex and the equation of the line of symmetry.

    a $\quad y = x^2 + 2x - 11$ **b** $\quad y = x^2 - 4x + 7$ **c** $\quad y = 4 - 6x - x^2$

    d $\quad y = x^2 - 8x + 1$ **e** $\quad y = 2x^2 + 12x - 3$ **f** $\quad y = 1 + 8x - 4x^2$

    g $\quad y = x^2 - 3x + 5$ **h** $\quad y = 3x^2 + 2x - 1$

4 Find, by completing the square, the maximum or minimum values of each of these. State the value of x for which each maximum or minimum occurs.

    a $\quad x^2 - 2x + 3$ **b** $\quad x^2 + 6x - 7$ **c** $\quad 8 + 6x - x^2$

    d $\quad 4 - 2x - x^2$ **e** $\quad x^2 + 3x - 1$ **f** $\quad x^2 - 5x - 7$

    g $\quad 3 - x - x^2$ **h** $\quad 9 + 3x - x^2$ **i** $\quad 2x^2 + 8x + 1$

    j $\quad 3x^2 - x + 4$ **k** $\quad 6 - x - 2x^2$ **l** $\quad 1 + 12x - 6x^2$

5 Find p, q and r such that

a $6x^2 + 12x + 1 \equiv p(x - q)^2 + r$ **b** $5 - 4x - 2x^2 \equiv p(x - q)^2 + r$

c $3x^2 - 2x + 4 \equiv p(x - q)^2 + r$ **d** $9x^2 + 12x + 8 \equiv (px + q)^2 + r$

6 Use the method of completing the square to find the maximum or minimum values of each of these functions and then sketch the curves, $y = f(x)$, marking the coordinates of the vertex, the intercept on the y-axis and the axis of symmetry.

a $f(x) = x^2 - 2x - 1$ **b** $f(x) = x^2 + 4x + 9$ **c** $f(x) = 3 + 6x - x^2$

3.5   Use of the discriminant

The discriminant of the quadratic function $y = ax^2 + bx + c$ is the value of $b^2 - 4ac$.

The value of the discriminant, $b^2 - 4ac$, can be used to determine whether the parabola cuts, touches or does not cut the x-axis. If they are real, the roots, α and β, of the equation $ax^2 + bx + c = 0$ show where the parabola cuts the x-axis; the constant, c, shows where it cuts the y-axis.

If α and β are real, the axis of symmetry of the parabola will be midway between α and β.

	Discriminant		
	$b^2 - 4ac > 0$	$b^2 - 4ac = 0$	$b^2 - 4ac < 0$
Number of roots	Two distinct real roots	One repeated real root (2 equal roots)	No real roots (2 complex roots)
Intersection with the x-axis	Two distinct points	Curve touches x-axis (or curve meets x-axis in two coincident points)	Curve and x-axis do not meet
Sketch for $a > 0$			
Sketch for $a < 0$			

Example 30 Find the number of distinct real roots of the given equations.

  a $3x^2 - 2x - 5 = 0$

  Compare $3x^2 - 2x - 5 = 0$ with $ax^2 + bx + c = 0$.

  $b^2 - 4ac = (-2)^2 - 4 \times 3 \times (-5) = 64 > 0$

  $a = 3, b = -2, c = -5$.

  Discriminant > 0.

  Therefore the equation $3x^2 - 2x - 5 = 0$ has two distinct real roots.

  b $x^2 - 2x + 1 = 0$

  $a = 1, b = -2, c = 1$.

  $b^2 - 4ac = (-2)^2 - 4 \times 1 \times 1 = 0$

  Discriminant $= 0$. Therefore the equation $x^2 - 2x + 1 = 0$ has a repeated root. That is, there is just one root.

  c $x^2 + 4 = 0$

  $a = 1, b = 0, c = 4$.

  $b^2 - 4ac = 0 - 4 \times 1 \times 4 = -16 < 0$

  Discriminant < 0. Therefore the equation $x^2 + 4 = 0$ has no real roots.

Example 31 By calculating the discriminant of $x^2 + 4x + 8 = 0$, or otherwise, show that $x^2 + 4x + 8$ is always positive.

  $x^2 + 4x + 8 = 0$

  $a = 1, b = 4, c = 8$.

  Discriminant $= b^2 - 4ac$

  $\qquad = 16 - 4 \times 1 \times 8 = -16 < 0$

  Alternatively, completing the square,
  $x^2 + 4x + 8 = (x + 2)^2 + 4$
  $(x + 2)^2 \geqslant 0$
  $\therefore \ x^2 + 4x + 8$ is always positive.

  Since the discriminant is less than zero, the graph of $y = x^2 + 4x + 8$ does not meet the x-axis. $a > 0$ so the parabola is $\cup$-shaped.

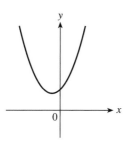

  The graph lies above the x-axis so $x^2 + 4x + 8$ is always positive.

Example 32   Find the values of k for which $2(k-1)x^2 + 2kx + k - 1 = 0$ has equal roots.

Equal roots $\Rightarrow$ discriminant $= 0$

> $a = 2(k-1)$, $b = 2k$, $c = k - 1$.

$$b^2 - 4ac = 4k^2 - 4 \times 2(k-1)(k-1)$$
$$= 4k^2 - 8(k^2 - 2k + 1)$$
$$= 4k^2 - 8k^2 + 16k - 8$$
$$= -4k^2 + 16k - 8$$

> The discriminant, $b^2 - 4ac$, is a quadratic expression in k so $b^2 - 4ac = 0$ introduces a second quadratic equation.

But $b^2 - 4ac = 0$, so

$$-4k^2 + 16k - 8 = 0$$

> Divide by -4.

$$k^2 - 4k + 2 = 0$$

> $a = 1$, $b = -4$, $c = 2$.

$$\Rightarrow \qquad k = \frac{4 \pm \sqrt{4^2 - 4 \times 1 \times 2}}{2}$$
$$= \frac{4 \pm \sqrt{8}}{2}$$
$$= \frac{4 \pm 2\sqrt{2}}{2}$$
$$= 2 \pm \sqrt{2}$$

Therefore the values of k for which $2(k-1)x^2 + 2kx + k - 1 = 0$ has equal roots are $k = 2 \pm \sqrt{2}$.

Exercise 3D

1   Calculate the value of the discriminant and hence state the number of distinct real roots of each of these equations.

    a $x^2 + 4x - 5 = 0$ **b** $x^2 + 4x + 5 = 0$ **c** $2x^2 + x - 5 = 0$

    d $x^2 + 4x + 4 = 0$ **e** $x^2 - 4x - 4 = 0$ **f** $3x^2 + 4x + 2 = 0$

    g $4x^2 - 25 = 0$ **h** $6 - 5x + x^2 = 0$

2   Find the value(s) of k if these equations have equal roots.

    a $x^2 - 2x + k = 0$ **b** $4x^2 + kx + 9 = 0$

    c $3x^2 + kx - 12 = 0$ **d** $kx^2 + 4x + 5 = 0$

    e $2x^2 + 5x - k = 0$ **f** $x^2 + (2k + 10)x + k^2 + 5 = 0$

3   Find the relation between a and b if $ax^2 + bx + 1 = 0$ has equal roots.

4   If $x^2 + mx + n$ is a perfect square, show that $m^2 = 4n$.

5   Find an equation whose roots are

    a $2, 3$ **b** $3, -5$ **c** $0, -6$ **d** p, q **e** $-\frac{2}{3}, \frac{1}{4}$

6   Calculate the possible values of k if $(k+1)x^2 + kx + k + 1 = 0$ has a repeated root.

7   By completing the square, show that $x^2 + 2x + 5$ is positive for all real values of x.

8   Show that $x^2 + 2px + 2p^2$ is positive for all real values of x.

**Exercise 3E
(Review)**

1 Solve for x.

a $3x + 4 = 6 - 7x$

b $2(x + 1) - (3x - 4) = 8 - (x + 2)$

c $x^2 - 4 = 0$

d $x^2 - 4x = 0$

e $x^3 - 4x = 0$

f $x^3 = 7x^2$

g $x^2 - 6 = x$

h $x^2 - 5x = 6$

i $x^4 - 17x^2 + 16 = 0$

j $2x^2 - 7x + 3 = 0$

k $x^5 - x^3 = 0$

l $6x^2 + 5x - 6 = 0$

2 Sketch the graphs of these parabolas, showing where the curves cut the x- and y-axes, the equation of the line of symmetry and the coordinates of each vertex.

a $y = x^2 + 6x$

b $y = 3(2 - x)(4 + x)$

c $y = (x - 5)(x + 2)$

d $y = 10 - 9x - x^2$

3 Calculate the value of the discriminant and hence state the number of distinct real roots of each of these equations.

a $x^2 - 5 = 0$

b $2x^2 + x + 1 = 0$

c $x^2 + 6x + 9 = 0$

d $2x^2 + 7x + 6 = 0$

4 Show, either by calculating the discriminant or by completing the square, that

a $x^2 - 8x + 20$ is never zero

b $12 - 4x - x^2 = 0$ has two distinct real roots

5 Use the method of completing the square to sketch

a $y = x^2 + 10x + 4$

b $y = 6 + x - x^2$

c $y = 2x^2 - 4x + 1$

d $y = 4 - 2x - 4x^2$

6 Find p, q and r such that $5x^2 - 2x + 1 \equiv p(x - q)^2 + r$. Hence, find the minimum value of $5x^2 - 2x + 1$ and the value of x for which it occurs.

7 Find the value of k for which $6x^2 + 2x + k = 0$ has equal roots.

8 Solve for x.

a $ax - b = 2x$ **b** $x^2 + y^2 = r^2$ **c** $\dfrac{ax + b}{cx + d} = k$ **d** $l = \frac{1}{2}\sqrt{\dfrac{2}{x}}$

9 Solve for x.

a $x^2 + kx = 0$ **b** $x^2 - a^2 = 0$ **c** $x^3 + kx^2 = 0$ **d** $(x - a)^2 = b$

10 Multiply $x^{\frac{2}{3}} + 2x^{\frac{1}{3}} + 1$ by $x^{\frac{1}{3}} - 2$. Check the answer by substituting $x = 8$.

A 'Test yourself' exercise (Ty3), extension examples (E3.31) and an 'Extension exercise' (Ext3) can be found on the CD-ROM.

Key points

Cross-multiplying

$\dfrac{a}{b} = \dfrac{c}{d} \Leftrightarrow ad = bc \quad (b \neq 0,\ d \neq 0)$

Solving equations

Linear equations

- Do the same to both sides.
- Collect all the terms in the unknown on one side.

Quadratic equations

- Factorisation $(x - \alpha)(x - \beta) = 0 \Rightarrow x = \alpha$ or $x = \beta$

- Formula: $ax^2 + bx + c = 0 \Rightarrow x = \dfrac{-b \pm \sqrt{b^2 - 4ac}}{2a}$

Completing the square

- $x^2 - kx + c = 0 \Rightarrow \left(x - \dfrac{k}{2}\right)^2 - \dfrac{k^2}{4} + c = 0$

Given that $ax^2 + bx + c \equiv p(x + q)^2 + r$, compare (equate) terms to find p, q and r.

Sketching a quadratic function

For $y = ax^2 + bx + c\ (a \neq 0)$

- $a > 0 \Rightarrow$ curve is $\cup$-shaped and has a minimum point.
- $a < 0 \Rightarrow$ curve is $\cap$-shaped and has a maximum point.
- The curve cuts the y-axis at $(0,\ c)$.
- The axis of symmetry is $x = -\dfrac{b}{2a}$.
- The discriminant indicates in how many points the curve intersects the x-axis.

For the completed-square form $y = a(x + p)^2 + q$.

- The vertex is at $(-p, q)$. It is either a maximum or a minimum point.
- The axis of symmetry is $x = -p$.

Discriminant of a quadratic equation

$b^2 - 4ac$ is the discriminant of the equation $ax^2 + bx + c = 0$

- $b^2 - 4ac > 0 \Rightarrow$ 2 distinct real roots
- $b^2 - 4ac = 0 \Rightarrow$ 1 repeated real root (2 equal roots)
- $b^2 - 4ac < 0 \Rightarrow$ No real roots.

Inequalities

In Chapter 3 we solved algebraic equations, finding values which satisfied the equations. There are many situations where rather than quantities being equal, one quantity might be less than or more than another, leading to inequalities. In this chapter we revise some techniques you may have met at GCSE and extend the topic further.

After working through this chapter you should be able to solve

- *linear inequalities, such as $10x - 4 < 5 - 6x$*
- *quadratic inequalities, such as $3x^2 - 7x - 5 < 0$.*

4.1   Linear inequalities

Consider the true statements $3 < 4$ and $-3 < 2$.

- Certain operations carried out on both sides of an inequality will not affect the truth of the statement.

- For some other operations the direction of the inequality sign must be reversed for the statement to remain true.

- Other operations should not be carried out since the inequality may or may not still be true.

Exercise 4A

1   Perform these operations on the inequality $3 < 4$.

In each case, state if the resulting inequality is true.

    a Add 4 to both sides.   **b** Subtract 100 from both sides.

    c Multiply both sides by 6.   **d** Divide both sides by -1.

    e Square both sides.   **f** Take the reciprocal of both sides.

    g Cube both sides.

2   Repeat Question **1** for the inequality $-3 < 2$.

The results of this exercise can be summarised as follows.

Linear inequalities are solved by doing the same to both sides, bearing in mind the situations where the sign must be reversed and avoiding unsafe operations.

Inequality *unaffected*:

- When adding same quantity to (or subtracting from) both sides

- When multiplying (or dividing) both sides by the same *positive* quantity

- Providing they are positive, squaring both sides

Inequality *true if sign is reversed*:

- When multiplying (or dividing) both sides by the same *negative* quantity
- Providing they are positive, taking the reciprocal of both sides

Unsafe operations:

- If either of the quantities is negative, squaring both sides or taking their reciprocals

Example 1   *This example shows two methods of solving an inequality and that both methods give the same answer.*

Solve $7x + 13 < 2x - 7$.

$7x + 13 < 2x - 7$

As with linear equations, all x terms must be collected on one side, with the rest of the terms on the other side.

$7x - 2x < -7 - 13$

Do this first by subtracting $2x$ and 13 from both sides.

$5x < -20$

$x < -4$

Now, divide by 5. Division is by a positive quantity, so no need to reverse the sign.

Alternatively

$7x + 13 < 2x - 7$

$13 + 7 < 2x - 7x$

This time, subtract $7x$ and add 7. This will result in $-5x$ on RHS

$20 < -5x$

Now, divide by -5. Division is by a *negative* quantity, so the sign must be *reversed*.

$-4 > x$

$x < -4$

Note: x is written on the LHS.

Note   It is usually possible, and makes sense, to avoid multiplying or dividing by a negative number by keeping the coefficient of x positive, as in the first solution.

Example 2   Find the range of values for which $2 < 3x - 4 \leqslant 7$. Show the solution on a number line.

Consider this as two separate inequalities: $2 < 3x - 4$ and $3x - 4 \leqslant 7$.

$$2 < 3x - 4 \qquad 3x - 4 \leqslant 7$$

To isolate x, add 4 to both sides.

$$6 < 3x \qquad 3x \leqslant 11$$

Divide each side by 3.

$$2 < x \qquad x \leqslant \tfrac{11}{3}$$

Rewrite with x on the LHS.

For both inequalities to hold $x > 2$ and $x \leqslant \tfrac{11}{3}$.

The two resulting inequalities can be shown on two separate number lines.

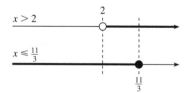

Note: The empty circle indicates that x cannot equal 2. The full circle indicates that x can equal $\tfrac{11}{3}$.

For both inequalities to hold, $2 < x \leqslant 3\tfrac{2}{3}$.

Combine this into a single number line.

Exercise 4B

1 Solve these inequalities.

a $3x + 4 < 2$ **b** $4 + 2x \geqslant 3$ **c** $2 \geqslant 7x - 5$

d $7 < 1 - 2x$ **e** $2a - 1 < 6a - 12$ **f** $3b + 4 > 6b + 5$

g $5c + 1 \geqslant 2c - 1$ **h** $2d + 9 \geqslant 5d - 2$ **i** $3 - 5e \leqslant 7 - 3e$

j $5f - 3 > 5 - 3f$ **k** $7 + 2(h - 3) < 1$ **l** $5 - (2 - 3x) \leqslant 9$

m $1 - (4x + 1) < 8$ **n** $7 + 2(5 - 3x) > 6x$

2 Solve these inequalities.

a $4(g - 2) > 3(g - 1)$ **b** $2(2x - 1) < 3(3x + 1)$

c $2 + 3(x - 4) < 3(2x - 5)$ **d** $x + 4(3 - 2x) \geqslant 2 - 2(4 - 3x)$

3 Solve these inequalities. Illustrate each solution on a number line.

a $-1 < 2x + 1 < 11$ **b** $4 \leqslant 7x - 3 \leqslant 11$

c $6 < 4 - x < 7$ **d** $-1 < 5x + 3 \leqslant 7$

e $3x < 5x + 1 \leqslant 2x + 9$ **f** $2 - x < 5 + 2x < 8$

4.2 Quadratic inequalities

Consider the inequalities $x^2 < 4$ and $x^2 > 9$ and the numbers on the number line that would satisfy them.

$x^2 < 4 \Leftrightarrow -2 < x < 2$

$x^2 > 9 \Leftrightarrow x < -3$ or $x > 3$

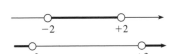

> The square of any number between -2 and 2 will be less than 4

Note $x < -3$ or $x > 3$. These inequalities represent two *separate* parts of the number line; they must *not* be combined in a single inequality.

These solutions can be illustrated graphically. (The bold lines show where the inequality holds.)

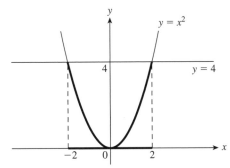

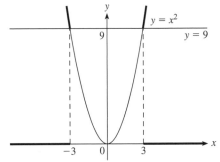

x^2 is less than 4 when the curve $y = x^2$ is *below* the line $y = 4$.
This occurs when $-2 < x < 2$.

x^2 is greater than 9 when the curve $y = x^2$ is *above* the line $y = 9$.
This occurs when $x < -3$ or $x > 3$.

So $\quad x^2 < a^2 \Leftrightarrow -a < x < a \quad (a > 0)$

and $\quad x^2 > a^2 \Leftrightarrow x < -a \quad$ or $\quad x > a \quad (a > 0)$

For inequalities such as $x^2 - x - 2 > 0$, a graphical approach can be used. Solving the inequality, in this example, means finding the values of x for which $y = x^2 - x - 2 > 0$, i.e. values of x for which the curve is above the x-axis.

To sketch $y = x^2 - x - 2$ first factorise the right-hand side.

$y = x^2 - x - 2 = (x - 2)(x + 1)$

When $y = 0$, $x = -1$ or $x = 2$. So the curve cuts the x-axis at $x = -1$ and $x = 2$.

The coefficient of x^2 is +ve, so the graph is ∪-shaped.

The values of y are *greater than zero*, i.e. $x^2 - x - 2 > 0$, when the curve is *above* the x-axis. That is, when $x < -1$ and $x > 2$.

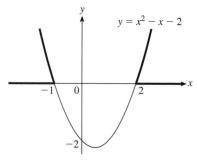

Note If the inequality had been reversed, i.e. $x^2 - x - 2 < 0$, the solution would be $-1 < x < 2$.

Any solution to a quadratic inequality, $f(x) > 0$ or $f(x) < 0$, is either the range of values *between* the roots of $f(x) = 0$ or the range of values *outside* the roots. To determine which, either draw a sketch or test one value of x.

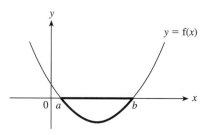

f(x) < 0 between the roots: $a < x < b$

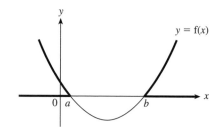

f(x) > 0 outside the roots: $x < a$ and $x > b$

> **To solve a quadratic inequality**
> - **Arrange the inequality with $ax^2 + bx + c$ on one side and zero on the other.**
> - **Solve $ax^2 + bx + c = 0$.**
> - **Sketch $y = ax^2 + bx + c$, marking the intercepts on the x-axis.**
> - **Look at the inequality sign to decide on the range of values of x required to make the inequality true. (This range will either be between the roots or be outside the roots of $ax^2 + bx + c = 0$.)**

Example 3   Solve $6 - x - x^2 < 0$.

Consider $6 - x - x^2 = 0$

$(3 + x)(2 - x) = 0$

$\therefore$ $x = -3$ or $x = 2$

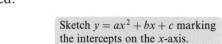
A sketch of $y = 6 - x - x^2$ is required.
Solve $ax^2 + bx + c = 0$.

Coefficient of x^2 term is $-$ve, so curve is $\cap$-shaped.

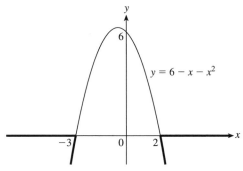

Sketch $y = ax^2 + bx + c$ marking the intercepts on the x-axis.

So $6 - x - x^2 < 0$ when $x < -3$ or $x > 2$.

Alternatively put $x = 0$ in the inequality. This gives $6 < 0$ which is not true, so the inequality is true *outside* the roots.

Example 4   Solve $(3x - 1)(x + 1) \leqslant x(2x - 3) - 5$.

$(3x - 1)(x + 1) \leqslant x(2x - 3) - 5$

Remove the brackets.

$3x^2 + 2x - 1 \leqslant 2x^2 - 3x - 5$

Rearrange with $ax^2 + bx + c$ on LHS and zero on RHS.

$x^2 + 5x + 4 \leqslant 0$

A sketch of $y = x^2 + 5x + 4$ is required. Solve $ax^2 + bx + c = 0$

Consider $x^2 + 5x + 4 = 0$

$(x + 1)(x + 4) = 0$

$\therefore$ $x = -1$ or $x = -4$

Coefficient of x^2 term is $+$ve, so curve is $\cup$-shaped.

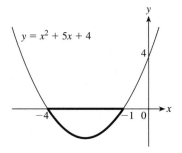

Sketch $y = ax^2 + bx + c$ marking the intercepts on the x-axis.

$y \leqslant 0$ when the curve is below the x-axis, i.e. for values of x between the roots.

So $x^2 + 5x + 4 \leqslant 0$

when $-4 \leqslant x \leqslant -1$

So $(3x - 1)(x + 1) \leqslant x(2x - 3) - 5$ when $-4 \leqslant x \leqslant -1$.

Alternatively put $x = 0$ in the inequality. This gives $-1 \leqslant -5$ which is not true, so the inequality is true *between* the roots.

Note   For an inequality with sign $\geqslant$ or $\leqslant$ the roots *are* included, so use $\geqslant$ or $\leqslant$ in the answer. For sign $>$ or $<$ the roots are excluded so use $>$ or $<$.

Extension

Polynomial inequalities with more than two linear factors can also be solved by a sketching method.

Example 5

Solve $(x-2)(x+1)(x+4) > 0$.

Sketching $y = (x-2)(x+1)(x+4)$

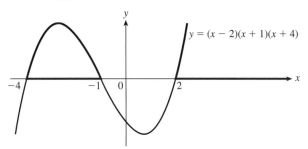

> Sketch $y = (x-2)(x+1)(x+4)$ marking the intercepts on the x-axis. $y > 0$ when the curve is above the x-axis.

So $(x-2)(x+1)(x+4) > 0 \Rightarrow -4 < x < -1$ or $x > 2$.

Exercise 4C

1 Solve these inequalities.

a $(x-2)(x-3) < 0$ **b** $(x+4)(x-5) > 0$ **c** $(2x+1)(x-1) \geqslant 0$

d $(2x-5)(3-2x) \leqslant 0$ **e** $x^2 - x > 0$ **f** $6 + x - x^2 < 0$

g $x^2 - 3x + 2 < 0$ **h** $x^2 - 7x - 18 > 0$ **i** $x^2 - x - 6 \geqslant 0$

j $4x - 3x^2 < 0$ **k** $y^2 - 7y + 12 > 0$ **l** $y^2 - 4y - 5 \geqslant 0$

m $6y^2 - 7y + 2 \leqslant 0$ **n** $3 - 5y - 8y^2 > 0$ **o** $2u^2 + 7u + 3 \leqslant 0$

p $2v^2 + 7v - 4 < 0$ **q** $4x^2 + 4x + 1 \geqslant 0$ **r** $9 + 6x + x^2 < 0$

2 Solve these inequalities.

a $3x^2 - 2 < 5x$

b $4x + 5 < 9x^2$

c $5x - x^2 > 4$

d $4(x-1) \geqslant x^2$

e $(x+1)^2 \leqslant 6x^2 + x + 1$

f $(x-5)^2 + 5(2x-3) > 2x(x+3) - 6$

3 By considering the discriminant, or otherwise, find the range of values of k that give each of these equations two distinct real roots.

a $x^2 - 2x + k = 0$ **b** $2x^2 + kx + 7 = 0$ **c** $kx^2 = 2x - k$

4 Find the range of values of k that give each of these equations no real roots.

a $3x^2 - x + k = 0$ **b** $3x^2 + kx + 7 = 0$ **c** $(k-1)x^2 = 3kx - 12$

5 Solve

a $(x-1)(x-2)(x+3) \leqslant 0$

b $(x-1)(2x-1)(4-x) > 0$

c $(5x-1)(x-1)^2 < 0$

d $(x^2-1)(x^2-4) \leqslant 0$

Exercise 4D (Review)

1 Solve these inequalities.

a $4x + 3 \geqslant 10$ **b** $2 - 7x < 9$

c $3x + 5 > 6x - 7$ **d** $8 + 2(3 - x) \leqslant 6 - x$

e $3(x + 5) - 7 > 8 - 4(2 - 3x)$ **f** $3y + 7 < 2 - 4(1 - y)$

2 Solve these inequalities.

a $x^2 > 4$ **b** $x^2 \leqslant 100$

c $(x - 1)(x + 2) > 0$ **d** $x^2 - 3x - 10 < 0$

e $(x - 5)^2 > 3$ **f** $2x^2 + 7x - 4 \geqslant 0$

3 By considering the discriminant, or otherwise, find the range of values of k that give each of these equations two distinct real roots.

a $x^2 + 3x + k = 0$ **b** $3x^2 + kx + 2 = 0$ **c** $k(x^2 + 1) = x - k$

4 Find the range of values of k that give these equations no real roots.

a $x^2 + 6x + k = 0$ **b** $2x^2 + kx + 1 = 0$ **c** $(k + 1)x^2 + 4kx + 9 = 0$

5 Show that $x^2 + 2kx + 9 \geqslant 0$ for all real values of x, if $k^2 \leqslant 9$.

There is a 'Test yourself' exercise (Ty4) and an 'Extension exercise' (Ext4) on the CD-ROM.

➤ Key points

Solving inequalities

Linear inequalities
Do the same to both sides (e.g. add, subtract, multiply, divide) but reverse the inequality sign if multiplying or dividing by a negative number.

Quadratic inequalities and those involving other polynomials
Sketch the graph of the quadratic or other polynomial and use the sketch to decide on the range of values of x required to make the inequality true.

To solve a quadratic inequality
- Arrange the inequality with $ax^2 + bx + c$ on one side and zero on the other.
- Solve $ax^2 + bx + c = 0$.
- Sketch $y = ax^2 + bx + c$, marking the intercepts on the x-axis.
- Look at the inequality sign to decide on the range of values of x required to make the inequality true. (This range will either be between the roots or be outside the roots of $ax^2 + bx + c = 0$.)

5 Simultaneous equations

Chapters 3 and 4 were concerned with solving equations and inequalities which involved one variable (one unknown quantity). Real-life problems, however, involve many unknown variables with many connections between them. In this chapter we revise some techniques you have met at GCSE and see how to deal with more complicated equations where there are two unknowns.

After working through this chapter you should be able to

■ solve simultaneous equations where both equations are linear
■ solve simultaneous equations where one equation is linear and one is quadratic
■ understand the connection between solving simultaneous equations and finding the intersections of graphs.

A graphical calculator should not be used in this chapter to find solutions. However it is a useful tool to check a solution once it has been obtained or to illustrate the solution graphically.

5.1 Simultaneous equations: Linear equations

This chapter deals with sets of equations with two (or more) unknowns.

Consider pairs of values which satisfy $x + y = 8$.

x	-1	0	1.2
y	9	8	6.8

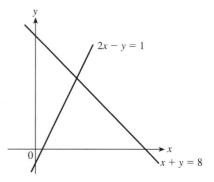

There is an infinite number of such pairs, x and y. These pairs can be plotted and in this case a straight line is obtained.

Similarly there is an infinite number of pairs of values of x and y which satisfy the equation $2x - y = 1$.

Solving equations simultaneously means finding the values of the unknowns that satisfy both (or all) equations at the same time.

The point of intersection of the lines is the solution of the equations. These values of x and y satisfy *both* equations simultaneously.

In this case there are *two* unknowns, x and y, and *two* equations are needed to find a solution. With *three* unknowns, *three* equations would be needed.

Two methods for solving simultaneous linear equations are explained:

■ Elimination

■ Substitution (see page 75).

Solving two linear simultaneous equations: Elimination method

Example 1   Solve the simultaneous equations $x + y = 8$ and $2x - y = 1$.

$$x + y = 8 \quad ①$$
$$2x - y = 1 \quad ②$$

> Notice that the y terms in both equations have matching coefficients with *different* signs.

$① + ②$ $3x = 9$

> *Add* the equations to eliminate the y terms.

$\therefore$ $x = 3$

Substituting in ① $3 + y = 8$

> Substitute $x = 3$ in equation ① to give the value of y.

$\therefore$ $y = 5$

Checking in ② $\text{LHS} = 2x - y$

> Check that the solution is correct by substituting $x = 3$ and $y = 5$ in equation ②.

$$= 2 \times 3 - 5$$
$$= 1 = \text{RHS}$$

So the solution is $x = 3$, $y = 5$.

Note   Always substitute and check in *different* equations. If one equation is used for substitution, use the other for checking.

Example 2   Solve the simultaneous equations $2x + 5y = 12$ and $2x - 3y = -4$.

$$2x + 5y = 12 \quad ①$$
$$2x - 3y = -4 \quad ②$$

> Notice that the x terms have matching coefficients with the *same* sign

$① - ②$ $5y - (-3y) = 12 - (-4)$

> *Subtract* the equations to eliminate the x terms.

$$8y = 16$$

$\therefore$ $y = 2$

Substituting in ② $2x - 6 = -4$

> Substitute $y = 2$ in equation ② to give the value of x.

$$2x = 2$$

$\therefore$ $x = 1$

Checking in ① $\text{LHS} = 2x + 5y$

> Check that the solution is correct by substituting $x = 1$ and $y = 2$ in equation ①.

$$= 2 \times 1 + 5 \times 2$$
$$= 12 = \text{RHS}$$

So the solution is $x = 1$, $y = 2$.

Example 3   Find the point of intersection of the lines $3x + 2y = 2$ and $x - 3y = -14$.

$$3x + 2y = 2 \qquad ①$$
$$x - 3y = -14 \qquad ②$$

> The solution of the simultaneous equations will give the point of intersection of the lines.

$② \times 3$ $3x - 9y = -42 \qquad ③$

$① - ③$ $2y - (-9y) = 2 - (-42)$

> Neither the x terms nor the y terms have matching coefficients. To match up the x coefficients multiply all the terms in ② by 3.

$$11y = 44$$

$\therefore \qquad\qquad y = 4$

> The matching coefficients in ① and ③ have the *same* sign so *subtract* ③ from ① (or vice versa) to eliminate the x terms.

Substituting in ② $x - 12 = -14$

> Substitute $y = 4$ in equation ② to give the value of x.

$\therefore \qquad\qquad x = -2$

Checking in ① $\text{LHS} = 3x + 2y$

> Check that the solution is correct by substituting $x = -2$ and $y = 4$ in equation ①.

$$= 3 \times (-2) + 2 \times 4$$
$$= 2 = \text{RHS}$$

So the solution is $x = -2$, $y = 4$ and the point of intersection of the lines is $(-2, 4)$.

Example 4   Solve the simultaneous equations $3x = -7 + 2y$ and $5y = 8 - 2x$.

$$3x = -7 + 2y \qquad ①$$
$$5y = 8 - 2x \qquad ②$$

> Rearrange the equations to line up x and y terms on the LHS.

$$3x - 2y = -7 \qquad ③$$
$$2x + 5y = 8 \qquad ④$$

$③ \times 5$ $15x - 10y = -35 \qquad ⑤$

$④ \times 2$ $4x + 10y = 16 \qquad ⑥$

> To match the y coefficients, multiply equation ③ by 5 and equation ④ by 2. (Or match up x coefficients by multiplying equation ③ by 2 and equation ④ by 3.)

$⑤ + ⑥$ $19x = -19$

$\therefore \qquad\qquad x = -1$

> The y coefficients in ⑤ and ⑥ now match with *different* signs, so *add* equations ⑤ and ⑥ to eliminate the y terms.

Substituting in ① $-3 = -7 + 2y$

> Substitute $x = -1$ in ① to give the value of y.

$$2y = 4$$

$\therefore \qquad\qquad y = 2$

Checking in ② $\text{LHS} = 5y = 10$

$\qquad\qquad \text{RHS} = 8 - 2x$

> Check that the solution is correct by substituting $x = -1$, $y = 2$ in equation ②. Use the original equations for checking as they will be correct (unless miscopied!).

$$= 8 - (-2)$$
$$= 10$$
$$\text{LHS} = \text{RHS}$$

So the solution is $x = -1$, $y = 2$.

➤ **Match the coefficients of terms with the same letter.**
If matching terms have the *same* sign then *subtracting* the equations will eliminate those terms.
If matching terms have *different* signs then *adding* the equations will eliminate those terms.

Exercise 5A

1 Solve these simultaneous equations by elimination.

a $x + y = 6$
$x - y = 2$

b $2c + d = 23$
$c - d = 1$

c $5x - 3y = 23$
$2x + 3y = 26$

d $-g + 6h = 15$
$g - 4h = -9$

e $x + 3y = 8$
$x - 2y = 3$

f $5r - 2s = 14$
$3r - 2s = 6$

g $5t - 3u = 9$
$5t + 2u = 19$

h $-x + 5y = 39$
$-x + 2y = 18$

i $5x + 6y = 14$
$x - 6y = 10$

j $3x + 2y = 7$
$5x + y = 7$

k $n - 6p - 1 = 0$
$3n + 2p - 13 = 0$

l $3x - 4y = 4$
$2x + 2y = 5$

2 Solve these simultaneous equations, by elimination.

a $s + 2t = 8$
$5s - 3t = 1$

b $3a - y = 5$
$2a - 10y = 1$

c $b + 2x = 3$
$4b + 3x = 2$

d $3z - 2y - 5 = 0$
$4z + 3y - 1 = 0$

e $4x + 3y = 5$
$3x + 2y = 4$

f $4r - 3q = 11$
$3r + 2q = 4$

g $3p - 2q = 5$
$2p - 3q = 0$

h $4r - 5s = 1$
$2r - 3s = 1$

i $7x - 11y = 21$
$5x - 3y = 15$

j $7x - 3y = 18$
$2x + 5y = 11$

Solving two linear simultaneous equations: Substitution method

Substitution is an alternative, more general, method of solving simultaneous equations.

Example 5 Solve the simultaneous equations $y = 3x + 4$ and $y = 2 - 7x$.

$$y = 3x + 4 \quad ①$$
$$y = 2 - 7x \quad ②$$

So
$$3x + 4 = 2 - 7x$$
$$10x = -2$$

$$\therefore \qquad x = -\tfrac{1}{5}$$

Substituting in ① $\quad y = 3 \times (-\tfrac{1}{5}) + 4 = 3\tfrac{2}{5}$

Checking in ② $\quad$ LHS $= 3\tfrac{2}{5}$

$\qquad\qquad$ RHS $= 2 + \tfrac{7}{5} = 3\tfrac{2}{5}$

$\qquad\qquad$ LHS $=$ RHS

So the solution is $x = -\tfrac{1}{5}$, $y = 3\tfrac{2}{5}$.

> For the solution, the values of y must be equal. Hence the expressions for y can be put equal to each other.

> Choose the equation that makes for the easiest substitution. Substitute $x = -\tfrac{1}{5}$ in equation ①.

> Check by substituting $x = -\tfrac{1}{5}$ and $y = 3\tfrac{2}{5}$ in equation ②.

Example 6   Solve the simultaneous equations $y - 2x = 5$ and $6y - 5x = 23$.

$$y - 2x = 5 \quad ①$$
$$6y - 5x = 23 \quad ②$$

> Rearrange one of the equations to express one unknown in terms of the other.

Rearranging ①
$$y = 2x + 5$$

> In this case ① can easily be arranged to express y in terms of x.

Substituting in ②
$$6(2x + 5) - 5x = 23$$
$$12x + 30 - 5x = 23$$
$$7x = -7$$
$$\therefore \qquad x = -1$$

> Substitute this expression for y in the other equation, to solve for x.

Substituting in ①
$$y - 2 \times (-1) = 5$$
$$y + 2 = 5$$
$$\therefore \qquad y = 3$$

> Substitute $x = -1$ in ① to give the value of y.

Checking in ②
$$\text{LHS} = 6 \times 3 - 5 \times (-1)$$
$$= 18 + 5$$
$$= 23 = \text{RHS}$$

> Check by substituting $x = -1$ and $y = 3$ in equation ②.

So the solution is $x = -1$, $y = 3$.

Before solving simultaneous equations, look at the coefficients.

■ **If there are fractional or decimal coefficients in any equation, multiply all terms in the equation by a suitable number, so that all coefficients become whole numbers.**

■ **If there is a common factor in all the coefficients of an equation, divide through by that factor before proceeding.**

Further examples of simultaneous equations

Example 7   *This example involves fractional coefficients.*

Solve the simultaneous equations $\dfrac{x}{3} - \dfrac{y}{4} = 0$ and $\dfrac{x}{2} + \dfrac{3y}{10} = 5\dfrac{2}{5}$.

$$\frac{x}{3} - \frac{y}{4} = 0 \quad ①$$

$$\frac{x}{2} + \frac{3y}{10} = 5\frac{2}{5} \quad ②$$

> Eliminate fractions by multiplying each equation by the LCM of its denominators: multiply equation ① by 12 and equation ② by 10.

① × 12	$4x - 3y = 0$	③
② × 10	$5x + 3y = 54$	④
③ + ④	$9x = 54$	
$\therefore$	$x = 6$	

> Coefficients of y match, but signs are *different* so *add* to eliminate the y terms.

Substituting in ③ $4 \times 6 - 3y = 0$

$\hspace{5cm} 3y = 24$

$\therefore \hspace{4.5cm} y = 8$

Checking in ② $\text{LHS} = \dfrac{6}{2} + \dfrac{3 \times 8}{10}$

$\hspace{4cm} = 3 + \tfrac{12}{5}$

$\hspace{4cm} = 5\tfrac{2}{5} = \text{RHS}$

Substitute $x = 6$ in equation ③ to give the value of y.

Since the substitution was in ③ (which was derived from ①), check by substituting $x = 6$ and $y = 8$ in the other equation, i.e. ②.

So the solution is $x = 6$, $y = 8$.

Example 8   *This example includes a common factor. When solving money problems, express all sums of money in the same units. Pence are used here to avoid decimals.*

In a book shop, 23 items were bought. Some were magazines costing £2.50 each, and the rest were books costing £3.00 each. The total cost was £61. Find how many magazines and how many books were bought.

Let the numbers of magazines and books bought be x and y respectively.

Then $\hspace{2.5cm} x + y = 23$ ①

and $\hspace{1.2cm} 250x + 300y = 6100$ ②

② ÷ 50 $\hspace{1.3cm} 5x + 6y = 122$ ③

① × 5 $\hspace{1.5cm} 5x + 5y = 115$ ④

③ − ④ $\hspace{3cm} y = 7$

Substituting in ① $x + 7 = 23$

$\therefore \hspace{3.5cm} x = 16$

Checking in ② $\text{LHS} = 250 \times 16 + 300 \times 7$

$\hspace{4cm} = 4000 + 2100$

$\hspace{4cm} = 6100 = \text{RHS}$

23 articles were bought.

Divide ② by 50, the HCF of the coefficients. (It is easier to work with smaller numbers.)

Or check in equation ③, which is a simpler form of equation ②.

So 16 magazines and 7 books were bought.

Example 9   Here are the tariffs of two mobile phone companies:

Company 1: £6 per month plus 2 pence per minute on a call

Company 2: No monthly fee but 5 pence per minute on a call.

a   Find the total length of calls (in minutes) during one month for which the charge would be the same for both companies.

A uses her mobile phone for 150 minutes per month on average. **B**'s monthly total is always more than 250 minutes.

b   Find which company would provide the better deal for each of these potential customers.

Solution **a** Let C pence be the charge of the service in one month.

Let t minutes be the time of calls in one month.

Choose units for C and convert all sums to the chosen units. In this case 'pence' have been used.

For Company 1	$C = 600 + 2t$
For Company 2	$C = 5t$
The charges are the same, so	$5t = 600 + 2t$
	$3t = 600$
$\therefore$	$t = 200$

To find the value of t for which the charges are the same, solve the equations simultaneously.

So the charge would be the same for calls totalling 200 minutes.

b For $t > 200$ Company 1 has a lower charge than Company 2.

For $t = 200$ charges are the same.

For $t < 200$ Company 2 has a lower charge than Company 1.

So for **A**, who uses the mobile on average for 150 minutes a month, Company 2 offers the better deal.

For **B**, who uses the mobile for more than 250 minutes a month, Company 1 offers the better deal.

One approach is to draw graphs for the two companies showing charge C, against time, t.

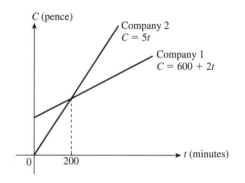

A method of solving three simultaneous equations is shown on the CD-ROM (E5.1).

Exercise 5B

1 Solve these simultaneous equations by the most appropriate method, elimination or substitution.

a $y = 2x + 1$
    $3y + 10x = 7$

b $x + y = 11$
    $3x - 5y = 1$

c $2x + y = 1$
    $3x + 2y = 3$

d $x = 3 - 7y$
    $2x + 9y = 11$

e $5x + 7y = 52$
    $3x - 8y = 19$

f $2s + 3x = 31$
    $3s + 4x = 44$

g $2x - 3y = 4$
    $5x + 2y = 1$

h $3x + 4y = 2$
    $5x - 3y = 1$

2 Solve these simultaneous equations for x and y.

a $7x - 8y = 9$
    $11x + 3y = -17$

b $7x + 4y = 9$
    $2x + 3y = 1$

c $3x + 2y = 5x + 2y = 7$

d $3x + 2y - 5 = 0$
    $5x - 3y - 21 = 0$

e $ax + by = m$
    $cx + dy = n$

3 Find the point of intersection for each of these pairs of lines.

a $y = 5x$
    $y = 3 - x$

b $y = 4x + 5$
    $y = 3x + 6$

c $2x - 3y + 5 = 0$
    $4x + 5y - 1 = 0$

d $4y - 2x = 7$
    $5y - 3x = 6$

If you want to apply these results to more real-life problems see the CD-ROM (A5.1).

5.2 Solving simultaneous equations: One linear and one quadratic

Plotting two *linear* equations on the same axes usually gives a single point of intersection.

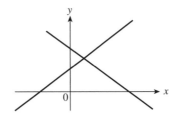

Plotting one *linear* and one *quadratic* equation will lead to zero, one or two point(s) of intersection.

Consider the possible points of intersection of a parabola and a general straight line:

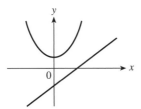

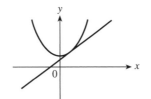

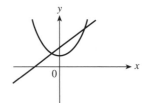

No point of intersection
The line misses the curve altogether.

One point of intersection (or two coincident points)
The line is a tangent to the curve.

Two points of intersection
The line cuts the curve in two distinct points.

Simultaneous equations with one linear and one quadratic equation are usually solved by

■ rearranging the linear equation to give one variable in terms of the other

■ substituting in the quadratic equation.

Note It is sometimes easier to rearrange the quadratic equation and substitute in the linear (e.g. Exercise 5C Question 2e).

Most of the examples in this section are illustrated graphically. It is not necessary to follow the details of the sketches, nor to know the names of the curves; it does, however, give geometrical meaning to solving the equations.

Example 10 Solve the simultaneous equations $y = x^2$ and $y = x + 2$.

$$y = x + 2 \qquad ①$$

> Equation ① is linear.

$$y = x^2 \qquad ②$$

> Equation ② is quadratic.

$$\Rightarrow \qquad x^2 = x + 2$$

> For the solution, the values of y must be equal. Hence the expressions for y can be put equal to each other.

$$x^2 - x - 2 = 0$$

$$(x-2)(x+1) = 0$$
$$\Rightarrow \qquad x = 2 \text{ or } x = -1$$

When $x = 2$, $y = 4$.

When $x = -1$, $y = 1$.

So solution is $x = -1$, $y = 1$ or $x = 2$, $y = 4$.

Substitute values of x in the linear equation ① to obtain corresponding values of y, and then check in ②.

Corresponding values of the unknowns must be given in pairs.

Note Illustrating the solution graphically:

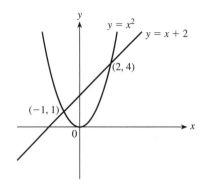

The points of intersection are $(-1, 1)$ and $(2, 4)$.

Note The linear equation should be used for substitution when solving simultaneous equations (one linear, one quadratic). If the quadratic equation is used, some solutions may be obtained which do not satisfy the linear equation.

The parabola is only one of the forms of quadratic curves.

Equations of the form $xy = c$ or $ax^2 \pm by^2 = c$ give hyperbolas, circles and ellipses.

Examples 11–13 illustrate lines intersecting with hyperbolas. Exercise 5C (on page 83) contains examples of lines intersecting with circles and ellipses, as well as parabolas and hyperbolas.

Example 11 Find the coordinates of the points of intersection of $x - 2y = 8$ and $xy = 24$.

$$x - 2y = 8 \qquad ①$$
$$xy = 24 \qquad ②$$

Equation ① is linear.

Equation ② is quadratic.

Rearranging ① $\qquad x = 8 + 2y$

Rearrange the linear equation to give one unknown in terms of the other. In this case it is easier to give x in terms of y. Then substitute in the quadratic equation.

Substituting in ② $\quad (8 + 2y)y = 24$
$$2y^2 + 8y - 24 = 0$$
$$y^2 + 4y - 12 = 0$$
$$(y + 6)(y - 2) = 0$$
$$\Rightarrow \qquad y = -6 \text{ or } y = 2$$

Multiply out and put all terms on one side, and then divide through by 2.

Substituting in ①

Substitute values of y in the linear equation to obtain corresponding values of x.

When $y = -6$, $x = -4$.

When $y = 2$, $x = 12$.

So the solutions are $x = -4$, $y = -6$ and $x = 12$, $y = 2$.

Corresponding values of the unknowns must be given in pairs.

Checking in ② $(-4) \times (-6) = 24$

$$12 \times 2 = 24$$

So the graphs intersect at $(-4, -6)$ and $(12, 2)$.

Note   Illustrating the solution graphically:

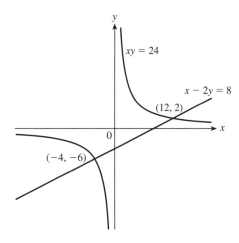

Example 12   Solve the simultaneous equations $2x + y = 5$ and $x^2 - y^2 = 3$.

$$2x + y = 5 \qquad ①$$

Equation ① is linear.

$$x^2 - y^2 = 3 \qquad ②$$

Equation ② is quadratic.

Rearranging ① $y = 5 - 2x$

Rearrange the linear equation to give one variable in terms of the other. Express y in terms of x to avoid using fractions, and then substitute in the quadratic equation ②.

Substituting in ② $x^2 - (5 - 2x)^2 = 3$

Multiply out $(5 - 2x)(5 - 2x)$.

$$x^2 - (25 - 20x + 4x^2) = 3$$

Keep the result in a bracket since it has to be subtracted.

$$3x^2 - 20x + 28 = 0$$

Put all terms on one side.

$$(3x - 14)(x - 2) = 0$$

Factorise or use the formula.

$$\Rightarrow \qquad \qquad x = \tfrac{14}{3} \text{ or } x = 2$$

Substitute values of x in the linear equation to obtain corresponding values of y.

Substituting in ①

When $x = \tfrac{14}{3}$, $y = 5 - \tfrac{28}{3} = -\tfrac{13}{3}$. When $x = 2$, $y = 1$.

So the solutions are $x = \tfrac{14}{3}$, $y = -\tfrac{13}{3}$
or $x = 2$, $y = 1$.

Give answers in pairs of x and y. Check by substituting in ②.

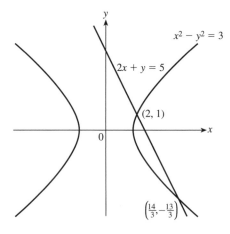

Note   Illustrating the solution graphically, points of intersection are $(2, 1)$ and $\left(\frac{14}{3}, -\frac{13}{3}\right)$.

Example 13   *This example shows two methods: using fractions, and avoiding fractions. The standard method of rearranging the linear equation and substituting in the quadratic equation may involve substituting algebraic fractions (as in part **a**). Sometimes algebraic fractions can be avoided by squaring the linear equation (it may need to be rearranged first) and matching terms (as in part **b**).*

Solve the simultaneous equations $3x - 2y = 1$ and $3x^2 - 2y^2 + 5 = 0$.

a   *Using fractions*

$$3x - 2y = 1 \qquad ① \qquad \boxed{\text{Linear}}$$

$$3x^2 - 2y^2 + 5 = 0 \qquad ② \qquad \boxed{\text{Quadratic}}$$

Rearranging ①
$$x = \frac{1 + 2y}{3} \qquad \boxed{\begin{array}{l}\text{Rearrange to give } x \text{ in terms} \\ \text{of } y \text{ or } y \text{ in terms of } x.\end{array}}$$

Substituting in ② $3\left(\dfrac{1 + 2y}{3}\right)^2 - 2y^2 + 5 = 0$ $\boxed{\begin{array}{l}\text{Substituting for } x \text{ in } ② \text{ avoids} \\ \text{problems with the minus sign.}\end{array}}$

$$\frac{\overset{1}{\cancel{3}}(1 + 2y)(1 + 2y)}{\underset{1}{\cancel{3}} \times 3} - 2y^2 + 5 = 0$$

$$\frac{1 + 4y + 4y^2}{3} - 2y^2 + 5 = 0 \qquad \boxed{\begin{array}{l}\text{Multiply through by 3} \\ \text{to eliminate the fraction.}\end{array}}$$

$$1 + 4y + 4y^2 - 6y^2 + 15 = 0$$

$$2y^2 - 4y - 16 = 0 \qquad \boxed{\text{Divide through by 2.}}$$

$$y^2 - 2y - 8 = 0$$

$$(y - 4)(y + 2) = 0$$

$$\Rightarrow \qquad\qquad y = 4 \text{ or } y = -2$$

Substituting in ① $\boxed{\begin{array}{l}\text{Substitute values of } y \text{ in the} \\ \text{linear equation to obtain} \\ \text{corresponding values of } x.\end{array}}$

When $y = 4$, $3x - 8 = 1$, and so $x = 3$.

When $y = -2$, $3x - (-4) = 1$, and so $x = -1$.

So the solutions are $x = 3$, $y = 4$ or $x = -1$, $y = -2$.

b *Avoiding fractions* $3x - 2y = 1$ ①   Linear

$$3x^2 - 2y^2 + 5 = 0 \quad ②$$ Quadratic

Rearranging ① $3x = 1 + 2y$ Rearrange the linear equation to give one unknown in terms of the other.

Squaring $9x^2 = (1 + 2y)^2$ Square both sides to give an expression for $9x^2$.

$$= 1 + 4y + 4y^2$$

② × 3 $9x^2 - 6y^2 + 15 = 0$ Multiply ② by 3 to give another expression for $9x^2$.

$$9x^2 = 6y^2 - 15$$

So $6y^2 - 15 = 1 + 4y + 4y^2$ Put the expressions for $9x^2$ equal to each other.

$$2y^2 - 4y - 16 = 0$$

$$y^2 - 2y - 8 = 0$$ Divide through by 2.

$$(y - 4)(y + 2) = 0$$

$\Rightarrow$ $y = 4$ or $y = -2$ Now proceed as in part **a**.

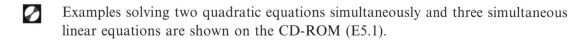 Examples solving two quadratic equations simultaneously and three simultaneous linear equations are shown on the CD-ROM (E5.1).

Exercise 5C

1 Solve these simultaneous equations.

  a $y = x^2$
     $y = x + 6$

  b $y = x^2$
     $y = 3 - 2x$

  c $y + x^2 = 8x$
     $y = 2x$

  d $y + 1 = x$
     $y = x^2 - 6x + 5$

  e $x - y = 0$
     $x^2 + y^2 = 18$

  f $x^2 + y^2 = 20$
     $2y - x = 0$

2 Find the coordinates of the points of intersection of these curves and lines.

  a $y = \dfrac{8}{x}$
     $y = 7 + x$

  b $xy = 4$
     $y = 2x + 2$

  c $x^2 + y^2 = 25$
     $y = x + 1$

  d $x^2 + 4y^2 = 4$
     $x = 2 - 2y$

  e $y^2 = x + 4$
     $2x + 5y = 4$

  f $4y^2 - 3x^2 = 1$
     $x - 2y = 1$

3 Solve

  a $x^2 + xy + y^2 = 7$
     $2x + y = 1$

  b $x^2 + 5x + y = 4$
     $x + y = 8$

  c $x^2 - xy - y^2 = -11$
     $2x + y = 1$

  d $x^2 + y^2 + 4x + 6y - 40 = 0$
     $x - y = 10$

5.3 Intersection of linear and quadratic curves: Three cases of the discriminant

The discriminant can be used to test for the number of points of intersection of a linear and a quadratic graph.

When the linear equation is rearranged and substituted into the quadratic equation a further quadratic equation is formed. The number of roots of this equation will correspond to the number of points in which the line and curve intersect.

Discriminant	Number of roots	Intersection of curve and line	
$b^2 - 4ac > 0$	Two distinct real roots.	Two distinct points.	
$b^2 - 4ac = 0$	One real root (or two equal roots).	Line is tangent to curve. Line meets curve in two coincident points.	
$b^2 - 4ac < 0$	No real roots.	Line and curve do not intersect.	

Example 14

*To find points of intersection, the equations in this example are solved simultaneously. Parts **a**–**c** relate to the three alternatives for the discriminant: > 0, < 0 and $= 0$. Part **d** is a special case.*

Find the coordinates of the points of intersection with $y = x^2$ of these lines.

a $x + y = 12$ **b** $x + y = -8$ **c** $x + y = -\frac{1}{4}$ **d** $x = 3$

Solution

a

$$x + y = 12 \quad \text{①}$$ ① is linear

$$y = x^2 \quad \text{②}$$ ② is quadratic

From ① $y = 12 - x$ Rearrange the linear equation.

So $x^2 = 12 - x$ The two expressions for y are equal.

$$x^2 + x - 12 = 0 \quad \text{③}$$ Solve the quadratic. Note that $b^2 - 4ac > 0$.

$$(x - 3)(x + 4) = 0$$

$\Rightarrow \quad x = 3 \quad \text{or} \quad x = -4$

When $x = 3$, $y = 9$.   Substitute in the linear equation for values of y.

When $x = -4$, $y = 16$.

There are two distinct points of intersection: $(3, 9)$ and $(-4, 16)$.

b

$$x + y = -8 \quad \text{①}$$

Linear

$$y = x^2 \quad \text{②}$$

Quadratic

From ① $y = -8 - x$

Rearrange the linear equation.

So $x^2 = -8 - x$

The two expressions for y are equal.

$$x^2 + x + 8 = 0$$

Note: $b^2 - 4ac < 0$.

∴ $x = \dfrac{-1 \pm \sqrt{1 - 32}}{2}$

There are no real roots of the equation $x^2 + x + 8 = 0$ so there are no points of intersection.

c

$$x + y = -\tfrac{1}{4} \quad \text{①}$$

Linear

$$y = x^2 \quad \text{②}$$

Quadratic

From ① $y = -\tfrac{1}{4} - x$

Rearrange the linear equation.

So $x^2 = -\tfrac{1}{4} - x$

The two expressions for y are equal.

$$4x^2 = -1 - 4x$$

Multiply through by 4.

$$4x^2 + 4x + 1 = 0$$

Note: $b^2 - 4ac = 0$.

$$(2x + 1)(2x + 1) = 0$$

∴ $x = -\tfrac{1}{2}$

One repeated root.

Substituting in ①, when $x = -\tfrac{1}{2}$, $y = \tfrac{1}{4}$.

Substitute in the linear equation.

There is one repeated root, so there is one point of intersection between the curve and the line. As the root is repeated, the line is a tangent to the curve at $\left(-\tfrac{1}{2}, \tfrac{1}{4}\right)$.

d $x = 3$

Linear

$y = x^2$

Quadratic

When $x = 3$, $y = 3^2 = 9$. There is one point of intersection $(3, 9)$.

This is not a repeated root, so line is not a tangent to the curve.

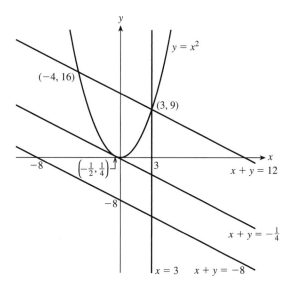

The four lines and their intersections with $y = x^2$ can be illustrated graphically.

Example 15   Find the values of k for which $kx + y = 4$ is a tangent to the curve $y = x^2 + 8$.

$$kx + y = 4 \qquad ①$$ Linear

$$y = x^2 + 8 \qquad ②$$ Quadratic

From ① $$y = 4 - kx$$ Rearrange the linear equation.

Graphs intersect when $x^2 + 8 = 4 - kx$ Put the two expressions for y equal to each other.

$$x^2 + kx + 4 = 0 \qquad ③$$

For the line to be a tangent to the curve, ③ must have equal roots, that is $b^2 - 4ac = 0$.

So $$k^2 - 16 = 0$$

$\therefore$ $$k = \pm 4$$ Substitute $k = \pm 4$ in equation ①.

So $y + 4x = 4$ and $y - 4x = 4$ are tangents to the curve $y = x^2 + 8$.

Note   The solution can be illustrated graphically:

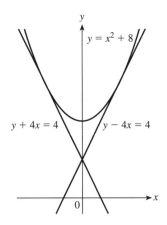

Example 16   Find the range of values of k for which $y = 2x + k$ meets $2x^2 + y^2 = 3$ in two distinct points.

$$y = 2x + k \qquad ①$$ To find the points of intersection, solve these equations simultaneously.

$$2x^2 + y^2 = 3 \qquad ②$$

Substituting ① in ② $$2x^2 + (2x + k)^2 = 3$$

$$2x^2 + 4x^2 + 4kx + k^2 = 3$$

$$6x^2 + 4kx + k^2 - 3 = 0$$

For two distinct points of intersection, $b^2 - 4ac > 0$ For the line and the curve to intersect in two distinct points, the discriminant of this equation must be +ve.

So $$16k^2 - 4 \times 6(k^2 - 3) > 0$$

($\div 8$) $$2k^2 - 3k^2 + 9 > 0$$

$$k^2 < 9$$ Solve this inequality.

$\therefore$ $-3 < k < 3$.

So the range of values required is $-3 < k < 3$.

Exercise 5D

1 Find the number of points of intersection of these straight lines and curves.

  a $y = 2x + 1$
     $y = x^2$

  b $y = 4 - x$
     $x^2 + y^2 = 4$

  c $y = \frac{1}{3}x + 1$
     $y = x^2 - 4$

  d $y = \frac{x}{4} + 1$
     $y^2 = x$

  e $2y + 3x + 1 = 0$
     $y = 2x^2$

  f $y = 2x + 1$
     $5y^2 + 5x^2 = 1$

2 Find the value of k for which $y = 3x + 1$ is a tangent to the curve $x^2 + y^2 = k$.

3 Find the range of values of k for which $y = x - 3$ meets $x^2 - 3y^2 = k$ in two distinct points.

4 Show that the line $y = x - 4$ touches the curve $x^2 + y^2 = 8$.

5 Find the value of k for which $y = kx + 1$ is a tangent to the curve $y^2 = 8x$.

6 A rectangular photograph has an area of $600\,\text{cm}^2$ and a perimeter of $1\,\text{m}$.
  Find its dimensions.

7 The hypotenuse of a right-angled triangle measures $17\,\text{cm}$.
  The shortest side is $7\,\text{cm}$ shorter than the middle side.
  Find the lengths of the sides.

8 The sides of a cuboid measure $1\,\text{cm}$, $a\,\text{cm}$ and $b\,\text{cm}$.
  Its volume is $12\,\text{cm}^3$ and its surface area is $38\,\text{cm}^2$.
  Find a and b.

9 Show that $y = x + c$ is a tangent to $x^2 + y^2 = 4$ if, and only if, $c^2 = 8$.
  (Show that if $y = x + c$ is a tangent then $c^2 = 8$, and also that if $c^2 = 8$ then $y = x + c$ is a tangent.)

Exercise 5E (Review)

1 Solve these.

  a $4x + 3y = 1$
     $3x + 2y = 1$

  b $17x - 2y = 28$
     $9x - 5y = 3$

  c $x + y = 7$
     $x^2 + y^2 = 25$

  d $3x - y = 5$
     $2x^2 + y^2 = 129$

  e $5x - 3y = 6$
     $3x - 4y = 1$

  f $3x - 5y = 16$
     $xy = 7$

2 Find the point of intersection of $y = 4x - 1$ and $3y - 8x + 2 = 0$.

3 Find the points of intersection of $y = x - 1$ and $x^2 + y^2 = 1$.

4 Prove that the line $y = 2x + \frac{5}{4}$ touches the curve $y^2 = 10x$.

5 Find the points of intersection of $x - y = 2$ with $x^2 - 4y^2 = 5$.

6 Find the possible values of k if $y = 2x + k$ meets $y = x^2 - 2x - 7$

  a in two distinct real points

  b in just one point

7 Find the range of values of k for which $kx + y = 3$ meets $x^2 + y^2 = 5$ in two distinct points.

8 Find the values of k for which $y = kx - 2$ is a tangent to the curve $y = x^2 - 8x + 7$.

9 Find the range of values of k for which $y = 2x + k$ and $x^2 + y^2 = 4$ do not intersect.

There are more review questions on the CD-ROM (see A5.3). Also on the CD-ROM is a 'Test yourself' exercise (Ty5) and an 'Extension exercise' (Ext5).

➤ Key points

Solving simultaneous equations

Two methods of solving simultaneous equations: elimination or substitution.

For two linear equations *either* method can be used.
For one linear and one quadratic equation, the substitution method is usually more suitable. The linear equation is usually the one to be rearranged, but not in every case.

- **Elimination**
  Match the coefficients of terms with the same letter.
  If matching terms have the *same* sign then *subtracting* the equations will eliminate those terms.
  If matching terms have *different* signs then *adding* the equations will eliminate those terms.

- **Substitution**
  Rearrange one of the equations to express one unknown in terms of the other. Substitute in the other equation.

Before solving simultaneous equations, look at the coefficients.

- If there are fractional or decimal coefficients in any equation, multiply all terms in the equation by a suitable number, so that all coefficients are whole numbers.
- If there is a common factor in all the coefficients of an equation, divide through by that factor before proceeding.

Always check the answer if possible.

1 Given that $(2 + \sqrt{7})(4 - \sqrt{7}) = a + b\sqrt{7}$, where a and b are integers,

  a find the value of a and the value of b.

  Given that $\dfrac{2 + \sqrt{7}}{4 + \sqrt{7}} = c + d\sqrt{7}$, where c and d are rational numbers,

  b find the value of c and the value of d.

Edexcel Jan 2001

2 Given that $2^x = \dfrac{1}{\sqrt{2}}$ and $2^y = 4\sqrt{2}$,

  a find the exact value of x and the exact value of y,

  b calculate the exact value of 2^{y-x}.

Edexcel Jan 2002

3 a Given that $8 = 2^k$, write down the value of k.

  b Given that $4^x = 8^{2-x}$, find the value of x.

Edexcel June 2001

4 Given that $27^{3x+1} = 9^y$,

  a obtain an expression for y in the form $y = ax + b$, where a and b are constants,

  b solve the equation $27^{3x+1} = 9$, giving your answer as an exact fraction.

London June 1998

5 Express each of the following in the form $m + n\sqrt{5}$ where m and n are rational numbers

  a $\dfrac{\sqrt{5} + 1}{\sqrt{5} - 2}$ **b** $(3\sqrt{5} + 2)(2\sqrt{5} - 2)$

6 a Prove, by completing the square, that the roots of the equation $x^2 + 2kx + c = 0$, where k and c are constants, are $-k \pm \sqrt{(k^2 - c)}$.

  The equation $x^2 + 2kx + 81 = 0$ has equal roots.

  b Find the possible values of k.

Edexcel Jan 2001

7 By substituting $t = x^{\frac{1}{2}}$, or otherwise, find the values of x for which $4x + 8 = 33x^{\frac{1}{2}}$.

London June 1997

8 a Find, as surds, the roots of the equation $2(x + 1)(x - 4) - (x - 2)^2 = 0$.

  b Hence find the set of values of x for which $2(x + 1)(x - 4) - (x - 2)^2 > 0$.

London Jan 1998

9 **a** Use algebra to solve $(x - 1)(x + 2) = 18$.

   b Hence, or otherwise, find the set of values of x for which $(x - 1)(x + 2) > 18$.

London June 1998

10 **a** Solve the equation $\dfrac{x - 6}{3} - \dfrac{3x - 2}{2} = \dfrac{3}{4}$.

   b Given that $ax + by + c = 0$, find an expression for y in terms of a, b, c and x.

11 **a** By completing the square, find in terms of k the roots of the equation

$$x^2 + 2kx - 7 = 0$$

   b Prove that, for all real values of k, the roots of $x^2 + 2kx - 7 = 0$ are real and different.

   c Given that $k = \sqrt{2}$, find the exact roots of the equation.

Edexcel May 2002

12 The equation $x^2 + 5kx + 2k = 0$, where k is a constant, has real roots.

   a Prove that $k(25k - 8) \geqslant 0$.

   b Hence find the set of possible values of k.

   c Write down the values of k for which the equation $x^2 + 5kx + 2k = 0$ has equal roots.

Edexcel June 2001

13 **a** Given that $3^x = 9^{y-1}$, show that $x = 2y - 2$.

   b Solve the simultaneous equations

$$x = 2y - 2,$$
$$x^2 = y^2 + 7.$$

Edexcel Jan 2003

14 **a** Express $4x^2 - 12x + 15$ in the form $(px + q)^2 + r$, where p, q and r are integers.

   b Find the coordinates of the vertex of the curve $y = 4x^2 - 12x + 15$ and sketch the curve.

15 Find the set of values of x for which

$$(2x + 1)(x - 2) > 2(x + 5).$$

Edexcel Mock paper

16 Express

$$\frac{2\sqrt{2}}{\sqrt{3} - 1} - \frac{2\sqrt{3}}{\sqrt{2} + 1},$$

in the form $p\sqrt{6} + q\sqrt{3} + r\sqrt{2}$, where the integers p, q and r are to be found.

Edexcel Mock paper

17 **a** Solve the inequality

$$3x - 8 > x + 13.$$

   b Solve the inequality

$$x^2 - 5x - 14 > 0.$$

Edexcel Nov 2002

18 Show that the elimination of x from the simultaneous equations

$$x - 2y = 1,$$
$$3xy - y^2 = 8,$$

produces the equation

$$5y^2 + 3y - 8 = 0.$$

Solve this quadratic equation and hence find the pairs (x, y) for which the simultaneous equations are satisfied.

London May 1995

19 a Expand $(1 + x^2)(1 + x^3)$, arranging your answer in ascending powers of x.

b Find, as a decimal number, the exact value of $(1 + x^2)(1 + x^3)$ for $x = 10^{-3}$.

London June 1996

20 The specification for a new rectangular car park states that the length x m is to be 5 m more than the breadth. The perimeter of the car park is to be greater than 32 m.

a Form a linear inequality in x.

The area of the car park is to be less than 104 m².

b Form a quadratic inequality in x.

c By solving your inequalities, determine the set of possible values of x.

Edexcel Specimen paper

21 Given that $y = 10^x$, show that

a $y^2 = 100^x$

b $\dfrac{y}{10} = 10^{x-1}$

c Using the results from **a** and **b** write the equation

$$100^x - 10\,001(10^{x-1}) + 100 = 0$$

as an equation in y.

d By first solving the equation in y, find the values of x which satisfy the given equation in x.

London June 1996

22 Given that $t^{\frac{1}{3}} = y$, $y \neq 0$,

a express $6t^{-\frac{1}{3}}$ in terms of y.

b Hence, or otherwise, find the values of t for which $6t^{-\frac{1}{3}} - t^{\frac{1}{3}} = 5$.

London Jan 1996

23 Find the set of values for x for which

a $6x - 7 < 2x + 3$,

b $2x^2 - 11x + 5 < 0$,

c both $6x - 7 < 2x + 3$ and $2x^2 - 11x + 5 < 0$.

Edexcel June 2003

6 Coordinate geometry and the straight line

Before the invention of coordinates, it was difficult to define the location of a place, to describe a complex shape or curve or to explain how to get from A to B. Using coordinates makes it so simple that it is hard to imagine life without them.

After working through this chapter you should know

- *equations of straight lines, including the forms $y - y_1 = m(x - x_1)$ and $ax + by = c$*
- *how to find the equation of a straight line through two given points*
- *how to find the equation of a straight line through a given point, with a given gradient*
- *that parallel lines have the same gradient*
- *that lines are perpendicular if, and only if, $m_1 m_2 = -1$.*

6.1 Introduction

The position of a point can be specified by an origin and a number of coordinates.

- On a line (1D – '1 dimensional'), an origin and one coordinate are needed to identify a point.
  The number line is an example of this.

- On a plane (2D), an origin and two coordinates are needed to identify a point.
  The Cartesian system, named after René Descartes (1596–1650), draws two axes at right angles to each other through the origin. Two coordinates are used: one in each of the two directions of the axes.

- Similarly in three dimensions (3D), an origin and three coordinates are needed to identify a point.
  The most frequently used system is three perpendicular axes drawn through the origin.

- In n dimensions, an origin and n coordinates are needed to identify a point.

There are other systems for identifying the position of points in 2D and 3D, e.g. polar coordinates, and position vectors.

In 1998, Richard Borcherds was awarded a Fields medal for creating a new area of algebra. He proved the so-called 'moonshine conjectures' of the 'Monster Group' – an abstract symmetrical snowflake that lives in 196 883-dimensional space and has 808 017 424 794 512 875 886 459 904 961 710 757 005 754 368 000 000 000 symmetries. That's more symmetries than the sun has atoms.

Superstring theory suggests that our universe may be part of a 26-dimensional space but this chapter is restricted to coordinate geometry in two dimensions!

6.2   Coordinate geometry of a plane

- The horizontal axis (across the page) is the x-axis. This can be written $\mathrm{O}x$.
- The vertical axis (up and down the page) is the y-axis. This can be written $\mathrm{O}y$.
- The point O, the origin, where the axes meet, is the zero point on both axes.
- The x-coordinate (or **abscissa**) of a point, P, is its directed distance, +ve or −ve, from the y-axis.
- The y-coordinate (or **ordinate**) of a point, P, is its directed distance, +ve or −ve, from the x-axis.

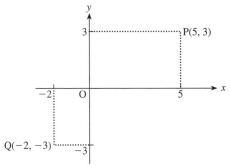

In the diagram, P is the point (5, 3). Q is the point (−2, −3).

6.3   The length of a line segment

In mathematics, a line is understood to be a straight line extending indefinitely in both directions.

A finite part of a line is called a **line segment**. AB is a line segment of the line l.

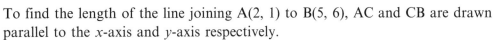

The length of a line segment can be calculated

- using numerical coordinates, or
- algebraically.

To find the length of the line joining A(2, 1) to B(5, 6), AC and CB are drawn parallel to the x-axis and y-axis respectively.

Pythagoras' theorem can then be applied to the right-angled triangle ABC.

$$AB^2 = AC^2 + BC^2$$
$$= (5-2)^2 + (6-1)^2$$
$$= 9 + 25$$
$$\therefore \quad AB = \sqrt{34}$$

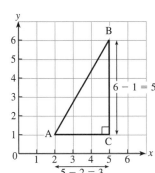

A similar method is used to find the length of the line joining $A(x_1, y_1)$ to $B(x_2, y_2)$.

$$AC = x_2 - x_1$$

$$BC = y_2 - y_1$$

Using Pythagoras' theorem

$$AB^2 = (x_2 - x_1)^2 + (y_2 - y_1)^2$$

$$\therefore \quad AB = \sqrt{(x_2 - x_1)^2 + (y_2 - y_1)^2}$$

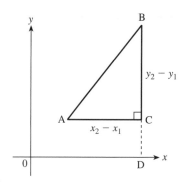

Suffixes are often used for points whose coordinates are unknown. (x_1, y_1) is read as 'x one, y one'.

$$BC = BD - CD$$
$$= y_2 - y_1$$

➤ The length of the line segment joining (x_1, y_1) to (x_2, y_2) is $\sqrt{(x_2 - x_1)^2 + (y_2 - y_1)^2}$

Example 1   Find the length of the line segment joining $A(2, -4)$ to $B(-6, -8)$.

$$AB = \sqrt{(-6 - 2)^2 + (-8 - (-4))^2}$$

$$= \sqrt{(-8)^2 + (-4)^2}$$

$$= \sqrt{80} = 4\sqrt{5}$$

The formula can be used with

$$\begin{array}{cccc} x_1 & y_1 & x_2 & y_2 \\ A(2, & -4) & B(-6, & -8) \end{array}$$

The same result is obtained if the coordinates of A are (x_2, y_2) and of B are (x_1, y_1).

$$\sqrt{80} = \sqrt{16} \times \sqrt{5} = 4\sqrt{5}$$

Example 2   Prove that the triangle with vertices $A(3, 7)$, $B(1, -4)$ and $C(-2, -3)$ is isosceles.

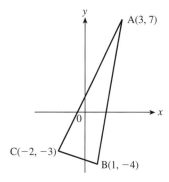

One method of proving that a triangle is isosceles is to show that two sides have equal length. A sketch can often help to decide which sides to test for equality. The sketch shows that AB and AC are likely to be the equal sides.

$$AB^2 = (3 - 1)^2 + (7 - (-4))^2$$

$$= 2^2 + 11^2 = 125$$

$$AC^2 = (3 - (-2))^2 + (7 - (-3))^2$$

$$= 5^2 + 10^2 = 125$$

Since $AB^2 = AC^2$

$$AB = AC$$

so $\triangle ABC$ is isosceles.

Substitute in the formula $d^2 = (x_2 - x_1)^2 + (y_2 - y_1)^2$.

$$y_2 - y_1 = 7 - (-4) = 7 + 4 = 11$$
Note: Take care with the signs.

There is no need to take square roots. AB and AC are both +ve, so $AB^2 = AC^2 \Rightarrow AB = AC$.

6.4   The mid-point of a line segment

To find the mid-point of the line segment joining A(2, 1) to B(8, 11), let M(x, y) be the mid-point of AB.

Draw ML parallel to BC and MN parallel to AC.

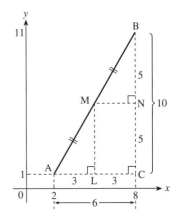

Then L and N are the mid-points of AC and BC respectively, so AL = LC and BN = NC.

AL = LC

∴   AL = 3

BN = NC

∴   NC = 5

From the diagram

$x = 2 + 3 = 5$ and $y = 1 + 5 = 6$

so M is the point (5, 6).

> An intuitive geometrical approach:
> Since M is halfway between A and B, its coordinates must be the arithmetic means of the coordinates of A and B
> $$x = \frac{2+8}{2} = 5 \quad y = \frac{1+11}{2} = 6$$

A similar method is used to find the mid-point, M(x, y), of the line segment joining A(x_1, y_1) to B(x_2, y_2). AL = LC, so:

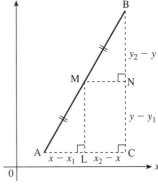

$x - x_1 = x_2 - x$

$\quad 2x = x_1 + x_2$

$\quad\quad x = \dfrac{x_1 + x_2}{2}$

Similarly, BN = NC, so

$y_2 - y = y - y_1$

$\quad 2y = y_1 + y_2$

$\quad\quad y = \dfrac{y_1 + y_2}{2}$

> **The mid-point of the line joining (x_1, y_1) to (x_2, y_2) is** $\left(\dfrac{x_1 + x_2}{2}, \ \dfrac{y_1 + y_2}{2} \right)$.

Example 3   Find the mid-point of the line segment joining (−9, 6) and (3, −1).

$\quad \overset{x_1 \quad y_1}{(-9, \ 6)} \quad \overset{x_2 \quad y_2}{(3, \ -1)}$

Using the formula above, the mid-point is $\left(\dfrac{-9+3}{2}, \ \dfrac{6+(-1)}{2} \right)$, i.e. $\left(-3, 2\tfrac{1}{2} \right)$.

Exercise 6A

1 Find the lengths of the straight lines joining these pairs of points.

    a A(3, 2) and B(8, 14)   **b** C(−1, 3) and D(4, 7)   **c** E(4, 2) and F(6, −10)

    d G(p, q) and H(r, s)   **e** J(−6, −1) and K(3, −4)   **f** L(6, 9) and M(6, 13)

    g N(1, 2) and P(5, 2)   **h** Q(−4, −2) and R(3, −7)   **i** S(−6, 1) and T(6, 6)

2 Find the coordinates of the mid-points of the line segments in Question 1.

3 Find the distance of the point (−15, 8) from the origin.

4 P, Q, R are the points (5, −3), (−6, 1) and (1, 8) respectively.
  Show that triangle PQR is isosceles, and find the coordinates of the mid-point of the base.

5 Repeat Question 4 for the points L(4, 4), M(−4, 1) and N(1, −4).

6 For each triangle ABC, show that it is right-angled, state which side is the hypotenuse, find the area, and find tan A.

    a A(13, −4), B(7, 4), C(3, 1)   **b** A(−1, 2), B(0, −11), C(3, 1)

7 Find h if the point (h, 0) is equidistant from the points (−1, 7) and (−3, −2).

8 Three of these four points lie on a circle whose centre is at the origin.

    A(−1, 7)   B(5, −5)   C(−7, 5)   D(7, −1)

  Which are they, and what is the radius of the circle?

9 Find a and b if the point (6, 3) is the mid-point of the line joining ($2a$, $2a - b$) and ($a - 2b$, $4a + 3b$).

10 The point (3, k) is a distance of 5 units from (0, 1).
  Find the two possible values of k.

11 Find the perimeter of triangle ABC where A is (7, 0), B is (4, 3) and C is (10, 1), giving the answer as simply as possible.

12 One test that a quadrilateral is a parallelogram is to show that its diagonals bisect each other. Use this test to show that ABCD is a parallelogram where A is (0, 5), B is (7, 7), C is (4, 3) and D is (−3, 1).

13 The points A, B, C, D, E and F are (−1, −2), (2, 0), (3, 5), (7, −1), (8, 4) and (11, 6) respectively.
  By considering the distances between the points, state which four points form

    a a square   **b** a rhombus   **c** a kite (two possibilities)

14 A and B are the points (−1, −6) and (5, −8) respectively.
  Which of these points lie on the perpendicular bisector of AB?

    P(3, −4)   Q(4, 0)   R(5, 2)   S(6, 5)

15 A and B are the points (12, 0) and (0, −5) respectively. Find the length of AB, and the length of the median, through the origin O, of the triangle OAB.

6.5 The gradient of a line

The gradient of a straight line measures how steep the line is. The gradient is defined as

$$\frac{\text{increase in } y}{\text{increase in } x}$$

in moving between two points on the line. The gradient of a line can be positive, negative, zero or infinite.

■ Positive gradient

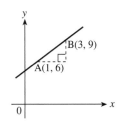

The increase in x from A to B $= 3 - 1 = 2$.
The increase in y from A to B $= 9 - 6 = 3$.

So the gradient of AB $= \frac{3}{2}$.

As x increases, so does y, so the gradient is +ve.

> A +ve gradient line goes from bottom left to top right.

■ Negative gradient

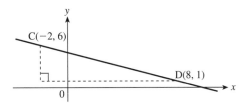

> A −ve gradient line goes from top left to bottom right.

The increase in x from C to D $= 8 - (-2) = 10$.
The 'increase' in y from C to D $= 1 - 6 = -5$.

Note   As y is actually decreasing, the 'increase' from C to D is −ve.

So the gradient of CD $= \frac{-5}{10} = -\frac{1}{2}$.

As x increases, y decreases, so the gradient is −ve.

In general, this diagram shows how to find the gradient, m, of the line through A(x_1, y_1) and B(x_2, y_2).

$$\boxed{\text{Gradient, } m = \frac{y_2 - y_1}{x_2 - x_1}}$$

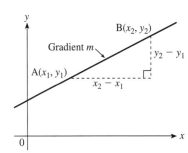

Example 4   Find the gradient of the line through the points $(2, -1)$ and $(4, -5)$.

$$x_1 \quad y_1 \qquad x_2 \quad y_2$$
$$(2, -1) \quad (4, -5)$$

$$\text{Gradient} = \frac{-5 - (-1)}{4 - 2} = \frac{-4}{2} = -2$$

The same result is obtained with
$$x_1 \quad y_1 \qquad x_2 \quad y_2$$
$$(4, -5) \text{ and } (2, -1)$$
i.e. the *order* of points does not matter because $\dfrac{y_2 - y_1}{x_2 - x_1} = \dfrac{y_1 - y_2}{x_1 - x_2}$

Any pair of points on the line will give the same result for the gradient because the resulting triangles are all similar.

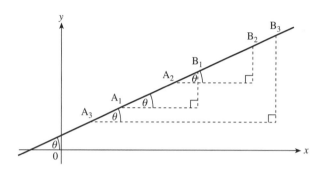

Gradient as tan θ

The gradient of a line is the tangent of the angle made by the line and the +ve direction of the x-axis.

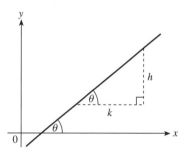

Gradient $= \tan \theta = \dfrac{h}{k}$

When θ is acute, $\tan \theta$ and the gradient are both +ve.

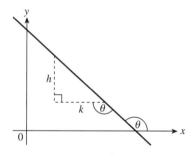

Gradient $= \tan \theta = -\dfrac{h}{k}$

When θ is obtuse, $\tan \theta$ and the gradient are both −ve.

Note   To avoid distorting the angle θ, the same scale must be used on both axes.

Special cases: zero and infinite gradients

Zero gradient: line parallel to x-axis

$$\text{Gradient} = \frac{3 - 3}{8 - 4} = 0$$

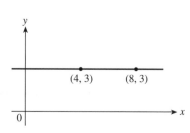

Infinite gradient: line parallel to y-axis

$$\text{Gradient} = \frac{8-3}{4-4} = \frac{5}{0}$$

Division by zero is undefined.
The gradient of the line is infinite.

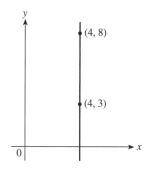

6.6   Parallel and perpendicular lines

- **Parallel lines**: lines are parallel $\Leftrightarrow$ lines have equal gradient
- **Perpendicular lines**: lines at right angles to each other

Imagine rotating the rectangle OAPB through $90°$ anticlockwise to OA$'$P$'$B$'$ (see diagram).

When the line segment OP, where P is the point (a, b), is rotated through $90°$ anticlockwise to OP$'$, the point P$'$ will have coordinates $(-b, a)$.

The gradient, m_1, of OP $= \dfrac{b}{a}$

The gradient, m_2, of OP$'$ $= -\dfrac{a}{b}$

$$m_1 m_2 = \frac{b}{a} \times -\frac{a}{b} = -1$$

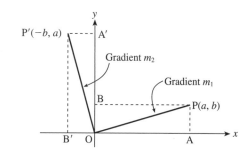

> **If two lines have gradients m_1 and m_2 then**
>
> $m_1 m_2 = -1 \quad \Leftrightarrow$ **the lines are perpendicular, $m_1 \neq 0$, $m_2 \neq 0$**
>
> $\quad m_1 = m_2 \quad \Leftrightarrow$ **the lines are parallel**

Example 5   Find the gradient of the line perpendicular to the line joining A(3, 4) to B(7, -8).

Gradient of AB $= \dfrac{-8-4}{7-3} = -3$

So, the gradient of the line perpendicular to AB is $\frac{1}{3}$.

$\boxed{-3 \times \frac{1}{3} = -1}$

Note   Given a gradient $\frac{a}{b}$, to find the perpendicular gradient, take the reciprocal $\frac{b}{a}$, and then reverse the sign $\left(-\frac{b}{a}\right)$.

Examples of gradients of pairs of lines which are perpendicular:
2 and $-\frac{1}{2}$; $\frac{p}{q}$ and $-\frac{q}{p}$; $\frac{7}{4}$ and $-\frac{4}{7}$.

6.7 Equation of a curve

The Cartesian equation of a curve is a relationship connecting the x and y values of points on the curve.

For example, $y = 3x + 1$, $x^2 + y^2 = 9$, $5x + 2y^7 = 7$.

These equations are satisfied by an infinite number of pairs of values of x and y, so there are an infinite number of points on each curve.

For an equation, pairs of values of x and y can be tabulated. These pairs can be plotted as points and the points joined by a smooth curve to give a graph.

Curves can also be plotted on a graphical calculator.

Consider $y = 3x + 1$.

$y = 3x + 1$ is the equation of a straight line.

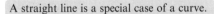

A straight line is a special case of a curve.

To plot the line, it is both necessary and sufficient to calculate two points on it, since two points define a straight line. It is sensible to calculate a third point as a check.

x	−2	0	2
y	−5	1	7

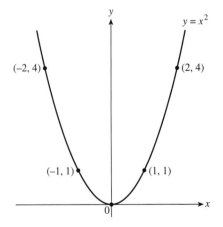

The points $(-2, -5)$, $(0, 1)$ and $(2, 7)$ are plotted and a straight line through the points, extended at both ends, is drawn.

Now consider $y = x^2$.

$y = x^2$ is the equation of a curve.
Many points are needed for an accurate graph.
Five points will give a rough idea of the shape.

x	−2	−1	0	1	2
y	4	1	0	1	4

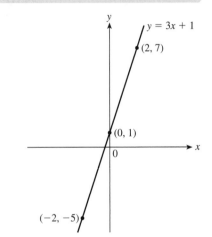

The points $(-2, 4)$, $(-1, 1)$, $(0, 0)$, $(1, 1)$ and $(2, 4)$ are plotted and a smooth curve drawn through the points.

For an accurate graph, more points would need to be worked out.

To test if a point lies on a curve

To test if a point lies on a curve, substitute the values of its coordinates into the equation. The point lies on the curve if, and only if, the equation is satisfied.

Example 6 Does $(3, -4)$ lie on the line $y = -2x + 3$ or on the curve $x^2 + y^2 = 25$?

$$y = -2x + 3$$
When $x = 3$ $y = -6 + 3$
$$= -3$$

> If $(3, -4)$ lies on the line then when $x = 3$ is substituted in the equation, $y = -4$ should result.

> The result is $y = -3$, so the point does *not* lie on the line.

$\therefore$ $(3, -4)$ does not lie on $y = -2x + 3$.

For the curve $x^2 + y^2 = 25$, when $x = 3$ and $y = -4$.

$$\text{LHS} = x^2 + y^2$$
$$= 3^2 + (-4)^2$$
$$= 9 + 16 = 25 = \text{RHS}$$

> Start with one side of the equation (LHS in this case) and show that it equals the other side.

$\therefore$ $(3, -4)$ does lie on the curve $x^2 + y^2 = 25$.

6.8 Equation of a straight line

A straight line can be defined by either one point on the line and its gradient or two points on the line.

To find the equation of a line given one point on the line and its gradient

Consider the equation of a line passing through A(3, 4) with gradient 2.

Let P(x, y) be any point on the line.

Gradient of AP $= \dfrac{y - 4}{x - 3}$.

But gradient $= 2$

So $\dfrac{y - 4}{x - 3} = 2$

$y - 4 = 2(x - 3)$

$y - 4 = 2x - 6$

$y = 2x - 2$

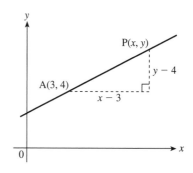

Now consider the equation of a line passing through A(x_1, y_1) with gradient m.

Let P(x, y) be any point on the line.

Gradient of AP $= \dfrac{y - y_1}{x - x_1}$

But gradient $= m$

So $\dfrac{y - y_1}{x - x_1} = m$

$\therefore$ $y - y_1 = m(x - x_1)$

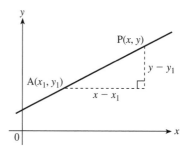

➤ | The equation of the line through (x_1, y_1) with gradient m is $y - y_1 = m(x - x_1)$.

To find the equation of a line given two points on the line

Consider the equation of the line passing through A(−2, 7) and B(6, 3).
Let P(x, y) be any point on the line.

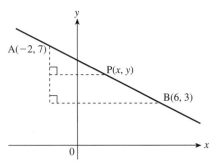

Gradient of AB $= \dfrac{3-7}{6-(-2)} = \dfrac{-4}{8} = -\dfrac{1}{2}$

Gradient of AP $= \dfrac{y-7}{x-(-2)} = \dfrac{y-7}{x+2}$

But, because A, B and P lie on the same line, the gradients of AB and AP are equal, so

$$\frac{y-7}{x+2} = -\frac{1}{2}$$

$$2y - 14 = -x - 2$$

$$2y + x = 12$$

Now consider the equation of the line through A(x_1, y_1) and B(x_2, y_2).
Let P(x, y) be any point on the line.

Gradient of AB $= \dfrac{y_2 - y_1}{x_2 - x_1}$

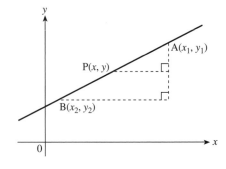

Gradient of AP $= \dfrac{y - y_1}{x - x_1}$

But the gradients are equal, so

$$\frac{y - y_1}{x - x_1} = \frac{y_2 - y_1}{x_2 - x_1}$$

Rearranging

$$\frac{y - y_1}{y_2 - y_1} = \frac{x - x_1}{x_2 - x_1}$$

> **The equation of the line through (x_1, y_1) and (x_2, y_2) is** $\dfrac{y - y_1}{y_2 - y_1} = \dfrac{x - x_1}{x_2 - x_1}$.

Use of this formula can be avoided. Given two points on a line, the equation of the line can be found by

- finding the gradient, $m = \dfrac{y_2 - y_1}{x_2 - x_1}$ and

- using either $y - y_1 = m(x - x_1)$, or $y = mx + c$ and substituting the coordinates of a point to find c.

Example 7 *Several methods are illustrated in this example.*

Find the equation of the line passing through $(3, -4)$ and $(-1, -2)$.

Method 1

Using $y - y_1 = m(x - x_1)$

$\begin{array}{cccc} x_1 & y_1 & x_2 & y_2 \\ (3, & -4) & (-1, & -2) \end{array}$

Gradient $= \dfrac{y_2 - y_1}{x_2 - x_1}$

> Alternatively use $y = mx + c$ so $y = -\frac{1}{2}x + c$. Substitute $x = 3$, $y = -4$ to find c.

$\qquad = \dfrac{-2 - (-4)}{-1 - 3} = \dfrac{2}{-4} = -\dfrac{1}{2}$

So equation is

$\qquad y - (-4) = -\frac{1}{2}(x - 3)$

$\qquad 2y + 8 = 3 - x$

> Multiply both sides by 2 to eliminate the fractions, taking care with signs.

So the equation of the line is $2y + x + 5 = 0$.

Method 2

Using $\dfrac{y - y_1}{y_2 - y_1} = \dfrac{x - x_1}{x_2 - x_1}$

Equation is

$\dfrac{y - (-4)}{-2 - (-4)} = \dfrac{x - 3}{-1 - 3}$

$\qquad \dfrac{y + 4}{2} = \dfrac{x - 3}{-4}$

> Multiply both sides by 4.

$\qquad 2y + 8 = 3 - x$

So the equation of the line is $2y + x + 5 = 0$.

Note If the form of the equation is specified in the question be sure to give the answer in the correct form. In any case all like terms must be confirmed.

Forms of equations of straight lines

An equation gives a straight line if, and only if, it can be expressed in the form $ax + by + c = 0$.

The only possible terms are x terms, y terms and a constant. There must be no squares, reciprocals, etc.

$ax + by + c = 0$ is the general form of the equation of a straight line.

One frequently used form of the equation of a straight line is $y = mx + c$.

Consider the line through $(0, c)$ with gradient m.

Using $y - y_1 = m(x - x_1)$

$y - c = mx$

$y = mx + c$

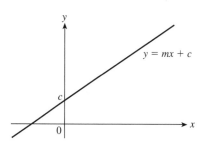

> The equation of a line with gradient m and intercept c on the y-axis is $y = mx + c$.

By arranging the equation of any straight line in the form $y = mx + c$, the gradient and intercept on the y-axis can be found directly.

Example 8 Find the gradient and intercept on the y-axis of the line with equation $3y - 2x + 7 = 0$.

$$3y - 2x + 7 = 0$$

Arrange the equation with y alone on the LHS.

$$3y = 2x - 7$$

Compare with $y = mx + c$

$(\div 3) \qquad y = \frac{2}{3}x - \frac{7}{3}$

Gradient   Intercept on y-axis

So, the line has gradient $\frac{2}{3}$ and intercept on the y-axis $-\frac{7}{3}$.

Example 9 Find the equation of the line with gradient $\frac{1}{7}$ and intercept on the y-axis -5.

$m = \frac{1}{7}$ and $c = -5$, so the equation is

$$y = \frac{1}{7}x - 5$$

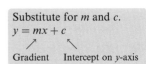

Substitute for m and c.
$y = mx + c$
Gradient   Intercept on y-axis

or

$(\times 7) \quad 7y = x - 35$

Forms of equations of straight lines	
$y = mx + c$	Line gradient m; intercept on y-axis, c
$y = mx$	Line gradient m, passes through the origin
$y = x + c$	Line gradient 1, makes an angle of $45°$ with the x-axis; intercept on y-axis, c
$y = k$	Line parallel to the x-axis through $(0, k)$
$y = 0$	x-axis
$x = k$	Line parallel to the y-axis through $(k, 0)$
$x = 0$	y-axis
$ax + by + c = 0$	General form of the equation of a straight line

Sketching straight lines

A sketch of a straight line should show where the line cuts the axes.

■ To find the intercept on the x-axis, put $y = 0$.

■ To find the intercept on the y-axis, put $x = 0$.

Example 10   Sketch $y = 2x - 8$.

When $x = 0$, $y = -8$, so the line intercepts the y-axis at $(0, -8)$.

When $y = 0$, $x = 4$, so the line intercepts the x-axis at $(4, 0)$.

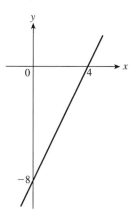

Exercise 6B

1   Find the gradients of the lines joining these pairs of points.
   a   (4, 3) and (8, 12)   **b**   (−2, −3) and (4, 6)   **c**   (5, 6) and (10, 2)
   d   (−3, 4) and (8, −6)   **e**   (−5, 3) and (−2, 3)   **f**   (p, q) and (r, s)
   g   (0, a) and (a, 0)   **h**   (0, 0) and (a, b)
   i $\left(-2\frac{1}{2}, -\frac{1}{2}\right)$ and $\left(4\frac{1}{2}, -1\right)$ **j**   (−7, −3) and (−1, −4)

2   Write down the gradients of the lines perpendicular to lines of gradient
   a 3   **b** $\frac{1}{4}$ **c** −6   **d** $-\frac{2}{3}$ **e** $2m$ **f** $-\frac{b}{a}$ **g** $-\frac{m}{2}$

3   Find if AB is parallel or perpendicular (or neither) to PQ in these cases.
   a   A(1, 4), B(6, 6), P(2, −1), Q(12, 3)
   b   A(−1, −1), B(0, 4), P(−4, 3), Q(6, 1)
   c   A(0, 3), B(7, 2), P(6, −1), Q(−1, −2)
   d   A(4, 3), B(8, 4), P(7, 1), Q(6, 5)
   e   A(3, 1), B(7, 3), P(−3, 2), Q(1, 0)
   f   A(−6, −1), B(−6, 3), P(2, 0), Q(2, −5)
   g   A(−6, −1), B(−6, 3), P(3, 1), Q(6, 1)

4   Determine whether these points lie on the given curves.
   a $y = 6x + 7$, (1, 13)   **b** $y = 2x + 2$, (13, 30)   **c** $3x + 4y = 1$, $\left(-1, \frac{1}{2}\right)$
   d $y = x^3 - 6$, (2, −2)   **e** $xy = 36$, (−9, −4)   **f** $x^2 + y^2 = 25$, (3, 4)

5   Find the coordinates of the points on the curve $y = 2x^2 - x - 1$ for which
   a $x = 2$ **b** $x = -3$ **c** $x = 0$

6 Find the x-coordinates of the points on the line $y = 2x + 3$ for which the y-coordinates are

   a 7   **b** 3   **c** -2

7 Find the points at which these lines cut the x-axis and the y-axis.

   a $y = 3x - 6$ **b** $2y + 5x = 8$ **c** $\dfrac{x}{3} - \dfrac{y}{4} = 1$ **d** $y - 2x = 0$

8 Sketch these lines on the same axes, labelling each line.

   a $y = 2x$ **b** $y = 3$ **c** $y = -x$ **d** $x + 2 = 0$ **e** $y = \frac{1}{2}x$

9 Sketch these straight lines on separate diagrams, showing the intercepts with the axes.

   a $3y = 2x + 6$ **b** $x - 4y + 2 = 0$ **c** $3x + y + 6 = 0$ **d** $7x = 3y + 5$

10 Write down the equations of the straight lines through the origin having gradients

   a $\frac{1}{3}$ **b** -2 **c** m

11 Rearrange these equations in the form $y = mx$ and hence write down the gradients of the lines they represent.

   a $4y = x$ **b** $5x + 4y = 0$ **c** $3x = 2y$ **d** $\dfrac{x}{4} = \dfrac{y}{7}$ **e** $\dfrac{x}{p} - \dfrac{y}{q} = 0$

12 Find the equations of the straight lines with

   a gradient 3 and intercept on the y-axis 2

   b gradient 3 and passing through $(0, -1)$

   c gradient $\frac{1}{5}$ and intercept on the y-axis 2

   d gradient $\frac{1}{5}$ and passing through $(0, 4)$.

13 Arrange these equations in the form $y = mx + c$. Hence write down the gradient and the intercept on the y-axis for each line.

   a $3y = 2x + 6$ **b** $x - 4y + 2 = 0$ **c** $3x + y + 6 = 0$

   d $7x = 3y + 5$ **e** $y + 4 = 0$ **f** $lx + my + n = 0$

14 Find the equations of the straight lines of given gradients, passing through the given points.

   a $4, (1, 3)$ **b** $3, (-2, 5)$ **c** $\frac{1}{3}, (2, -5)$

   d $-\frac{3}{4}, (7, 5)$ **e** $\frac{1}{2}, \left(\frac{1}{3}, -\frac{1}{2}\right)$ **f** $a, (3, 2a)$

15 Find the equations of the straight lines joining these pairs of points.

   a $(1, 6)$ and $(5, 9)$ **b** $(3, 2)$ and $(7, -3)$ **c** $(-3, 4)$ and $(8, 1)$

   d $(-1, -4)$ and $(4, -3)$ **e** $\left(\frac{1}{2}, 2\right)$ and $\left(3, \frac{1}{3}\right)$ **f** $(k, 3h)$ and $(3k, h)$

16 Find the equation of the straight line through

   a (5, 4), parallel to $3x - 4y + 7 = 0$

   b (−2, 3), parallel to $5x - 2y - 1 = 0$

   c (4, 0), perpendicular to $x + 7y + 4 = 0$

   d (−2, −3), perpendicular to $4x + 3y - 5 = 0$.

17 Calculate the area of the triangle formed by the line $3x - 7y + 4 = 0$ and the axes.

18 Find a if the line joining (3, a) to (−6, 5) has gradient $\frac{1}{3}$.

19 Given that $ax + 3y + 2 = 0$ and $2x - by - 5 = 0$ are perpendicular lines, find the ratio $a{:}b$.

20 Show that A(−3, 1), B(1, 2), C(0, −1), D(−4, −2) are the vertices of a parallelogram.

21 Show that P(1, 7), Q(7, 5), R(6, 2), S(0, 4) are the vertices of a rectangle. Calculate the lengths of the diagonals, and find their point of intersection.

22 Show that D(−2, 0), E$\left(\frac{1}{2}, 1\frac{1}{2}\right)$ and F$\left(3\frac{1}{2}, -3\frac{1}{2}\right)$ are the vertices of a right-angled triangle, and find the length of the shortest side, and the mid-point of the hypotenuse.

6.9 Applications

The intersection of two graphs consists of the points common to both graphs. To find these points, the equations of the graphs have to be solved simultaneously (see Chapter 5).

Example 11 Find the point of intersection of the two lines $y = 1 - x$ and $y = 2x - 8$.

$$y = 1 - x \quad \text{①}$$
$$y = 2x - 8 \quad \text{②}$$

> Solve the equations of the two lines simultaneously.

The two lines intersect where

$$1 - x = 2x - 8$$

> The two expressions for y must be equal.

$$3x = 9$$

$$\therefore \qquad x = 3$$

Substituting in ① $y = 1 - 3 = -2$

> Check $x = 3$, $y = -2$ in ②.

The values $x = 3$, $y = -2$ satisfy both equations, so (3, −2) lies on both lines.

So the point of intersection is (3, −2).

Example 12   A and B are the points (2, 5) and (8, 4) respectively. Find the equation of the median through the origin, O, of the triangle OAB.

The required median is OM where M is the mid-point of AB.

> The medians of a triangle are the lines joining the vertices to the mid-points of the opposite sides.

The mid-point, M, is at $\left(\dfrac{2+8}{2}, \dfrac{5+4}{2}\right) = \left(5, 4\tfrac{1}{2}\right)$.

OM passes through (0, 0) and $\left(5, 4\tfrac{1}{2}\right)$.

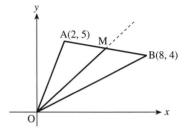

So the gradient of OM $= \dfrac{4\tfrac{1}{2} - 0}{5 - 0} = \dfrac{9}{10}$

> $\begin{array}{cc} x_1 \ y_1 & x_2 \ y_2 \\ (0, \ 0) & \left(5, \ 4\tfrac{1}{2}\right) \end{array}$ using gradient $= \dfrac{y_2 - y_1}{x_2 - x_1}$

∴   Equation of OM is $y = \tfrac{9}{10}x$.

> Using $y - y_1 = m(x - x_1)$.

Example 13   Find the equation of the perpendicular bisector of the line joining A(−2, 5) to B(−8, −3).

The perpendicular bisector of AB

■ passes through the mid-point of AB

■ is perpendicular to AB, so the product of the gradients of AB and the bisector is −1.

Mid-point of AB $= \left(\dfrac{-2 + (-8)}{2}, \dfrac{5 + (-3)}{2}\right) = (-5, 1)$

> $\begin{array}{cc} x_1 \ y_1 & x_2 \ y_2 \\ A(-2, \ 5) & B(-8, \ -3) \end{array}$
> Using $\left(\dfrac{x_1 + x_2}{2}, \dfrac{y_1 + y_2}{2}\right)$

Gradient of AB $= \dfrac{-3 - 5}{-8 - (-2)} = \dfrac{-8}{-6} = \dfrac{4}{3}$

> Using $\dfrac{y_2 - y_1}{x_2 - x_1}$

So, the gradient of a line perpendicular to AB is $-\tfrac{3}{4}$.

> $m_1 m_2 = -1$

So, the perpendicular bisector passes through (−5, 1) with gradient $-\tfrac{3}{4}$.

The equation is $y - 1 = -\tfrac{3}{4}(x - (-5))$

> Using $y - y_1 = m(x - x_1)$

$4y - 4 = -3x - 15$

$4y + 3x + 11 = 0$

Example 14   The points A, B, C and D are (7, 10), (5, 9), (2, 3) and (8, 6) respectively. Prove that ABCD is a trapezium with AB parallel to CD. Show that CD = 3AB.

Gradient of CD $= \dfrac{6-3}{8-2} = \dfrac{3}{6} = \dfrac{1}{2}$

Gradient of AB $= \dfrac{10-9}{7-5} = \dfrac{1}{2}$

Gradient AB = Gradient CD, so AB ∥ CD.

$$CD = \sqrt{(8-2)^2 + (6-3)^2}$$
$$= \sqrt{36+9}$$
$$= \sqrt{45}$$
$$= 3\sqrt{5}$$
$$AB = \sqrt{(7-5)^2 + (10-9)^2}$$
$$= \sqrt{2^2 + 1^2}$$
$$= \sqrt{5}$$

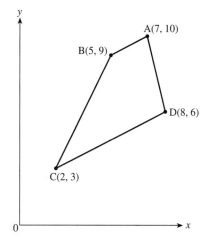

Note: This question can be solved using vector geometry (covered later in the A-level course)

∴ CD = 3AB. So ABCD is a trapezium with AB parallel to CD and CD = 3AB.

Exercise 6C

1 Find the points of intersection of these pairs of straight lines.

  a $x + y = 0$
      $y = -7$

  b $y = 5x + 2$
      $y = 3x - 1$

  c $3x + 2y - 1 = 0$
      $4x + 5y + 3 = 0$

  d $5x + 7y + 29 = 0$
      $11x - 3y - 65 = 0$

2 Find the equation of the straight line joining the origin to the mid-point of the line joining A(3, 2) and B(5, −1).

3 For the line segments joining these pairs of points, find the mid-point, the gradient, the perpendicular gradient, and the equation of the perpendicular bisector.

  a   (3, 5) and (7, 7)
  b   (−2, 4) and (−4, 8)
  c   (−1, 3) and (−2, −5)
  d   (p, q) and (7p, 3q)

4 Given the points A(−4, 8) and B(0, −2), find the equation of the perpendicular bisector of AB.

5 Find the equation of the perpendicular bisector of PQ where P is (7, 3) and Q is (−6, 1).

6 **a** Find the equation of the straight line *l* through P(7, 5) perpendicular to the straight line AB whose equation is $3x + 4y - 16 = 0$.

  b Find the point of intersection of the line AB with the line *l* in part **a**.

  c Hence find the perpendicular distance from P to AB.

7   Find the perpendicular distance from (2, 4) to the line $y = 2x + 10$.

8   **a**   Find the equation of the straight line joining the points A(7, 0) and B(0, 2).

    b   Find the equation of the line AC such that the x-axis bisects angle BAC.

9   A line is drawn through the point (2, 3) making an angle of 45° with the positive direction of the x-axis, and it meets the line $x = 6$ at P. Find the distance of P from the origin O, and the equation of the line through P perpendicular to OP.

10   Prove that the points $(-5, 4)$, $(-1, -2)$ and $(5, 2)$ lie at three of the corners of a square. Find the coordinates of the fourth corner, and the area of the square.

11   The vertices of a quadrilateral ABCD are A(4, 0), B(14, 11), C(0, 6) and D(-10, -5). Prove that the diagonals AC, BD bisect each other at right angles, and that the length of BD is four times that of AC.

12   The coordinates of the vertices A, B, C of the triangle ABC are $(-3, 7)$, $(2, 19)$, $(10, 7)$ respectively.

    a   Prove that the triangle is isosceles.

    b   Calculate the length of the perpendicular from B to AC, and use it to find the area of the triangle.

13   A triangle ABC has A at the point (7, 9), B at (3, 5), C at (5, 1).
Find the equation of the line joining the mid-points of AB and AC.
Find also the area of the triangle enclosed by this line and the axes.

14   One side of a rhombus is the line $y = 2x$, and two opposite vertices are the points (0, 0) and $\left(4\frac{1}{2}, 4\frac{1}{2}\right)$. Find the equations of the diagonals, the coordinates of the other two vertices and the length of the side.

15   Here are two possible methods of proving that three points A, B and C are collinear (i.e. lie on the same line).

    a   Prove that $AB + BC = AC$ (assuming B lies between A and C).
Use this method to show that P, Q and R are collinear where P is (2, 6), Q is (4, 9) and R is $(-6, -6)$.

    b   Prove that AB and BC are parallel. Then since B is a point in common, A, B and C must be collinear. Use this method to show that L, M and N are collinear where L is (0, 4), M is (3, 13) and N is $(-1, 1)$.

16   Find all possible values of a and b given that $y = ax + 14$ is the perpendicular bisector of the line joining (1, 2) to $(b, 6)$.

17   Find a and b given that $y + 4x = 11$ is the perpendicular bisector of the line joining $(a, 2)$ to $(6, b)$.

Exercise 6D (Review)

1 Find these equations.
   a The line joining the points (2, 4) and (−3, 1)
   b The line through (3, 1) parallel to the line $3x + 5y = 6$
   c The line through (3, −4) perpendicular to the line $5x − 2y = 3$

2 a Find the equation of the line joining A(−3, 2) and B(6, 8).
   b The line AB meets the line $3x + 2y − 21 = 0$ at M.
      Find the coordinates of M and show that M divides AB in the ratio 2:1.

3 a Find the equation of the line passing through the point A(4, −2) and perpendicular to the line l whose equation is $2x − y − 5 = 0$.
   b Find the coordinates of the foot of the perpendicular from A to the line l.
   c Hence find the perpendicular distance from A to l.

4 The coordinates of P, Q and R are (5, 9), (14, −3) and (2, 3) respectively.
   a Show that triangle PQR is a right-angled triangle.
   b Find angle PQR correct to 3 significant figures.

5 The area of any quadrilateral with perpendicular diagonals of lengths d_1 and d_2 is given by $\frac{1}{2}d_1d_2$.
   Show that the area of the quadrilateral ABCD where the points A, B, C and D have coordinates (−1, 4), (3, 7), (5, 2) and (−5, −17) respectively is 80 units^2.

6 The points D, E and F have coordinates (5, −2), (2, 9) and (9, 2) respectively.
   a Find the equation of l, the perpendicular bisector of EF.
   b Find the coordinates of the point where l meets DE.

7 The points A and B have coordinates (h, k) and $(3h, −5k)$, respectively.
   a Find the coordinates of the mid-point of AB.
   b Find the gradient of AB.
   c Hence find the equation of the perpendicular bisector of AB.

8 Given that P and Q are the points $(s, 2s)$ and $(3s, 8s)$ respectively, find the equation of the perpendicular bisector of PQ.

9 The line l has equation $4y + 3x = 12$.
   a Find the coordinates of the intercepts of l on the axes.
   b Hence or otherwise find the equation of the reflection of l in the x-axis, the y-axis, and the line $y = x$.
   c Find the area of the triangle enclosed by the line l and the axes.

10 Find the coordinates of the points at which these curves cut the y-axis and the x-axis.
    a $y = x^2 − x − 12$ b $y = 6x^2 − 7x + 2$ c $y = x^2 − 6x + 9$
    d $y = x^3 − 9x^2$ e $y = (x + 1)(x − 5)^2$ f $y = (x^2 − 1)(x^2 − 9)$

11 The parabola $y = x^2 + 4$ meets the straight line $y = 5x + 10$ at the points A and B.
    a Find the coordinates of A and B.
    b Find the length AB.

12 The parabola $y = x^2 + 2x - 7$ meets the line $y = 17 - 3x$ at the points P and Q. Find the length of PQ.

13 Given that a triangle has vertices (2, 3), (4, 9) and (5, 2)

   a find the perimeter of the triangle, leaving surds in the answer

   b show that the triangle is right-angled.

Extension material on coordinate geometry can be found on the CD-ROM (E6.10). There is also a 'Test yourself' exercise (Ty6) and an 'Extension exercise' (Ext6).

➤ Key points

Coordinate geometry

■ For the points $A(x_1, y_1)$ and $B(x_2, y_2)$

$$AB = \sqrt{(x_2 - x_1)^2 + (y_2 - y_1)^2}$$

Mid-point of AB is $\left(\dfrac{x_1 + x_2}{2}, \dfrac{y_1 + y_2}{2} \right)$

Gradient of $AB = \dfrac{y_2 - y_1}{x_2 - x_1}$

Equation of line through A with gradient m: $y - y_1 = m(x - x_1)$

Equation of AB: $\dfrac{y - y_1}{y_2 - y_1} = \dfrac{x - x_1}{x_2 - x_1}$

■ The equation of a line with gradient m and intercept c on the y-axis is $y = mx + c$.

■ Forms of equations of straight lines

$y = mx + c$	Line gradient m; intercept on y-axis, c
$y = mx$	Line gradient m, passes through the origin
$y = x + c$	Line gradient 1, makes an angle of 45° with the x-axis; intercept on y-axis, c
$y = k$	Line parallel to the x-axis through $(0, k)$
$y = 0$	x-axis
$x = k$	Line parallel to the y-axis through $(k, 0)$
$x = 0$	y-axis
$ax + by + c = 0$	General form of the equation of a straight line

■ For lines with gradients m_1, m_2

   $m_1 = m_2 \Leftrightarrow$ the lines are parallel

  $m_1 m_2 = -1 \Leftrightarrow$ the lines are perpendicular, $m_1 \neq 0$, $m_2 \neq 0$

■ The Cartesian equation of a curve is a relationship connecting the x and y values of points on the curve.

Graphs of functions

Drawing graphs of equations gives them a physical meaning, and, conversely, curves, paths (of planets, for example), and designs can be expressed as equations. This chapter introduces some graphs and ways of transforming them.

After working through this chapter you should be able to

■ *sketch the graphs of simple functions, including cubics and reciprocals. (For quadratics see Chapter 3.)*

■ *sketch curves defined by simple equations*

■ *understand the term 'asymptote'*

■ *apply to the graph of $y = f(x)$, the transformations $y = af(x)$, $y = f(x) + a$, $y = f(x + a)$ and $y = f(ax)$*

■ *give a geometrical interpretation of the algebraic solution of equations*

■ *use the intersection points of graphs of functions to solve equations.*

7.1 Graphs of standard functions

The x-axis is the line $y = 0$ and the y-axis is the line $x = 0$.

Lines parallel to the x-axis have equation $y = k$, where k is constant.

Lines parallel to the y-axis have equation $x = k$, where k is constant.

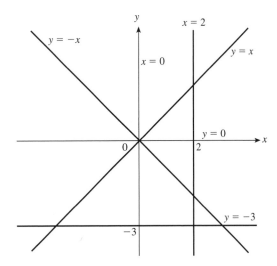

Three useful standard curves:

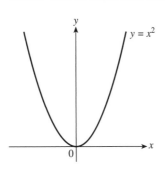 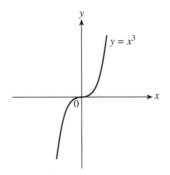 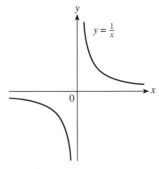

$y = x^2$ is a
quadratic function.

$y = x^3$ is a
cubic function.

$y = \frac{1}{x}$ is a
reciprocal function.

Other interesting curves include:

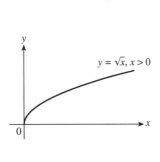 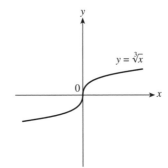 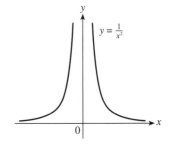

Continuous and discontinuous curves; asymptotes

Curves without a break, such as $y = x^2$ and $y = x^3$ are called **continuous**.

The graphs of $y = \frac{1}{x}$ and $y = \frac{1}{x^2}$ have breaks in the curves at $x = 0$. Curves such as these are called **discontinuous**. Discontinuous curves cannot be drawn without taking the pencil off the paper.

When $y = \frac{1}{x}$ and $y = \frac{1}{x^2}$, the graphs shoot off to plus or minus infinity when $x = 0$. The line $x = 0$ (the y-axis) is an **asymptote** to each curve. For very large positive and negative values of x, the curve approaches the x-axis but *never* reaches it. The x-axis is also an asymptote to each curve.

Note   As a curve approaches an asymptote it becomes closer and closer to the asymptote but never reaches it. An asymptote which is not the x- or y-axis is shown with a dashed line. (See Example 8 on page 119.)

A graphical calculator is a useful tool when studying curve sketching. Asymptotes are not usually shown on graphical calculators. On certain calculators what appear to be asymptotes are drawn. Be wary! The calculator may be incorrectly joining up two branches of the graph.

Asymptotes are discussed more fully on the CD-ROM (see E7.1).

7.2 Transformation of graphs

In this section, curves are translated (moved), reflected in an axis and stretched.

When studying these transformations, try drawing the graphs on a graphical calculator or with a graph-plotting program on a computer.

Example 1 Consider $y = x^2$ and $y = x^2 + 3$.

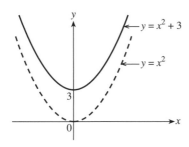

For $y = x^2 + 3$ all points of $y = x^2$ have been moved up 3 units.

The graph of $y = x^2$ has been translated by $\binom{0}{3}$.

➤ $y = \mathbf{f}(x) + a$ **is a translation** $\binom{0}{a}$ **of** $y = \mathbf{f}(x)$.

> **Add** 3 to the function ⇒ move **up** 3.
> **Subtract** 3 from the function ⇒ move **down** 3.

Example 2 Consider $y = x^2$ and $y = (x - 3)^2$.

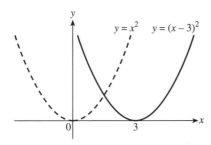

> Check by putting in a value for x in $y = (x - 3)^2$. For example $x = 3$ gives $y = 0$.

For $y = (x - 3)^2$ all points of $y = x^2$ have been moved 3 units to the right.

The graph of $y = x^2$ has been translated by $\binom{3}{0}$.

➤ $y = \mathbf{f}(x - a)$ **is a translation** $\binom{a}{0}$ **of** $y = \mathbf{f}(x)$.

> Replace x by $x - 3$ ⇒ move **right** 3.
> Replace x by $x + 3$ ⇒ move **left** 3.

**Example 3
Extension**

This example combines the two previous results. Combinations of transformations will be covered later in the A-level course. A general translation, as illustrated by this example, is useful for transforming circles (Chapter 13).

Consider $y = x^2$ and $y = (x + 1)^2 - 4$.

$(x + 1)^2 - 4$ is a quadratic expression in completed square form.

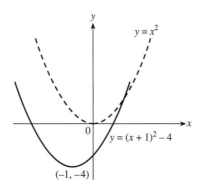

Since x has been replaced by $x + 1$, the graph of $y = x^2$ moves 1 unit to the left.

Since 4 has been subtracted, the graph then moves down 4.

So $y = x^2$ is translated by $\binom{-1}{-4}$ to give $y = (x + 1)^2 - 4$.

There will be a minimum point at $(-1, -4)$.

> $y = \mathrm{f}(x - a) + b$ **is a translation** $\binom{a}{b}$ **of** $y = \mathrm{f}(x)$.

Note $y = \mathrm{f}(x - a) + b$ can be rewritten as $(y - b) = \mathrm{f}(x - a)$. Replacing x by $(x - a)$ and y by $(y - b)$ translates the graph by $\binom{a}{b}$.

Example 4   Consider $y = x^2$ and $y = -x^2$.

For $y = -x^2$ all points of $y = x^2$ have been reflected in the x-axis.

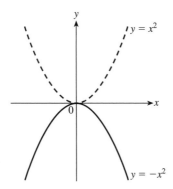

> Check by putting a value for x in $y = x^2$.
> For example $x = 3$ gives $y = -9$.

> $y = -\mathrm{f}(x)$ **is a reflection in the** x**-axis of** $y = \mathrm{f}(x)$.

Example 5   Consider $y = x^3 + 1$ and $y = (-x)^3 + 1 = -x^3 + 1$.

For $y = -x^3 + 1$ all points of $y = x^3 + 1$ have been reflected in the y-axis.
In the equation, x has been replaced by $-x$.

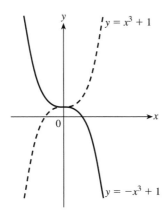

➤ | $y = \mathbf{f}(-x)$ is a reflection in the y-axis of $y = \mathbf{f}(x)$. |

Although the sine curve is not studied until later (in Chapter 16, section 16.3) it is used for the next two examples because it provides a better illustration of stretches than other functions. If you have not met a sine curve before, try drawing it on a computer or graphical calculator.

Example 6   Consider $y = \sin x$ and $y = 2\sin x$.

For $y = 2\sin x$ all points of $y = \sin x$ have had their y-coordinate multiplied by 2.
This transformation is a stretch in the y direction of scale factor 2.

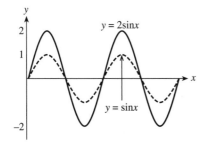

➤ | $y = a\mathbf{f}(x)$ is a stretch in the y direction of scale factor a. |

$a > 1 \Rightarrow$ curve is stretched away from x-axis.
$0 < a < 1 \Rightarrow$ curve is compressed towards x-axis.

Example 7 Consider $y = \sin x$ and $y = \sin 2x$.

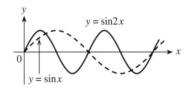

For $y = \sin 2x$ all the points of $y = \sin x$ have had their x-coordinate multiplied by $\frac{1}{2}$. This transformation is a stretch in the x direction of scale factor $\frac{1}{2}$.

Similarly, for $y = \sin\frac{1}{2}x$ all the points of $y = \sin x$ have had their x-coordinate multiplied by 2.

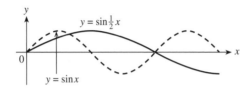

This transformation is a stretch in the x direction of scale factor 2.

➤ | $y = \mathrm{f}(ax)$ **is a stretch in the** x **direction of scale factor** $\frac{1}{a}$. |

$a > 1 \Rightarrow$ curve is compressed towards y-axis.
$0 < a < 1 \Rightarrow$ curve is stretched away from y-axis.

Exercise 7A

1 Sketch $y = x^2$ and use transformations of graphs to sketch, on the same axes
   a $y = x^2 + 3$ **b** $y = (x - 2)^2$

2 Sketch $y = x^2$ and use transformations of graphs to sketch, on the same axes
   a $y = -x^2$ **b** $y = 4x^2$ **c** $y = \frac{1}{2}x^2$ **d** $y = -\frac{3}{2}x^2$

   Check the answers by substituting suitable values of x.

3 Sketch $y = x^3$ and, on the same axes
   a $y = x^3 - 3$ **b** $y = (x + 2)^3$

4 Sketch $y = x^3$ and, on the same axes
   a $y = 2x^3$ **b** $y = 4 + x^3$

5 Sketch $y = x^3$ and, on the same axes
   a $y = -x^3$ **b** $y = \frac{1}{4}x^3$

6 Describe the transformations which change the graph of $y = \mathrm{f}(x)$ into
   a $y = \mathrm{f}(x) + 6$ **b** $y = \mathrm{f}(x) - 5$ **c** $y = \frac{1}{2}\mathrm{f}(x)$ **d** $y = \mathrm{f}(4x)$

7 Describe the transformations which change the graph of $y = \mathrm{f}(x)$ into
   a $y = \mathrm{f}(x + 3)$ **b** $y = -\mathrm{f}(x)$ **c** $y = -5\mathrm{f}(x)$ **d** $y = \mathrm{f}(-x)$

7.3 Sketching curves 1

Sketching a curve means showing its general shape and marking certain important features. Only if an accurate **plot** of a graph is required should a large number of points be tabulated.

A graphical calculator should be used with care; it can mislead. There is no substitute for knowing functions and their graphs and practising curve sketching techniques.

When a graph is to be sketched, consider these points:

- Is the graph a *standard function* or a *transformation* of one?
- Where does the curve *cross the axes*?
  Putting $x = 0$ will give the intercepts on the y-axis.
  Putting $y = 0$ will give the intercepts on the x-axis.

A second section on curve sketching, in Chapter 15, extends this work further.

Recall (See page 64)

> The graph of any equation of the form $y = ax^2 + bx + c$ $(a \neq 0)$ is a parabola.
> If $a > 0$ the parabola is $\cup$-shaped.
> If $a < 0$ the parabola is $\cap$-shaped.
> To sketch a parabola, either complete the square to find the coordinates of the vertex and the equation of the axis of symmetry, or, if the points of intersection with the x-axis are required, solve $ax^2 + bx + c = 0$.
> Put $x = 0$ to find the intercept on the y-axis.

Example 8 Sketch $y = f(x)$ given that $f(x) = \dfrac{1}{x} + 4$ for $x \neq 0$.

State the equations of any asymptotes.

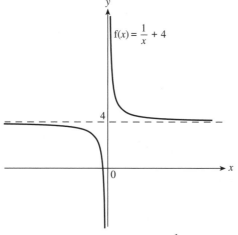

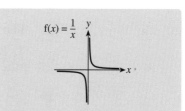

The graph of $\frac{1}{x}$ is translated 4 up.

The graph of $y = \frac{1}{x}$ has asymptotes $x = 0$ and $y = 0$. After the translation the asymptotes will be at $x = 0$ and $y = 4$. First draw the new asymptote $y = 4$ with a dashed line and then add the translated curve.

$f(x)$ is a translation $\begin{pmatrix} 0 \\ 4 \end{pmatrix}$ of $\dfrac{1}{x}$. The lines $x = 0$ (y-axis) and $y = 4$ are asymptotes.

Example 9   The diagrams show sketches of $y = f(x)$. The graph cuts the x-axis at A(1, 0) and has a maximum point at B(2, 3). The y-axis and the line $y = 1$ are asymptotes.

a   On the first diagram, sketch $y = -f(x)$.

b   On the second diagram sketch $y = 2f(x)$

Mark the new positions, A′ and B′, of the points A and B after the transformations, stating their coordinates. State the equations of the asymptotes of the transformed curve.

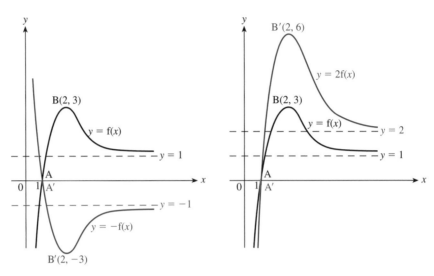

a $y = -f(x)$
The graph of $y = f(x)$ is reflected in the x-axis.

The point A(1, 0), which is on the line of reflection, will not be changed by the transformation.
So A′ has coordinates (1, 0).

> Any point on the line of reflection is unchanged by the reflection.

B is reflected in the x-axis so B′ has coordinates (2, −3).

The y-axis will still be an asymptote after the reflection. The asymptote $y = 1$ of the original graph will be reflected to $y = -1$.

> In a reflection in the x-axis $(a, b) \rightarrow (a, -b)$. Similarly for a reflection in the y-axis $(a, b) \rightarrow (-a, b)$

b $y = 2f(x)$
The graph of $y = f(x)$ is stretched in the y-direction, scale factor 2.
The point A(1, 0), which is on the x-axis, will not be changed by the transformation.
So A′ has coordinates (1, 0).

> Any point on the x-axis is unchanged by a stretch in the y direction. Similarly any point on the y-axis is unchanged by a stretch in the x direction.

B is stretched in the y direction, scale factor 2, so B′ has coordinates (2, 6).

The y-axis will still be an asymptote after the stretch. The asymptote $y = 1$ of the original graph will be moved to $y = 2$.

Example 10 $f(x) = x^2 + 2$

  a   Sketch $y = f(x)$ and $y = f(x + 3)$ on the same axes, marking the coordinates of the vertex of each parabola.

  b   Find an expression for the function $f(x + 3)$ in the form $ax^2 + bx + c$.

  c   Find the point of intersection of $y = f(x)$ and $y = f(x + 3)$

Solution   **a**

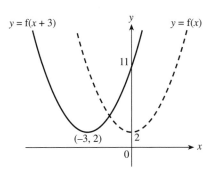

> To obtain the graph of $y = x^2 + 2$,
> $y = x^2$ is translated $\begin{pmatrix} 0 \\ 2 \end{pmatrix}$, vertex at $(0, 2)$.

> The vertex is at $(-3, 2)$.

> $f(x + 3)$ is $f(x)$ translated by $\begin{pmatrix} -3 \\ 0 \end{pmatrix}$, i.e. 3 units to the left.

  b $f(x + 3) = (x + 3)^2 + 2$
  $$= x^2 + 6x + 11$$

> To obtain the function $f(x + 3)$, substitute $(x + 3)$ for x in $f(x) = x^2 + 2$.

  c   The two curves intersect where $f(x) = f(x + 3)$.

  So $x^2 + 2 = x^2 + 6x + 11$

  $$6x = -9$$

  $$x = -\tfrac{3}{2}$$

  When $x = -\tfrac{3}{2}$, $y = \tfrac{9}{4} + 2 = \tfrac{17}{4}$, so the curves intersect at $\left(-\tfrac{3}{2}, \tfrac{17}{4}\right)$.

> *Note:* When $x = 0$, $f(x + 3) = 11$, $\therefore$ curve cuts y-axis at $(0, 11)$.

> *A graphical calculator can provide a useful double check when learning curve sketching techniques. However, resist the temptation to have a 'quick look first'. It is important to do the sketch before checking on a graphical calculator or computer. It is not always possible to use a calculator; see Questions 4–7 of Exercise 7B. You may not use a graphical calculator in the C1 exam.*

Exercise 7B   **1**   Sketch $y = \dfrac{1}{x}$ and, on the same axes

  a $y = \dfrac{1}{x} + 2$, $x \neq 0$ **b** $y = \dfrac{1}{x - 1}$, $x \neq 1$

  State the equations of any asymptotes.

2 Sketch $y = \dfrac{1}{x}$, $x \neq 0$ and, on the same axes

    a $y = \dfrac{2}{x}$ **b** $y = -\dfrac{3}{x}$

    State the equations of any asymptotes.

3 Sketch $y = \dfrac{1}{x}$, $x \neq 0$ and, on the same axes

    a $y = \dfrac{1}{2x}$ **b** $y = \dfrac{1}{3x}$

4 Copy this sketch of $y = g(x)$ and, on the same axes, sketch $y = -g(x)$ and $y = g(-x)$.

    Mark the images of points A and B under the transformations, stating their coordinates.

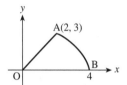

5 Copy this sketch of $y = f(x)$ and, on the same axes, sketch $y = -f(x)$, $y = f(x - 3)$, $y = f(-x)$, $y = f(2x)$.

    Mark the images of points O, A and B under the transformations, stating their coordinates.

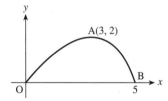

6 The graph shows a sketch of $y = g(x)$.

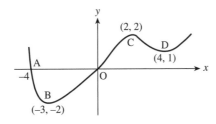

a Copy the sketch and, on the same axes, sketch $y = 2g(x)$ and $y = -g(x)$.

b Mark the images of points A, B, C, D and O under the transformations, stating their coordinates.

c Describe each transformation geometrically, i.e. in terms of translations, reflections and stretches.

7 a Copy the original sketch of $g(x)$ in Question 6 and, on the same axes, sketch $y = g(x - 1)$ and $y = g(x) + 3$.

b Mark the images of points A, B, C, D and O under the transformations, stating their coordinates.

c Describe each transformation geometrically, i.e. in terms of translations, reflections and stretches.

8 Express $x^2 + 2x + 8$ in the form $(x + a)^2 + b$, by completing the square or otherwise.

Hence sketch the parabola $y = x^2 + 2x + 8$, marking the coordinates of the vertex and the intercept on the y-axis.

9 Express these in completed square form and hence sketch each parabola, marking the coordinates of the vertex and the intercept on the y-axis.

a $y = x^2 - 2x + 4$ **b** $y = x^2 + 4x - 5$ **c** $y = x^2 - 6x + 3$

d $y = x^2 + x + 1$ **e** $y = 7 - 2x - x^2$

10 Given $f(x) = x^2$ sketch these graphs, each on a separate diagram, and find an expression, giving y in terms of x, for each function.

a $y = f(x) - 1$ **b** $y = f(x + 1)$ **c** $y = f(2x)$ **d** $y = -3f(x)$

11 Given $g(x) = x^3$ sketch these graphs, each on a separate diagram, and find expressions, giving y in terms of x, for each function.

a $y = 4g(x)$ **b** $y = g(-x)$ **c** $y = g(3x)$ **d** $y = g(x - 2)$

7.4 Geometrical interpretation of the solution of equations

Solving a pair of equations

In Chapter 5, on solving simultaneous equations, Examples 10, 11 and 12 on pages 79, 80 and 81 gave a geometrical interpretation of the algebraic solution of equations.

> ➤ When solving any pair of equations in x and y, the solution gives the point(s) of intersection, if they exist, of the graphs of the two equations. A repeated root in the solution indicates that the curves described by the two equations touch.
>
> Conversely, the points of intersection of graphs give approximate solutions to equations.

Example 11 Find the coordinates of the points of intersection of the parabola $y = 1 - (x + 4)^2$ with

a $y = -8$ **b** $y = 2x + 10$ **c** $y = x^2 + 3x - 10$

a $y = 1 - (x + 4)^2$ meets $y = -8$ where

$$1 - (x + 4)^2 = -8$$
$$(x + 4)^2 = 1 + 8$$
$$= 9$$
$$\therefore \quad x + 4 = \pm 3$$

$x = -7$ or -1

> Where the line and parabola meet, the two expressions for y must be equal.

> Take the square root of both sides. Remember the root can be +ve or −ve.

At the points of intersection $y = -8$.
So the coordinates of the points of intersection are $(-7, -8)$ and $(-1, -8)$.

> The quadratic equation could be solved by factorising or by using the formula, but the form in which it is given makes 'completing the square' the most efficient method.

b $y = 1 - (x + 4)^2$ meets $y = 2x + 10$ where

$$1 - (x + 4)^2 = 2x + 10$$
$$1 - (x^2 + 8x + 16) = 2x + 10$$
$$x^2 + 10x + 25 = 0$$
$$(x + 5)^2 = 0$$

$x = -5$ is a repeated root, so the line $y = 2x + 10$ is a tangent to the parabola.
When $x = -5$, $y = 0$, so $y = 2x + 10$ touches $y = 1 - (x + 4)^2$ at $(-5, 0)$.

> The discriminant $b^2 - 4ac = 10^2 - 4 \times 1 \times 25 = 0$
> So the equation $x^2 + 10x + 25 = 0$ has equal roots and hence the line is a tangent to the curve.

c $y = 1 - (x + 4)^2$ meets $y = x^2 + 3x - 10$ where
$$x^2 + 3x - 10 = 1 - (x + 4)^2$$
$$= 1 - (x^2 + 8x + 16)$$
$$= 1 - x^2 - 8x - 16$$
$$2x^2 + 11x + 5 = 0$$
$$(2x + 1)(x + 5) = 0$$

$\therefore x = -5$ or $-\frac{1}{2}$

When $x = -5$, $y = 1 - (-5 + 4)^2 = 0$

When $x = -\frac{1}{2}$,
$$y = 1 - (-\tfrac{1}{2} + 4)^2$$
$$= 1 - \frac{49}{4}$$
$$= -\frac{45}{4}$$

So $y = 1 - (x + 4)^2$ and $y = x^2 + 3x - 10$
intersect at $(-5, 0)$ and $(-\frac{1}{2}, -\frac{45}{4})$.

Solving an equation in one unknown

The equation $x^3 - 3x^2 = 2$ has one real solution. It is not easy to solve algebraically but an appropriate solution can be found using a graphical method. Plotting a graph will give a rough approximation to the solution; a graphical calculator or the numerical methods introduced later in the A-level course, will give a better one.

To solve the equation $x^3 - 3x^2 = 2$ by plotting graphs there are several options. For example,

i   Plot $y = x^3 - 3x^2$ and $y = 2$. Where the graphs intersect the values of y are equal, so $x^3 - 3x^2 = 2$. The root of the equation is the value of x where the graphs intersect.

or

ii Rewrite the equation as $x^3 - 3x^2 - 2 = 0$ and plot $y = x^3 - 3x^2 - 2$.
The root of the equation is the value of x where the graph cuts the x-axis ($y = 0$).

or

iii Rewrite the equation as $x^3 = 3x^2 + 2$ and plot $y = x^3$ and $y = 3x^2 + 2$.
The root of the equation is the value of x where the graphs intersect.

Example 12

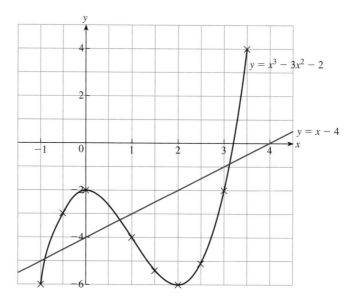

a The diagram shows the graph of $y = x^3 - 3x^2 - 2$ for $-1 \leqslant x \leqslant 3.5$. Use the graph to find an approximation to the root of $x^3 - 3x^2 - 2 = 0$, giving the answer correct to one decimal place.

b By drawing another graph on the same axes, find an approximate solution of the equation $x^3 - 3x^2 - 2 = x - 4$.

c State the equation of the graph which would need to be drawn to solve $x^3 - 3x^2 - 2x + 1 = 0$.

Solution **a** The root of the equation $x^3 - 3x^2 - 2 = 0$ is the value of x where the graph of $y = x^3 - 3x^2 - 2$ cuts the x-axis. Reading from the graph, an estimate of the root is 3.2, to one decimal place.

b The graph of $y = x - 4$ is plotted on the same axes.
The graphs intersect where $x = -0.9$, 0.7 and 3.1. So an approximate solution of the equation $x^3 - 3x^2 - 2 = x - 4$ is $x = -0.9$, 0.7 and 3.1.

c The equation, $x^3 - 3x^2 - 2x + 1 = 0$, has to be arranged so the LHS matches the graph, $y = x^3 - 3x^2 - 2$, which has already been drawn.

> Adding $2x - 3$ to both sides of $x^3 - 3x^2 - 2x + 1 = 0$ will make the LHS equal to $x^3 - 3x^2 - 2$ as required.

Rearranging gives $x^3 - 3x^2 - 2 = 2x - 3$.
If $y = 2x - 3$ is drawn on the same axes, the solution to the equation, $x^3 - 3x^2 - 2x + 1 = 0$, will be given by the value(s) of x where the graphs intersect.

Exercise 7C

1 Find the points of intersection, if any, of these lines and curves.

   a $y = x^2 + 2x - 12$
     $y = -4$

   b $y = 1 + x - 2x^2$
     $y = 9 - 7x$

   c $y = 3x^2 + 2x + 4$
     $y = 5$

   d $y = x^2 + 20x + 12$
     $y = 10x - 12$

   e $y = -2 - 4x - x^2$
     $y = -8 - 3x$

   f $y = x^2 + 6x + 3$
     $y = 7x + 2$

2 Find the point(s) of intersection of the following pairs of parabolas.

   a $y = x^2 + 20x + 100$
     $y = x^2 - 20x + 100$

   b $y = x^2 - 10x + 13$
     $y = x^2 - 4x + 7$

   c $y = x^2 - 7x + 8$
     $y = -x^2 + 9x - 6$

   d $y = x^2 - 2x + 4$
     $y = 2x^2 - 4x + 4$

   e $y = 2 - 2x - x^2$
     $y = 3x^2 + 10x + 11$

   f $y = (x + 1)^2 + 6$
     $y = 6 - 2(x + 1)^2$

3 Explain why these pairs of parabolas do not intersect, illustrating each answer with a sketch.

   a $y = x^2 + 2x + 4$
     $y = x^2 + 2x + 5$

   b $y = x^2 - 4x + 8$
     $y = -14 + 8x - x^2$

4 By solving the simultaneous equations, find the points of intersection of $y = x^2 + 2x + 2$ with

   a $y = 5$

   b $y = 7 - 2x$

   c $y = x^3 + 2$

5 The curve $y = x^2 + 4x + 2$ meets the line $y = 8 - x$ at two points. Find the distance between the two points.

6 Find the possible values of a for which $y = ax + 1$ is a tangent to $y = 3x^2 - 4x + 4$.

7 Find the possible values of k for which $y = x - 8$ is a tangent to $y = 4x^2 + kx + 1$.

8 Find the x coordinates of the points of intersection of this pair of parabolas.
   $y = 4x^2 - 12x - 3$
   $y = x^2 + 11x + 5$

9 Find the coordinates of the points of intersection of this pair of parabolas.
   $y = 4x^2 - 2x - 1$
   $y = -2x^2 + 3x + 5$

10 a Draw axes so that x can take values between -2 and 2 and y between -8 and 8. Complete the table.

x	-2	-1.5	-1	-0.5	0	0.5	1	1.5	2
$4 - x^2$			3						
x^3									8

   Draw the graphs of $y = 4 - x^2$ and $y = x^3$. Plot as many points as are needed to obtain a smooth curve, tabulating more points if necessary.

   b Use the graphs to find an approximate solution to the equation $4 - x^2 = x^3$.

11 a Draw axes so that x can take values between 0 and 2 and y between 0 and 10. Draw the graph of $y = x^3 + \frac{1}{x}$ for $0 < x \leqslant 2$.

b By drawing an appropriate straight line on the same axes, estimate the roots of the equation $x^3 + \frac{1}{x} = 5$.

c By drawing an appropriate straight line on the same axes, estimate the roots of the equation $x^3 + \frac{1}{x} - 2x - 1 = 0$.

12 a Draw axes so that x can take values between -3 and 3 and y between -10 and 10. Draw the graph of $y = x^3 - 6x + 1$ for $-3 \leqslant x \leqslant 3$.

b Use the graph to find the approximate roots of $x^3 - 6x + 1 = 0$.

c By drawing an appropriate straight line on the same axes, estimate the root of the equation $x^3 - 6x = 6$.

d State the line that should be drawn to find estimates for the roots of $x^3 - 8x + 2 = 0$.

13 By solving the equation $x^2 = x^3$, find the points of intersection of $y = x^2$ and $y = x^3$. Sketch, on the same axes, $y = x^2$ and $y = x^3$ showing where the curves intersect.

14 Sketch $y = x^2$ and $y = x^4$ showing where the curves intersect.

15 Sketch $y = x^3$ and $y = x^5$ showing clearly their points of intersection.

1 Find the points of intersection of

a $y = x^2 + 3x - 7$
  $y = 3$

b $y = x - 3x^2$
  $y = 2x - 2$

c $y = 5x^2 - 2x + 1$
  $y = 6 - 3x - x^2$

d $y = x^2 + 20x$
  $y = x^3$

2 Sketch $y = x^2$ and, on the same axes

a $y = x^2 - 2$

b $y = -x^2$

c $y = 4x^2$

3 Sketch $y = x^3$ and, on the same axes

a $y = (x + 2)^3$

b $y = 3 + x^3$

c $y = 10x^3$

4 Find the value of c for which $y = x + c$ is a tangent to $y = 3 - x - 5x^2$.

5 a Draw axes so that x can take values between -3 and 3 and y between -4 and 7. Draw the graph of $y = \frac{1}{10}(2x^3 + 3x^2 - 12)$ for $-3 \leqslant x \leqslant 3$.

b Use the graph to find an approximate root of $2x^3 + 3x^2 - 12 = -20$.

c By drawing an appropriate straight line on the same axes, estimate the root of the equation $2x^3 + 3x^2 = 42$.

d Show that the x coordinates of the points of intersection of $y = 6 - 3x$ and $y = \frac{1}{10}(2x^3 + 3x^2 - 12)$ would be the solutions of the equation $2x^3 + 3x^2 + 30x - 72 = 0$.

e Draw, on the same axes, the line $y = 6 - 3x$, and find, from the graph, an approximate solution of $2x^3 + 3x^2 + 30x - 72 = 0$.

A 'Test yourself' exercise can be found on the CD-ROM (Ty7).

Key points

Transformation of graphs

All these transformations apply to $y = f(x)$. Assume $a > 0$.

- $y = f(x) + a$ is a translation $\begin{pmatrix} 0 \\ a \end{pmatrix}$.
  Adding a to the function moves the graph a units up.
- $y = f(x) - a$ is a translation $\begin{pmatrix} 0 \\ -a \end{pmatrix}$.
  Subtracting a from the function moves the graph a units down.
- $y = f(x - a)$ is a translation $\begin{pmatrix} a \\ 0 \end{pmatrix}$.
  Subtracting a from x (i.e. replacing x by $x - a$) moves the graph a units to the right.
- $y = f(x + a)$ is a translation $\begin{pmatrix} -a \\ 0 \end{pmatrix}$.
  Adding a to x (i.e. replacing x by $x + a$) moves the graph a units to the left.
- $y = -f(x)$ is a reflection in the x-axis.
- $y = f(-x)$ is a reflection in the y-axis.
- $y = af(x)$ is a stretch parallel to the y-axis with scale factor a.
- $y = f(ax)$ is a stretch parallel to the x-axis with scale factor $\frac{1}{a}$.

Sketching curves

Many curves can be sketched from knowledge of standard curves (see page 114) and transformations.

The graph of any equation of the form $y = ax^2 + bx + c$ $(a \neq 0)$ is a parabola.

If $a > 0$ the parabola is ∪-shaped, if $a < 0$ the parabola is ∩-shaped.

To sketch a parabola, either complete the square to find the coordinates of the vertex and the equation of the axis of symmetry or, if the points of intersection with the x-axis are required, solve $ax^2 + bx + c = 0$.

Put $x = 0$ to find the intercept on the y-axis.

Geometrical interpretation of the algebraic solution of equations

When solving any pair of equations in x and y, the solution gives the point(s) of intersection, if they exist, of the graphs of the two equations. A repeated root in the solution indicates that the curves described by the two equations touch.

Asymptotes

As a curve approaches an asymptote it becomes closer and closer to the asymptote but never reaches it.

8 Sequences and series; arithmetic series

Since the advent of computers, sequences have become an exciting field of mathematics, producing, for example, fractals, such as those on the cover of this book. More practical examples of sequences include theoretical models of population growth or decay or the annual increase of a sum of money invested at compound interest.

After working through this chapter you should

■ *know both what a sequence is and what a series is*

■ *be able to generate a sequence from a formula or from a relation of the form* $x_{n+1} = f(x_n)$

■ *know what an arithmetic series is*

■ *be able to find the nth term and the sum to n terms of an arithmetic series*

■ *understand Σ notation*

■ *be able to prove and use the formulae $S_n = \frac{n}{2}(a + l)$ and $S_n = \frac{n}{2}(2a + (n-1)d)$*

■ *know that the sum of the first n natural numbers is $\frac{n}{2}(n+1)$.*

8.1 Sequences

A **sequence** is a set of terms, in a definite order, where the terms are obtained by some rule.

A **finite sequence** ends after a certain number of terms.

An **infinite sequence** is one that continues indefinitely.

Consider the infinite sequence 1, 3, 5, 7, ..., the sequence of odd numbers:

1st term $= 2 \times 1 - 1 = 1$

2nd term $= 2 \times 2 - 1 = 3$

3rd term $= 2 \times 3 - 1 = 5$

⋮   ⋮

nth term $= 2 \times n - 1 = 2n - 1$

> Notice there is a constant difference of 2 between the terms, leading to $2n$ in the formula.

Notation

The terms of a sequence can be expressed as $u_1, u_2, u_3, \ldots, u_n, \ldots$ where $u_1 =$ first term, $u_2 =$ second term and so on.

The terms can also be expressed as $u_0, u_1, u_2, \ldots, u_n, \ldots$ where $u_0 =$ first term and so on. In this book u_1 refers to the first term.

Finding the formula for the terms of a sequence

The formula for the terms in a sequence can be given either as a formula for the nth term or as a recurrence relation (or both). A **recurrence relation** defines the first term(s) in the sequence and the relation between successive terms.

For example, the sequence 5, 8, 11, 14, ... can be defined as $u_n = 3n + 2$ and/or as $u_1 = 5$, $u_{n+1} = u_n + 3$, $n \geqslant 1$.

> 1st term: $u_1 = 5$
> 2nd term: $u_2 = u_1 + 3 = 5 + 3 = 8$, etc.

When looking for the rule defining a sequence, these points may help:

■ Look at the difference between consecutive terms. Is it constant?

■ Compare the sequence with sequences such as the squares (1, 4, 9, 16, ...; nth term $= n^2$) and the cubes (1, 8, 27, 64, ...; nth term $= n^3$).

■ Look for powers of numbers, e.g. for the sequence 1, 2, 4, 8, ... the nth term is 2^{n-1}.

■ Do the signs of the terms alternate, e.g. $-1, +2, -3, +4, \ldots$? If so, use the fact that
$(-1)^k = -1$ when k is odd, and
$(-1)^k = +1$ when k is even.

If the first term is $-$ve, include $(-1)^n$ in the expression for the nth term. If the first term is $+$ve, include $(-1)^{n+1}$ or $(-1)^{n-1}$ in the expression for the nth term.

This technique is illustrated in Example 3.

Example 1 Find the next three terms in the sequence 5, 8, 11, 14, ... and the formula for the nth term.

Look at the difference between the terms.

The constant difference of 3 will lead to a formula containing $3n$.

The sequence with nth term $3n$ has terms 3, 6, 9, 12, ...

The terms of the sequence 5, 8, 11, 14, ... are each 2 more, so its nth term is $3n + 2$.

> This sequence could also be defined by the **recurrence relation**:
> $u_{n+1} = u_n + 3$, $n \geqslant 1$, $u_1 = 5$
> The first term is $5\,(u_1 = 5)$ and each term is 3 more than the previous one $(u_{n+1} = u_n + 3)$.

Check: *1st* term $= 3 \times 1 + 2 = 5$
             2nd term $= 3 \times 2 + 2 = 8$

Note Checking several terms does not prove that the formula is correct, but does give some reassurance.

Example 2   The nth term of a sequence is given by $x_n = \dfrac{1}{2^n}$.

    a   Find the first four terms of the sequence.

    b   Which term in the sequence is $\dfrac{1}{1024}$?

    c   Express the sequence as a recurrence relation.

Solution   **a** Putting $n = 1$ $x_1 = \dfrac{1}{2^1} = \dfrac{1}{2}$

                   $n = 2$ $x_2 = \dfrac{1}{2^2} = \dfrac{1}{4}$

                   $n = 3$ $x_3 = \dfrac{1}{2^3} = \dfrac{1}{8}$

                   $n = 4$ $x_4 = \dfrac{1}{2^4} = \dfrac{1}{16}$

      The first four terms are $\frac{1}{2}, \frac{1}{4}, \frac{1}{8}$ and $\frac{1}{16}$.

    b Let the rth term be $\dfrac{1}{1024}$.

      So $\dfrac{1}{2^r} = \dfrac{1}{1024}$

      $\therefore$ $2^r = 1024$

      But $2^{10} = 1024$

      $\therefore$ $r = 10$

      So $\dfrac{1}{1024}$ is the 10th term.

> To find which power of 2 is 1024, either
> - express 1024 as the product of prime factors ($1024 = 2^{10}$) or
> - use the calculator to find which power of 2 is 1024, or
> - use logs: $r \log_{10} 2 = \log_{10} 1024$
>
>   (Chapter 18) $r = \dfrac{\log_{10} 1024}{\log_{10} 2}$

    c $x_{n+1} = \frac{1}{2} x_n$ $n \geqslant 1$ $x_1 = \frac{1}{2}$ Each term is half the previous one. The first term is $\frac{1}{2}$.

Example 3   Find the nth term of the sequence $+1, -4, +9, -16, +25, \ldots$.

The terms of the sequence are the square numbers $1, 4, 9, 16, 25, \ldots$ but with alternating signs.

For the sequence $+1, -4, +9, -16, +25, \ldots$.

$u_n = (-1)^{n+1} n^2$

> Check:
> 1st term $= (-1)^{1+1} \times 1^2$
> $= (-1)^2 \times 1 = 1$
> 2nd term $= (-1)^{2+1} \times 2^2$
> $= (-1)^3 \times 4 = -4$ ✓

Example 4 A sequence is defined by a recurrence relation of the form $M_{n+1} = aM_n + b$

Given that $M_1 = 10$, $M_2 = 20$, $M_3 = 24$, find the value of a and the value of b and hence find M_4.

$M_1 = 10$, $M_2 = 20$, $M_3 = 24$ State the data.

Therefore $20 = 10a + b$ ① Substituting M_1 and M_2.

$24 = 20a + b$ ② Substituting M_2 and M_3.

Subtracting ① from ②

$$4 = 10a$$
$$a = \frac{4}{10}$$ Solve ① and ② to find a and b.
$$= \frac{2}{5}$$

Substituting in ①

$$20 = 10 \times \frac{2}{5} + b$$
$$= 4 + b$$
$$b = 16$$

Therefore $M_{n+1} = \frac{2}{5}M_n + 16$ Substitute for a and b in the recurrence relation.

$$M_4 = \frac{2}{5}M_3 + 16$$
$$= \frac{2}{5} \times 24 + 16$$ Substitute for M_3 to find M_4.
$$= 25.6$$

To find out more about the behaviour of sequences look on the **CD-ROM** (E8.1).

Exercise 8A *In this exercise, use of a spreadsheet or graphical calculator is recommended for Question 3.*

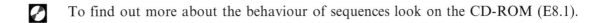

1 For each of these sequences, find the first four terms of the sequence, and which term of the sequence equals X.

   a $u_n = 3n$ $X = 303$ **b** $u_n = 3^n$ $X = 729$

   c $u_n = 4n - 3$ $X = 245$ **d** $u_n = 8 - 2n$ $X = -32$

   e $u_n = n^4$ $X = 14\,641$ **f** $u_n = n^2 - 1$ $X = 840$

2 For each of these sequences, find the next three terms, and a formula for the nth term.

   a $11, 16, 21, 26, \ldots$ **b** $1, \frac{1}{4}, \frac{1}{16}, \frac{1}{64}, \ldots$ **c** $20, 17, 14, 11, \ldots$

   d $2, 4, 6, 8, \ldots$ **e** $2, 4, 8, 16, \ldots$ **f** $\frac{2}{1}, \frac{3}{2}, \frac{4}{3}, \frac{5}{4}, \ldots$

   g $1, 4, 9, 16, \ldots$ **h** $1, 8, 27, 64, \ldots$ **i** $2, 16, 54, 128, \ldots$

3 For each of these sequences, find the first six terms.

  a $u_1 = 3 \quad u_{n+1} = u_n + 4 \quad n \geqslant 1$ **b** $u_1 = 64 \quad u_{n+1} = \dfrac{u_n}{2} \quad n \geqslant 1$

  c $u_1 = 1 \quad u_{n+1} = u_n + 2n + 1 \quad n \geqslant 1$ **d** $u_1 = 6 \quad u_{n+1} = \dfrac{u_n + 1}{2} \quad n \geqslant 1$

4 A sequence is defined by a recurrence relation of the form $x_{n+1} = 0.2(c - 3x_n)$, $n \geqslant 1$, where c is a constant.
  Given that $x_1 = 8$ and $x_2 = -3.2$, find the value of c and hence find x_3.

5 A recurrence relation is of the form $C_{n+1} = kC_n + 12$, where k is a constant.
  Given that $C_1 = 16$, $C_2 = 24$, find k and hence find C_3 and C_4.

6 A sequence is defined by the recurrence relation $p_{n+1} = 1.04p_n - q$, $p_0 = 1000$, where q is a constant

  a Given that $q = 30$, calculate p_1, p_2 and p_3.

  b Given instead that $q = 400$, calculate p_1, p_2 and p_3.

  c Find the value of q that would give the value of p_n unchanged.

8.2 Series and Σ notation

When the terms of a sequence are added, the sum of the terms is called a **series**. A finite series is one which ends after a finite number of terms. An infinite series continues indefinitely.

S_n is used to denote the sum of the first n terms. S_∞ is used to denote the sum of an infinite series, if it exists.

The greek capital letter, Σ (pronounced 'sigma'), is used to show that the terms of a sequence are to be added.

Last value of r in the sequence

$$\sum_{r=3}^{7} u_r = u_3 + u_4 + u_5 + u_6 + u_7$$

General term

First value of r in the sequence

$$\sum_{r=2}^{50} 3r^2 = 3 \times 2^2 + 3 \times 3^2 + 3 \times 4^2 + \cdots + 3 \times 50^2$$

$\cdots$ indicates more terms

The values of r go up in steps of 1 from the first to the last.

When there is no possibility of confusion, the notation can be simplified.

$$\sum_{1}^{10} r^3 = 1^3 + 2^3 + 3^3 + \cdots + 10^3$$

The notation can also be used for an infinite series.

$$\sum_{1}^{\infty} \frac{1}{r} = \tfrac{1}{1} + \tfrac{1}{2} + \tfrac{1}{3} + \cdots$$

Example 5 Write the terms of the series.

a $\displaystyle\sum_{2}^{5} m^2 = 2^2 + 3^2 + 4^2 + 5^2$

b $\displaystyle\sum_{1}^{k} m(m-1) = 1 \times 0 + 2 \times 1 + 3 \times 2 + 4 \times 3 + \cdots + k(k-1)$

Example 6 Write, using $\sum$ notation, $1 \times 4 + 2 \times 5 + 3 \times 6 + \cdots + 10 \times 13$.

Notice that there are two sequences within the series, labelled • and ▪, and that there are ten terms.

$$\overset{\bullet}{1} \times \overset{\blacksquare}{4} + \overset{\bullet}{2} \times \overset{\blacksquare}{5} + \overset{\bullet}{3} \times \overset{\blacksquare}{6} + \cdots + \overset{\bullet}{10} \times \overset{\blacksquare}{13}$$

• $1, 2, 3, \ldots 10, \Rightarrow u_r = r$

▪ $4, 5, 6, \ldots, 13 \Rightarrow u_r = r + 3$

So $1 \times 4 + 2 \times 5 + 3 \times 6 + \cdots + 10 \times 13 = \displaystyle\sum_{1}^{10} r(r+3)$

Here are some useful results. (The proof of these is left to the reader.)

➤

Summation results

$$\sum_{h}^{k} (a_r + b_r) = \sum_{h}^{k} a_r + \sum_{h}^{k} b_r$$

$$\sum_{h}^{j} a_r + \sum_{j+1}^{k} a_r = \sum_{h}^{k} a_r \quad h < j < k$$

$$\sum_{1}^{n} c = nc \quad \text{where } c \text{ is a constant}$$

$$\sum_{m}^{n} k a_r = k \sum_{m}^{n} a_r \quad \text{where } k \text{ is a constant}$$

Exercise 8B

1 Write in full

a $\displaystyle\sum_{1}^{4} m^3$

b $\displaystyle\sum_{2}^{n} m^2$

c $\displaystyle\sum_{1}^{n} (m^2 + m)$

d $\displaystyle\sum_{1}^{3} \frac{1}{m(m+1)}$

e $\displaystyle\sum_{2}^{5} 2^m$

f $\displaystyle\sum_{1}^{4} (-1)^m m^2$

g $\displaystyle\sum_{1}^{n} m^m$

h $\displaystyle\sum_{3}^{6} \frac{(-1)^m}{m}$

i $\displaystyle\sum_{n}^{n+2} m(m-1)$

j $\displaystyle\sum_{n-2}^{n} \frac{m}{m+1}$

2 Write in the Σ notation

a $1 + 2 + 3 + \cdots + n$

b $1^4 + 2^4 + \cdots + n^4 + (n+1)^4$

c $1 + \frac{1}{2} + \frac{1}{3} + \frac{1}{4} + \frac{1}{5}$

d $3^2 + 3^3 + 3^4 + 3^5$

Exercise 8B (continued)

e $1 + \frac{2}{3} + \frac{3}{9} + \frac{4}{27} + \frac{5}{81}$

f $\frac{1 \times 3}{4} + \frac{2 \times 5}{6} + \frac{3 \times 7}{8} + \frac{4 \times 9}{10} + \frac{5 \times 11}{12}$

g $-1 + 2 - 3 + 4 - 5 + 6$

h $2 \times 7 + 3 \times 8 + 4 \times 9 + 5 \times 10 + 6 \times 11$

i $1 - 2 + 4 - 8 + 16 - 32$

j $1 \times 3 - 2 \times 5 + 3 \times 7 - 4 \times 9 + 5 \times 11$

3 Write in full and evaluate

a $\sum\limits_{1}^{6} n$

b $\sum\limits_{1}^{5} (2r + 3)$

c $\sum\limits_{1}^{4} (-1)^m m^3$

d $\sum\limits_{1}^{5} \frac{n}{n + 1}$

e $\sum\limits_{1}^{6} \left(-\frac{2}{p}\right)$

f $\sum\limits_{5}^{9} n(n - 3)$

g $\sum\limits_{4}^{7} (-2)^r$

h $\sum\limits_{2}^{7} \frac{(-1)^r}{3(r - 1)}$

i $\sum\limits_{1}^{5} (5 + (-2)^r)$

j $\sum\limits_{n=1}^{2k} (-1)^n$

k $\sum\limits_{1}^{6} 4$

l $\sum\limits_{4}^{30} (2m - 1) - \sum\limits_{8}^{30} (2m - 1)$

4 Write in full and simplify

a $\sum\limits_{n=1}^{5} k^2 n$

b $\sum\limits_{k=1}^{5} k^2 n$

c $\sum\limits_{r=1}^{6} \frac{ar(r + 1)}{2}$

d $\sum\limits_{n=1}^{2k} (-1)^{n+1} n$

e $\sum\limits_{m=1}^{4} \frac{a}{m}$

f $\sum\limits_{r=1}^{4} \frac{a}{m}$

8.3 Arithmetic series

Consider the series $2 + 6 + 10 + 14 + \cdots$ and $2 + 6 + 18 + 54 + \cdots$

In the first case, each term has 4 *added* to it to obtain the next term.
The series is said to have a **common difference** of 4.
Such a series is called an **arithmetic series** or **progression** (often abbreviated to **AP**).

The terms of an arithmetic sequence or arithmetic series are said to be in *arithmetic progression*.

In the second series, each term is *multiplied* by 3 to obtain the next term.
The series is said to have a **common ratio** of 3.
Such a series is called a **geometric series** or **progression** (often abbreviated to **GP**).

Geometric series will be discussed in Chapter 20.

> An arithmetic series or arithmetic progression (AP)
> is a series whose consecutive terms have a common difference.

If the common difference is d, each term is obtained from the previous term by adding d. Expressed formally: $u_r = u_{r-1} + d$ or $u_r - u_{r-1} = d$.

Problems involving arithmetic series can be solved with or without using formulae. It is often helpful to consider intuitive methods first.

Examples 7 to 9 are solved without using the formulae for arithmetic series.

Example 7   Find the tenth term of the arithmetic series: $2 + 6 + 10 + 14 + \cdots$

The common difference is the difference between consecutive terms.
In this case, the common difference is $6 - 2 = 4$.

For the tenth term, nine lots of the common difference will have been added to the first term, 2.

So the tenth term is $2 + 9 \times 4 = 38$.

Example 8   Find the number of terms in the arithmetic series: $50 + 47 + 44 + 41 + \cdots - 34$

The common difference is $47 - 50 = -3$.

For the last term, -34, 84 will have been subtracted from the first term, 50
$(50 - (-34) = 84)$.

Since 3 is subtracted for each new term and 84 is subtracted in all, there must be $84/3 = 28$ terms after the first.

So there are 29 terms in all.

Example 9   Find the sum of the arithmetic series: $2 + 5 + 8 + 11 + \cdots + 32$.

Let $S = 2 + 5 + 8 + 11 + \cdots + 29 + 32$ ①

> The common difference is 3. $30 \, (= 32 - 2)$ is added to the first term to obtain the last term.

Writing the terms in reverse order

$S = 32 + 29 + 26 + 23 + \cdots + 5 + 2$ ②

Adding ① and ②

$2S = (2 + 32) + (5 + 29) + (8 + 26) + \cdots + (32 + 2)$

$ = 34 + 34 + 34 + \cdots + 34$

> 30 (the difference between the last and first terms) is 10 lots of the common difference, 3. So there are $10 + 1 = 11$ terms in all.

$ = 11 \times 34$

$S = \dfrac{11 \times 34}{2} = 187$

> When Carl Friedrich Gauss (1777–1855) was ten years old, he was told by his teacher to add up all the numbers from 1 to 100. The teacher, so the story goes, assumed the task would keep Gauss occupied for some time. In fact he gave the answer immediately, using, presumably, the method in Example 9.

➤

$$\text{Sum of an arithmetic series} = \frac{\text{Number of terms} \times \text{Sum of first and last terms}}{2}$$

Exercise 8C

1 Which of these series are arithmetic series?
Write down the common differences of those that are arithmetic series.

a $7 + 8\frac{1}{2} + 10 + 11\frac{1}{2}$

b $-2 - 5 - 8 - 11$

c $1 + 1.1 + 1.2 + 1.3$

d $1 + 1.1 + 1.11 + 1.111$

e $\frac{1}{2} + \frac{5}{6} + \frac{7}{6} + \frac{3}{2}$

f $1^2 + 2^2 + 3^2 + 4^2$

g $n + 2n + 3n + 4n$

h $1 + \frac{1}{2} + \frac{1}{3} + \frac{1}{4}$

i $1\frac{1}{8} + 2\frac{1}{4} + 3\frac{3}{8} + 4\frac{1}{2}$

j $19 + 12 + 5 - 2 - 9$

k $1 - 2 + 3 - 4 + 5$

l $1 + 0.8 + 0.6 + 0.4$

2 Write down the terms indicated in each of these arithmetic series.

a $3 + 11 + \cdots$; 10th, 19th

b $8 + 5 + \cdots$; 15th, 31st

c $\frac{1}{4} + \frac{7}{8} + \cdots$; 12th, nth

d $50 + 48 + \cdots$; 100th, nth

e $7 + 6\frac{1}{2} + \cdots$; 42nd, nth

f $3 + 7 + \cdots$; 200th, $(n + 1)$th

3 Find the number of terms in these arithmetic series.

a $2 + 4 + 6 + \cdots + 46$

b $50 + 47 + 44 + \cdots + 14$

c $2.7 + 3.2 + \cdots + 17.7$

d $6\frac{1}{4} + 7\frac{1}{2} + \cdots + 31\frac{1}{4}$

e $407 + 401 + \cdots - 133$

f $2 - 9 - \cdots - 130$

4 Find the sums of these arithmetic series.

a $1 + 3 + 5 + \cdots + 101$

b $-10 - 7 - 4 - \cdots + 50$

c $2.01 + 2.02 + 2.03 + \cdots + 3.00$

d $x + 3x + 5x + \cdots + 21x$

e $a + (a + d) + \cdots + \{a + (n - 1)d\}$

5 Find the sums of these arithmetic series as far as the terms indicated.

a $4 + 10 + \cdots + 12$th term

b $15 + 13 + \cdots + 20$th term

c $1 + 2 + \cdots + 200$th term

d $20 + 13 + \cdots + 16$th term

e $6 + 10 + \cdots + n$th term

f $1\frac{1}{4} + 1 + \cdots + n$th term

Formulae for arithmetic series

Let the first term be a, last term l, common difference d and sum of the first n terms S_n. Look at the first four terms:

First term $= a$

Second term $= a + d$

Third term $= a + 2d$

Fourth term $= a + 3d$

Formula for the nth term

➤
> nth term $= a + (n-1)d$
>
> If there are n terms in all, the last term, l is given by $l = a + (n-1)d$

Note   For an arithmetic series, $u_{n+1} = u_n + d$

Formula for the sum of an arithmetic series

$S_n = a + (a+d) + (a+2d) + (a+3d) + \cdots + (l-2d) + (l-d) + l$ ①

Writing the terms in reverse order

$S_n = l + (l-d) + (l-2d) + (l-3d) + \cdots + (a+2d) + (a+d) + a$ ②

Adding ① and ②

$2S_n = (a+l) + (a+l) + (a+l) + (a+l) + \cdots + (a+l) + (a+l) + (a+l)$

$ = n(a+l)$

$S_n = \dfrac{n}{2}(a+l)$ ③

If the last term is not known but the number of terms, n, is known, then since the last term, l, is the nth term

$l = a + (n-1)d$

Substituting this in ③

$S_n = \dfrac{n}{2}\Big(a + a + (n-1)d\Big)$

$ = \dfrac{n}{2}\Big(2a + (n-1)d\Big)$

➤
> $S_n = \dfrac{n}{2}(a+l)$
>
> $S_n = \dfrac{n}{2}\Big(2a + (n-1)d\Big)$

Examples 10 to 18 are solved using the formulae for arithmetic series.

Example 10   Find the fiftieth term of the arithmetic series: $\frac{1}{2} + 2 + 3\frac{1}{2} + 5 + \cdots$

$a = \frac{1}{2}, \quad d = 1\frac{1}{2}, \quad n = 50.$ $\boxed{\text{Use } u_n = a + (n-1)d}$

$u_{50} = \dfrac{1}{2} + 49 \times \dfrac{3}{2} = \dfrac{1 + 147}{2} = 74$

The fiftieth term is 74.

Example 11 Find the number of terms in the arithmetic series: $67 + 62 + 57 + \cdots - 13$

$a = 67, \quad d = -5, \quad u_n = -13.$ $\qquad$ Use $u_n = a + (n-1)d$

$-13 = 67 - 5(n - 1)$

$5(n - 1) = 80$

$n - 1 = 16$

$\therefore \quad n = 17$

The arithmetic series has 17 terms.

Example 12 *Examples 10 and 11 were quite simple. For more complex calculations, as in this example, assemble the data and then state which formula is used.*

Find the sum of these arithmetic series:

a $\quad k + 4k + 7k + 10k + \cdots + 91k$

b $\quad \frac{1}{4} - \frac{1}{4} - \frac{3}{4} - \frac{5}{4} - \cdots$ as far as the sixteenth term

Solution $\quad$ **a** $\quad a = k, \quad d = 3k, \quad u_n = l = 91k$ $\qquad$ Assemble the data.

$\qquad u_n = a + (n - 1)d$ $\qquad$ Before finding the sum, the number of terms must be calculated. Quote the formula to be used.

$\qquad 91k = k + 3k(n - 1)$

$\qquad 91 = 1 + 3n - 3$ $\qquad$ Divide through by k.

$\qquad 3n = 93$

$\qquad \therefore \quad n = 31$ $\qquad$ There are 31 terms.

$\qquad S_n = \dfrac{n}{2}(a + l)$ $\qquad$ Quote the next formula to be used.

$\qquad \quad = \frac{31}{2} \times (k + 91k)$

$\qquad \quad = \frac{31}{2} \times 92k$

$\qquad \quad = 31 \times 46k$

$\qquad \quad = 1426k$

b $\quad a = \frac{1}{4}, \quad d = -\frac{1}{2}, \quad n = 16$ $\qquad$ Assemble the data.

$\qquad S_n = \dfrac{n}{2}\left(2a + (n - 1)d\right)$ $\qquad$ Quote the formula to be used.

$\qquad S_{16} = \frac{16}{2}\left(2 \times \frac{1}{4} + 15 \times \left(-\frac{1}{2}\right)\right)$

$\qquad \quad = 8\left(\frac{1}{2} - 7\frac{1}{2}\right)$

$\qquad \quad = 8 \times (-7)$

$\qquad \quad = -56$

Example 13 Evaluate $\displaystyle\sum_{1}^{100}(3r+1)$.

> At first sight this may not look like an arithmetic series. Writing out the first few terms will show that it is one.

$$\sum_{1}^{100}(3r+1) = 4 + 7 + 10 + \cdots + 301$$

This is an arithmetic series with $a = 4$, $d = 3$ and $n = 100$.

$$S_n = \frac{n}{2}\Big(2a + (n-1)d\Big)$$

> Or use $S_n = \frac{n}{2}(a + l)$ with $l = 301$.

$$\sum_{1}^{100}(3r+1) = \frac{100}{2}(2 \times 4 + 99 \times 3)$$

$$= 50 \times 305 = 15\,250$$

> Alternatively, use
> $$\sum_{1}^{100}(3r+1) = 3\sum_{1}^{100}r + 100.$$

Example 14 Show that the sum of the first n natural numbers $\displaystyle\sum_{1}^{n}r = \frac{n}{2}(n+1)$

$$\sum_{1}^{n}r = 1 + 2 + \ldots + n$$

This is an arithmetic series of n terms with first term $a = 1$ and last term $l = n$.

$$\sum_{1}^{n}r = \frac{n}{2}(a + l)$$

> Quote the formula to be used.

$$= \frac{n}{2}(n+1)$$

> Substitute in the formula.

Example 15 **a** Find the sum of the integers 1 to 1000 inclusive

b Find the sum of the integers 1 to 1000 inclusive that are *not* divisible by 5.

Solution **a** The sum of the first 1000 integers is

$$S_{1000} = \sum_{1}^{1000}r = \frac{1000(1000 + 1)}{2}$$

> Use the formula $\displaystyle\sum_{1}^{n}r = \frac{n}{2}(n+1)$

$$= 500 \times 1001$$

$$= 500\,500$$

The sum of the integers from 1 to 1000 is $500\,500$.

b Let S be the required sum and S' be the sum of integers from 1 to 1000 divisible by 5.

> The sum of the integers *not* divisible by 5 equals the sum of the integers which are divisible by 5 subtracted from the sum of all the integers.

Then $S = S_{1000} - S'$

But $S' = 5 + 10 + 15 + \cdots + 1000$

$$= \frac{200}{2} \times (5 + 1000)$$

$$= 100 \times 1005$$

$$= 100\,500$$

> There are 200 multiples of 5. Find this either by dividing each term by 5 mentally; this would number the terms 1 to 200. Alternatively, use $a = 5$, $d = 5$, $u_n = 1000$ and $u_n = a + (n-1)d$:
> $$1000 = 5 + 5(n-1)$$
> $$= 5 + 5n - 5$$
> $$5n = 1000$$
> $$n = 200$$

So $S = 500\,500 - 100\,500$

$$= 400\,000$$

The sum of the integers from 1 to 1000 not divisible by 5 is $400\,000$.

Example 16   The third term of an arithmetic series is 13 and the sum of the first six terms is 96. Find the tenth term.

Let the first term be a and the common difference d.

3rd term $= 13$ so $a + 2d = 13$ ①

$S_6 = 96$ so $\frac{6}{2}(2a + 5d) = 96$

$$3(2a + 5d) = 96$$

$$2a + 5d = 32 \quad ②$$

① $\times 2$ $2a + 4d = 26 \quad ③$

② $-$ ③ $d = 6$

Substituting in ①

$$a + 12 = 13$$

$$a = 1$$

10th term $= a + 9d$

$$= 1 + 9 \times 6$$

$$= 55$$

> There are two unknowns so two equations are needed to find their values.

> S_6 is the sum of 6 terms, so $n = 6$.

> Divide both sides by 3.

> Check mentally by substituting $a = 1$ and $d = 6$ in ②.

Example 17   Three consecutive terms of an arithmetic series are $2x - 2$, $x - 3$ and $1 - x$. Find the next term.

Since the terms are consecutive terms of an arithmetic series

$$d = x - 3 - (2x - 2) = -x - 1$$

Also

$$d = 1 - x - (x - 3) = -2x + 4$$

So $-x - 1 = -2x + 4$

$\therefore$ $x = 5$

and $d = -6$

So the terms are 8, 2 and -4.

Next term is $-4 - 6 = -10$.

> The difference, d, between consecutive terms is constant. So for consecutive terms, a, b and c, $b - a = c - b$.

> Substitute for x in one of the expressions to find d.

> Substitute $x = 5$ in $2x - 2$, $x - 3$ and $1 - x$.

Example 18   The sum of the first n terms of a series is $3n^2 + 6n$.

a   Find the first three terms of the series.

b   Prove algebraically that the series is an arithmetic series.

Solution $S_n = 3n^2 + 6n$

a $S_1 = u_1 = 3 + 6 = 9$ Substitute $n = 1$ in S_n to find S_1

$$S_2 = 3 \times 2^2 + 6 \times 2 = 24$$

Substitute $n = 2$ in S_n for S_2
$S_2 = u_1 + u_2$

So $u_1 + u_2 = 24$

$$u_2 = 24 - 9 = 15$$

$$S_3 = 27 + 18 = 45$$

$\therefore$ $u_3 = 45 - 24 = 21$ $S_3 - S_2 = u_3$

So the first three terms are 9, 15 and 21.

b $u_n = S_n - S_{n-1}$

$$= 3n^2 + 6n - [3(n-1)^2 + 6(n-1)]$$

$$= 3n^2 + 6n - 3(n^2 - 2n + 1) - 6n + 6$$

$$= 3n^2 + 6n - 3n^2 + 6n - 3 - 6n + 6$$

$$= 6n + 3$$

$S_n = u_1 + u_2 + \cdots + u_{n-1} + u_n$
$S_{n-1} = u_1 + u_2 + \cdots + u_{n-1}$
so $S_n - S_{n-1} = u_n$

Take care with the signs when removing the brackets.

$d = u_n - u_{n-1}$

$$= 6n + 3 - [6(n-1) + 3]$$

$$= 6n + 3 - (6n - 6 + 3)$$

$$= 6$$

The common difference, d, is the difference between consecutive terms.

6 is the common difference of the terms found in **a**.

The difference between consecutive terms is constant.

So the series is arithmetic.

Exercise 8D   *For this exercise, use the formulae for arithmetic series.*

1   Find the number of terms in these arithmetic series.

   a $10 + 12 + 14 + \cdots + 80$ **b** $3 + 6 + 9 + \cdots + 51$

   c $3 + 7 + \cdots + 87$ **d** $13 + 8 + \cdots - 32$

   e $-7 - 5\frac{1}{2} - \cdots + 18\frac{1}{2}$ **f** $2 + 4 + \cdots + 4n$

   g $x + 2x + \cdots + nx$ **h** $a + (a+d) + \cdots + (a + (n-1)d)$

2   Write down the forty-first term of each arithmetic series in Question 1.

3   Find the sums of these arithmetic series.

   a $2 + 7 + 12 + \cdots + 77$ **b** $71 + 67 + 63 + \cdots - 53$

   c $1 + 1\frac{1}{6} + 1\frac{1}{3} + \cdots + 4\frac{1}{2}$ **d** $15 + 11 + \cdots - 25$

   e $a + (a+1) + \cdots + (a + n - 1)$ **f** $a + (a-d) + (a-2d) + \cdots + (a - (n-1)d)$

4 Find the sums of these arithmetic series to the terms indicated.

a $3 + 6 + \cdots$; 10th term   **b** $16 + 12 + \cdots$; 9th term

c $-3 - 2 - \cdots$; 14th term   **d** $2 + 6 + \cdots$; nth term

e $3\frac{1}{2} + 3 + \cdots$; 12th term   **f** $p + 3p + \cdots$; rth term

5 Evaluate

a $\displaystyle\sum_{3}^{12} r$ **b** $\displaystyle\sum_{1}^{6} (3r + 2)$ **c** $\displaystyle\sum_{1}^{8} (2r - 7)$

d $\displaystyle\sum_{1}^{11} (32 - 5r)$ **e** $\displaystyle\sum_{1}^{n} (2r + 1)$ **f** $\displaystyle\sum_{6}^{n} \left(4 - \frac{r}{2}\right)$

6 The third term of an arithmetic series is 5 and the fifth term is 9.
Find the first term and the common difference.

7 The second term of an arithmetic series is 16 and the fifth term is 37.
Find the first term and the sum of the first eight terms.

8 The first term of an arithmetic series is -12 and the last term is 40.
If the sum of the series is 196, find the number of terms and the common difference.

9 Show that the sum of the integers from 1 to n is $\frac{1}{2}n(n + 1)$.

10 The twenty-first term of an arithmetic series is 37 and the sum of the first twenty terms is 320. What is the sum of the first ten terms?

11 The sum of the first n terms of an arithmetic series is $n(n + 4)$.

a Find the first three terms.

b Find the sum to six terms.

12 The third term of an arithmetic series is 5 and the sixth term is 11.
Find the first term and the sum of the first eight terms.

13 The sixth term of an arithmetic series is 2 and the ninth term is 11.
Find the first term and the sum of the first twelve terms.

14 The first term of an arithmetic series is 14.
The common difference and the sum of the first n terms are both -6.
Find n.

15 The sum of the first n terms of a series is $n^2 + 5n$.
a Find the first three terms of the series.
b Prove algebraically that the series is an arithmetic series.

16 Find the sum of the odd numbers between 100 and 200.

17 Find the sum of the even numbers, divisible by three, lying between 400 and 500.

18 The twenty-first term of an arithmetic series is $5\frac{1}{2}$, and the sum of the first twenty-one terms is $94\frac{1}{2}$.
Find the first term, the common difference and the sum of the first thirty terms.

19 The first two terms of an arithmetic series are -3.1 and -2.8.

    a Which term is the first positive one, and what is its value?

    b How many terms must be summed to give a positive total?

20 The first three terms of an arithmetic series are $2x + 1$, $3x + 2$ and $5x - 1$. Find x.

21 The first three terms of an arithmetic series are $y - 1$, $2y$ and $4y - 2$.
Find the fourth and fifth terms. Do not include y in the answer.

22 Three consecutive terms of an arithmetic series are $z + 4$, $2z + 9$ and $6 - z$.
Find z and hence find the sum of the three terms.

23 A shop assistant is arranging a display of a triangular array of tins so as to have one tin in the top row, two in the second, three in the third and so on. If there are 100 tins altogether, how many rows can be completed and how many tins will be left over?

24 A farmer has a triangular orchard in which trees are planted in rows. The row along the base of the triangle has 64 trees. Each row has three fewer trees than the previous one until the row at the top of the triangle contains a single tree.

    a Find the total number of rows of trees.

    b Find the number of trees in **i** the fifth, and **ii** the twelfth row.

    c The seven shortest rows are removed to make way for a road.
       Find the percentage of trees lost.

25 A lecture theatre has a trapezium-shaped floor plan, so that the number of chairs in successive rows are in arithmetic series. The back row of chairs contains eight chairs and the front row contains thirty. There are twelve rows altogether.

    a Find the number of seats in the theatre.

    b Find the percentage of the seats that are in the rear half of the theatre.

26 A teacher illustrates arithmetic series by cutting a length of string into pieces so that the lengths of the pieces form an arithmetic sequence and the entire length of string is used up exactly.

    a On one occasion a 1 m length is cut so that the first piece measures 30 cm and the fourth 15 cm. Find how many pieces there are altogether.

    b On another occasion a 2 m length is cut so that the first piece measures 2 cm and there are eight pieces. Find the length of the longest piece.

1 Find the sum of the even numbers up to and including 100.

2 What is the sum of the integers from 1 to 100 which are not divisible by 6?

3 How many terms of the arithmetic series $15 + 13 + 11 + \cdots$ are required to make a total of -36?

4 Find the first five terms of these sequences.

    a $u_n = (-2)^{n+1}$ **b** $u_n = 3 - \dfrac{1}{2^n}$

    c $u_1 = 2,\ u_n = 3u_{n-1} - 1$ for $n \geqslant 2$ **d** $u_1 = 5,\ u_n = \dfrac{5}{u_{n+1}} + 1$ for $n \geqslant 2$

5 Evaluate these sums.

    a $\displaystyle\sum_{1}^{4} r^2$ **b** $\displaystyle\sum_{1}^{5} (-1)^r r^2$

    c $\displaystyle\sum_{4}^{8} (k+1)(k-6)$ **d** $\displaystyle\sum_{1}^{5} (k-1)(k-2)(k-3)(k-4)(k-5)$

6 The sum of the first six terms of an arithmetic series is 21, and the seventh term is three times the sum of the third and the fourth.
Find the first term and the common difference.

7 The sum of the first five terms of an arithmetic series is 30, and the third term is equal to the sum of the first two.
Find the first five terms of the series.

8 A recurrence relation is of the form $p_{n+1} = ap_n + b$, $n \geqslant 1$, where a and b are constants.
Given that $p_1 = 100$, $p_2 = 97$ and $p_3 = 94.15$, find a and b.

9 The sum of the integers from 1 to $n - 1$ is equal to the sum of the integers from $n + 1$ to 49. Find n.

10 If a and b are the first and last terms of an arithmetic series of $r + 2$ terms, find the second and the $(r + 1)$th term.

11 The third, fourth and fifth terms of an arithmetic series are $2x$, $x - 3$ and $11 - x$.

    a Find the first term.

    b Find the sum of the first fifteen terms.

There is a 'Test yourself' exercise on the CD-ROM (Ty8).

➤ Key points

Sequences and series

A **sequence** is a set of terms, in a definite order, where the terms are obtained by some rule.

A sequence can be defined by a formula for the nth term, such as $u_n = n^2 + 1$
or by recurrence relation, such as $x_{n+1} = x_n + 2$, $n > 1$, $x_1 = 5$.

A **series** is the sum of a sequence.

Arithmetic series or progression (AP)

An **arithmetic series** is a series whose consecutive terms have a common difference.

For first term, a, common difference, d

nth term $= a + (n - 1)d$

For n terms with last term, l, and sum to n terms, S_n

$$l = a + (n - 1)d$$

$$S_n = \frac{n}{2}(a + l)$$

$$S_n = \frac{n}{2}\left(2a + (n - 1)d\right)$$

For the sum of the first n natural numbers $\displaystyle\sum_1^n r = \frac{n}{2}(n + 1)$

For an arithmetic series with terms $u_1, u_2, u_3, \ldots, u_r, \ldots$

$$d = u_r - u_{r-1}$$

Differentiation

The slope of a roof, the acceleration of a car, the changing population of a virus can all be expressed as rates of change. Calculus, with its study of differentiation and integration, enables us to deal with many different types of rates of change and can be applied, in a very powerful way, to problems of mechanics, biology, electromagnetism, economic theory, . . .

After working through this chapter you should be able to

- *differentiate a function*
- *apply differentiation to gradients, tangents and normals*
- *interpret rates of change*
- *calculate second derivatives.*

9.1 Rates of change

Consider this distance–time graph.

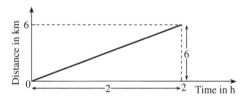

The gradient of a straight line measures how steep the line is. In this case, the gradient measures the rate at which distance varies with time.

Gradient $= \frac{6}{2} = 3$.

This corresponds to a speed of $3\,\text{km/h}$.

> **Speed** measures the rate of change of distance with respect to time.

There are numerous situations where the rate of change of one variable with respect to another is needed. In the above case, the relationship between the two variables is linear: the gradient is constant. In most cases, the variables are not related in a linear way.

The slope of this curve varies. From O to A the curve is not as steep as from A to B. From A to B the distance increases faster with respect to time than from O to A.

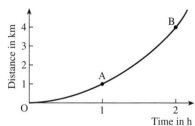

The speed would be greater during the motion from A to B than from O to A.

When a relationship is not linear, the slope of the curve (and therefore the gradient and rate of change) varies.

Tangent to a curve

The tangent to a curve at any point is the straight line which touches the curve at that point.

At P the gradient of the curve is defined as the gradient of the tangent at P.

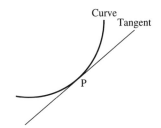

Gradient of a curve

The gradient of a curve at any point is the gradient of the tangent to the curve at that point.

To find an approximate value for the gradient of a curve at any point, the curve could be drawn, a tangent to the curve at the point drawn 'by eye' and its gradient calculated.

- Plot carefully the graph of $y = x^2$ for values of x from 0 to 4.
- Draw 'by eye' the tangents to the curve at $x = \frac{1}{2}, 1, 1\frac{1}{2}, 2, 3$.
- Find the gradient of each tangent.
- Compare the values of x and the gradient at each point.

This should have led to an interesting result. A more formal approach to find the gradient of a curve at any point is now given.

- To find the gradient of the tangent at P, first choose a point P_1 on the curve close to P.
- Find the gradient of the line PP_1.
  The gradient of PP_1 is an approximation to the gradient of the tangent at P.
- Choose a point P_2 on the curve nearer to P than P_1.
- Find the gradient of the line PP_2.
  The gradient of PP_2 is a better approximation to the gradient of the tangent.

Repeat this process. As the points P_3, P_4, ... approach closer and closer to P, so the gradients of PP_3, PP_4, ... approach closer and closer to the gradient of the tangent.

This approach is now used to find the gradient of $y = x^2$ at (1, 1).

Consider the points P_1, P_2, P_3, P_4, on the curve $y = x^2$.

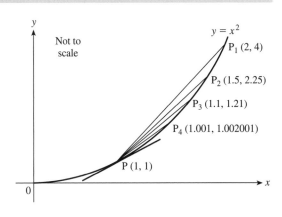

P_1 (2, 4) Gradient of $PP_1 = \dfrac{4-1}{2-1} = 3$

P_2 (1.5, 2.25) Gradient of $PP_2 = \dfrac{2.25-1}{1.5-1} = 2.5$

P_3 (1.1, 1.21) Gradient of $PP_3 = \dfrac{1.21-1}{1.1-1} = 2.1$

P_4 (1.001, 1.002001) Gradient of $PP_4 = \dfrac{1.002001-1}{1.001-1} = 2.001$

As the points P_1, P_2, P_3, P_4 approach closer and closer to P the gradients appear to approach closer and closer to 2.

To show that the gradients of the lines approach 2, which is called the **limit**, a sequence of points, P_k, can be taken closer and closer to P. A spreadsheet is an efficient way of carrying out the calculations.

The algebraic method given next illustrates a more general way of finding a gradient, in this case at the point (1, 1), for the function $y = x^2$.

Let P_k be the point $(1 + h, (1 + h)^2)$.

$$\text{Gradient of } PP_k = \frac{(1+h)^2 - 1}{(1+h) - 1}$$
$$= \frac{1 + 2h + h^2 - 1}{h}$$
$$= \frac{2h + h^2}{h}$$
$$= \frac{h(2 + h)}{h}$$
$$= 2 + h$$

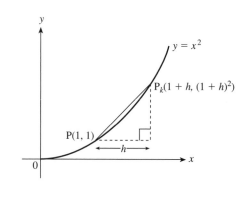

As P_k approaches closer and closer to P, h tends to 0, and so $2 + h$ tends to 2.

By taking appropriate values of h one can get as close to the value, 2, as one wishes.

Expressed formally:

The limit of $2 + h$ as $h \to 0$ is 2, or $\lim\limits_{h \to 0}(2 + h) = 2$.

$\lim\limits_{h \to 0}(2 + h)$ is read as 'limit as h tends to 0 of $2 + h$'.

So Gradient at $P(1, 1) = \lim\limits_{h \to 0}(\text{Gradient of } PP_k)$
$$= \lim\limits_{h \to 0}(2 + h)$$
$$= 2$$

This corresponds to the result found numerically (above) and should correspond to those found graphically (page 149).

Note Such a limit does not exist at all points for all functions. This chapter deals only with functions where the limit exists.

For a fuller discussion on limits, see the CD-Rom (E7.1).

The method can now be applied to finding the gradient at any point on $y = x^2$.

To find the gradient at $P(x, x^2)$, take a point $P_1(x + h, (x + h)^2)$ on the curve close to P.

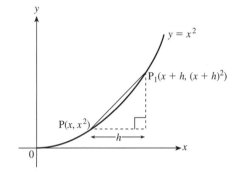

$$\text{Gradient of PP}_1 = \frac{(x + h)^2 - x^2}{(x + h) - x}$$

$$= \frac{x^2 + 2xh + h^2 - x^2}{h}$$

$$= 2x + h$$

$$\text{Gradient at P} = \lim_{h \to 0} (\text{Gradient of PP}_1)$$

$$= \lim_{h \to 0} (2x + h)$$

$$= 2x$$

At any point on $y = x^2$ the gradient of the curve is $2x$.

So for $y = x^2$ the **gradient function** is $2x$.

> This should have been the result suspected from drawing tangents by eye to the graph. See page 149.

This result enables the gradient at *any* point on $y = x^2$ to be found.

At the point (3, 9) $x = 3$, so the gradient is $2 \times 3 = 6$.

At the point (5, 25) $x = 5$, so the gradient is $2 \times 5 = 10$.

9.2 Differentiation

The process of finding the gradient function is called **differentiation**. There are many other terms used within calculus. This section introduces the notation and vocabulary of **differential calculus**.

The notation $\dfrac{dy}{dx}$

For any curve, $y = f(x)$, the gradient can be defined as follows.

Let $P(x, y)$ be any point on the curve and $P_1(x + \delta x, y + \delta y)$ be a point on the curve near P.

δx and δy are the small changes in x and y between P and P_1.

> Read 'δx' as 'delta x'.

So gradient of $PP_1 = \dfrac{\delta y}{\delta x}$

gradient at $P = \lim_{\delta x \to 0} (\text{Gradient of } PP_1)$

$= \lim_{\delta x \to 0} \dfrac{\delta y}{\delta x}$

$= \dfrac{dy}{dx}$

Read as 'dy by dx'.

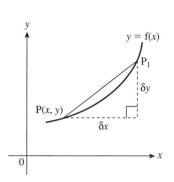

$\dfrac{dy}{dx}$ is the notation used for the gradient. It represents the limit of $\dfrac{\delta y}{\delta x}$ as $\delta x \to 0$.

$\dfrac{dy}{dx}$ is *not* a fraction. It must be considered as a single entity.

$\dfrac{dy}{dx}$ **is the gradient of the curve, i.e. the rate of change of y with respect to x.**

Function notation

It is usually easier to work with h in place of δx.

Consider $y = f(x)$. Let $P(x, f(x))$ be any point on the curve and $P_1(x + h, f(x + h))$ be a point near P.

Gradient of $PP_1 = \dfrac{f(x + h) - f(x)}{(x + h) - x}$

gradient at $P = \lim_{h \to 0} (\text{Gradient of } PP_1)$

$= \lim_{h \to 0} \dfrac{f(x + h) - f(x)}{h}$

In function notation, the gradient is expressed as $f'(x)$,

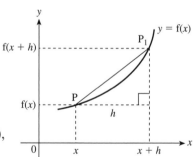

$\boxed{f'(x) = \lim_{h \to 0} \dfrac{f(x + h) - f(x)}{h}}$

Vocabulary

Consider $y = f(x) = x^2$.

The gradient function can be written as $\dfrac{dy}{dx}$ or $f'(x)$ or $\dfrac{d}{dx}(f(x))$ or $\dfrac{d}{dx}(x^2)$ or y'.

When a function of x, $y = f(x)$, is differentiated once, the result is called

- the (first) **derivative** with respect to x, or
- the (first) **differential coefficient** with respect to x, or
- the (first) **derived function** with respect to x, or
- the **gradient function**.

When a function of x, $y = f(x)$, is differentiated twice, the result is called the **second derivative** (or differential coefficient or derived function) with respect to x.

This is written as $\dfrac{d^2 y}{dx^2}$ or $f''(x)$ or $\dfrac{d}{dx}\left(\dfrac{dy}{dx}\right)$ or y''.

> Read as 'd 2y by d x squared'.

Note $\dfrac{d^2 y}{dx^2} = \dfrac{d}{dx}\left(\dfrac{dy}{dx}\right)$ is a function, which is the gradient of the gradient function.

> The first derivative (the gradient) is written as $\dfrac{dy}{dx}$, y' or $f'(x)$.
>
> The second derivative (the gradient of the gradient function) is written as $\dfrac{d^2 y}{dx^2}$, y'' or $f''(x)$.
>
> Similarly, the third derivative is written as $\dfrac{d^3 y}{dx^3}$, y''' or $f'''(x)$.

So far all differentiation has been with respect to x. One can differentiate with respect to any variable.

For example, if $s = t^2$, differentiating s with respect to t gives $\dfrac{ds}{dt} = 2t$.

Differentiating from first principles

The definition

$$f'(x) = \lim_{h \to 0} \frac{f(x + h) - f(x)}{h}$$

will now be used to differentiate x^3. This method is called 'differentiating from first principles'.

Example 1   Find, from first principles, the derivative of x^3.

$f(x) = x^3$

$f'(x) = \lim\limits_{h \to 0} \dfrac{f(x + h) - f(x)}{h}$ To differentiate from first principles, use this formula.

$\qquad = \lim\limits_{h \to 0} \dfrac{(x + h)^3 - x^3}{h}$

$\qquad = \lim\limits_{h \to 0} \dfrac{x^3 + 3x^2 h + 3xh^2 + h^3 - x^3}{h}$ The x^3 terms cancel.

$\qquad = \lim\limits_{h \to 0} (3x^2 + 3xh + h^2)$ As h tends to 0, $3xh$ and h^2 also tend to 0.

$\qquad = 3x^2$

So the derivative of x^3 is $3x^2$

Example 2 Find, from first principles, the differential coefficient of $\frac{1}{x}$.

$$f(x) = \frac{1}{x}$$

$$f'(x) = \lim_{h \to 0} \frac{f(x+h) - f(x)}{h}$$

$$= \lim_{h \to 0} \frac{\frac{1}{x+h} - \frac{1}{x}}{h}$$

> Multiply numerator and denominator by $x(x+h)$.

$$= \lim_{h \to 0} \frac{x - (x+h)}{hx(x+h)}$$

$$= \lim_{h \to 0} \frac{-h}{hx(x+h)}$$

$$= \lim_{h \to 0} \frac{-1}{x(x+h)}$$

> As h tends to 0, $(x+h)$ in the denominator tends to x.

$$= -\frac{1}{x^2}$$

So the differential coefficient of $\frac{1}{x}$ is $-\frac{1}{x^2}$.

Extension Exercise 9A

1 Find from first principles, the derived functions of these expressions.

    a x **b** $5x$ **c** $2x^2$ **d** $x^2 + 3$ **e** 5 **f** $\dfrac{1}{x^2}$ **g** x^4

Summary of results obtained by differentiating from first principles

y	5	x	x^2	x^3	x^4	$2x^2$	$\frac{1}{x} = x^{-1}$	$\frac{1}{x^2} = x^{-2}$
$\frac{dy}{dx}$	0	1	$2x$	$3x^2$	$4x^3$	$4x$	$-\frac{1}{x^2} = -x^{-2}$	$-\frac{2}{x^3} = -2x^{-3}$

Differentiation of polynomials

The results in the table above lead to these two rules:

➤
$$y = x^n \Rightarrow \frac{dy}{dx} = nx^{n-1} \quad n \in \mathbb{R}$$

$$y = kx^n \Rightarrow \frac{dy}{dx} = knx^{n-1}$$

> *The result for y = kx^n can be proved for positive integers using the binomial expansion. To prove it is true for all real values of n is beyond the scope of this book.*

Note To differentiate a power of x, multiply by the power and reduce the power by one.

Example 3   Find the derivatives.

a $\dfrac{d}{dx}(7x^5) = 5 \times 7x^{5-1} = 35x^4$

b $\dfrac{d}{dx}\left(x^{\frac{1}{2}}\right) = \frac{1}{2}x^{\frac{1}{2}-1} = \frac{1}{2}x^{-\frac{1}{2}}$

c $\dfrac{d}{dx}(3x^{-4}) = -4 \times 3x^{-4-1} = -12x^{-5}$

Differentiation of $y = c$

$y = c$ represents a line parallel to the x-axis. This has gradient zero.

Differentiating by the rule confirms this result.

$$y = c = cx^0$$

$$\Rightarrow \quad \frac{dy}{dx} = 0 \times cx^{-1} = 0$$

> Since $x^0 = 1$

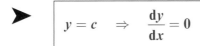

$$y = c \quad \Rightarrow \quad \frac{dy}{dx} = 0$$

Differentiation of $y = kx$

$y = kx$ represents a line with gradient k. Differentiating by the rule confirms this result.

$$y = kx^1$$

$$\frac{dy}{dx} = 1 \times kx^{1-1}$$

$$= kx^0$$

$$= k$$

$$y = kx \quad \Rightarrow \quad \frac{dy}{dx} = k$$

Example 4   Find the derivatives.

a $y = 3x \quad \Rightarrow \quad \dfrac{dy}{dx} = 3$

b $y = 7 \quad \Rightarrow \quad \dfrac{dy}{dx} = 0$

Differentiation of a number of terms

By differentiating from first principles it can be shown that, for the sum of a number of terms, such as

$$3x^3 - 2x^2 + \frac{4}{x} - \frac{1}{x^2}$$

the derived function is the sum of the derived functions of each term. For example

$$y = 3x^3 - 2x^2 + \frac{4}{x} - \frac{1}{x^2}$$

$$\frac{dy}{dx} = 9x^2 - 4x - \frac{4}{x^2} + \frac{2}{x^3}$$

The proofs of the next two general results are beyond the scope of this book, but the results are included for completeness.

➤

$$y = f(x) \pm g(x) \quad \Rightarrow \quad \frac{dy}{dx} = f'(x) \pm g'(x)$$

$$y = kf(x) \qquad \Rightarrow \quad \frac{dy}{dx} = kf'(x)$$

➤

Rewrite expressions as sums of terms with powers of x before differentiating, i.e. write

$$y = (x+1)(3x-1) \quad \text{as} \quad y = 3x^2 + 2x - 1$$

$$\text{or} \quad y = \frac{x^5 - x^2}{\sqrt{x}} \qquad \text{as} \quad y = x^{\frac{9}{2}} - x^{\frac{3}{2}}$$

Example 5 Differentiate with respect to x.

a $\quad y = ax^2 + bx + c$

$$\frac{dy}{dx} = 2ax + b$$

b $\quad y = \dfrac{6x + 2x^4}{x^2}$ Divide by x^2 before differentiating.

$$= 6x^{-1} + 2x^2$$

$$\frac{dy}{dx} = -6x^{-2} + 4x$$

c $\quad y = \dfrac{3}{x}$ Write in index notation.

$$= 3x^{-1}$$

$$\frac{dy}{dx} = -3x^{-2}$$ The power is reduced by 1.

$$= -\frac{3}{x^2}$$

d
$$y = \frac{1}{4\sqrt{x}}$$

$$= \frac{1}{4} \times \frac{1}{\sqrt{x}}$$

> Write $\frac{1}{\sqrt{x}}$ as $x^{-\frac{1}{2}}$. Note $\frac{1}{4}$ is unchanged.

$$= \frac{1}{4}x^{-\frac{1}{2}}$$

> The power is reduced by 1: $-\frac{1}{2} - 1 = -\frac{3}{2}$.

$$\frac{dy}{dx} = -\frac{1}{8}x^{-\frac{3}{2}}$$

> Alternatively, write $\frac{dy}{dx} = -\frac{1}{8\sqrt{x^3}} = -\frac{1}{8x\sqrt{x}}$.

e
$$y = \frac{2}{3x^{\frac{1}{3}}}$$

$$= \frac{2}{3} \times \frac{1}{x^{\frac{1}{3}}}$$

> Write $\frac{1}{x^{\frac{1}{3}}}$ as $x^{-\frac{1}{3}}$. *Note*: $\frac{2}{3}$ is unchanged.

$$= \frac{2}{3}x^{-\frac{1}{3}}$$

> The power is reduced by 1: $-\frac{1}{3} - 1 = -\frac{4}{3}$.

$$\frac{dy}{dx} = -\frac{2}{9}x^{-\frac{4}{3}}$$

> Alternatively, write $\frac{dy}{dx} = -\frac{2}{9\sqrt[3]{x^4}}$.

f
$$y = 2x^2 - 3x + 4 - \frac{5}{x}$$

$$\frac{dy}{dx} = 4x - 3 + \frac{5}{x^2}$$

Example 6 Find $\frac{d^2y}{dx^2}$ given $y = 10x^3 - 3x^2 + 4x - 1$.

$$y = 10x^3 - 3x^2 + 4x - 1$$

> Differentiating y once gives $\frac{dy}{dx}$.

$$\frac{dy}{dx} = 30x^2 - 6x + 4$$

$$\frac{d^2y}{dx^2} = 60x - 6$$

> Differentiating again gives $\frac{d^2y}{dx^2}$.

Example 7 Find the y-coordinate and the gradient of $y = (3x + 1)^2$ when $x = 2$.

When $x = 2$ $\quad y = (3 \times 2 + 1)^2 = 49$

$$y = (3x + 1)^2$$

> Multiply out the brackets before differentiating.

$$= 9x^2 + 6x + 1$$

$$\frac{dy}{dx} = 18x + 6$$

When $x = 2$ $\quad \frac{dy}{dx} = 18 \times 2 + 6 = 42$

So when $x = 2$, the y-coordinate is 49 and the gradient is 42.

Example 8 Find the points on $y = x + \frac{1}{x}$ where the gradient is $\frac{3}{4}$.

$$y = x + \frac{1}{x}$$

$$\frac{dy}{dx} = 1 - \frac{1}{x^2}$$

When gradient $= \frac{3}{4}$ $\qquad 1 - \frac{1}{x^2} = \frac{3}{4}$

$$\frac{1}{x^2} = 1 - \frac{3}{4} = \frac{1}{4}$$

$$\therefore \qquad\qquad\qquad x = \pm 2$$

When $x = 2$, $y = 2 + \frac{1}{2} = \frac{5}{2}$.

And when $x = -2$, $y = -2 - \frac{1}{2} = -\frac{5}{2}$.

So the gradient is $\frac{3}{4}$ at $(2, \frac{5}{2})$ and at $(-2, -\frac{5}{2})$.

Exercise 9B

1 Find $\dfrac{dy}{dx}$ for each of these.

a $y = x^4$ **b** $y = 3x^4$ **c** $y = 5x^4$ **d** $y = x^{12}$ **e** $y = 3x^7$

f $y = 5x$ **g** $y = 5x + 3$ **h** $y = 5x^2 - 3x$ **i** $y = 5$ **j** $y = x^{-5}$

k $y = 4x^{-3}$ **l** $y = \dfrac{1}{x^2}$ **m** $y = \dfrac{2}{x}$ **n** $y = -\dfrac{3}{x^2}$ **o** $y = \dfrac{1}{3x^3}$

p $y = -\dfrac{1}{x^4}$ **q** $y = \dfrac{3}{4x^5}$ **r** $y = x^{\frac{1}{3}}$ **s** $y = 3x^{-\frac{1}{2}}$ **t** $y = \sqrt{x}$

u $y = \sqrt[4]{x}$ **v** $y = \dfrac{1}{\sqrt{x}}$ **w** $y = -\dfrac{3}{\sqrt[3]{x}}$ **x** $y = 4\sqrt{x^3}$ **y** $y = \dfrac{3}{2\sqrt{x}}$

2 Find the derived function, $f'(x)$, for these functions.

a $f(x) = 3x^4 - 2x^3 + x^2 - x + 10$ **b** $f(x) = 2x^4 + \frac{1}{3}x^3 - \frac{1}{4}x^2 + 2$

c $f(x) = x^6 + \dfrac{1}{x} - \dfrac{3}{\sqrt{x}}$ **d** $f(x) = ax^3 + bx^2 + cx$

e $f(x) = 2x(3x^2 - 4)$ **f** $f(x) = \dfrac{10x^5 + 3x^4}{2x^2}$

g $f(x) = \dfrac{6x + 3 + \sqrt{x}}{x^2}$ **h** $f(x) = \left(\sqrt{x} + \dfrac{1}{\sqrt{x}}\right)^2$

3 Differentiate these expressions with respect to x.

a $-x$ **b** $+10$ **c** $4x^3 - 3x + 2$ **d** $\frac{1}{2}ax^2 - 2bx + c$

e $2(x^2 + x)$ **f** $\dfrac{3x(x+1)}{x^4}$ **g** $x^{\frac{1}{2}} + x^{\frac{1}{3}} + x^{\frac{1}{4}}$ **h** $3\sqrt{x} + \dfrac{4}{\sqrt{x}} + \dfrac{5}{x}$

i $\frac{1}{3}(x^3 - 3x + 6)$ **j** $(x+1)(x+2)$ **k** $\dfrac{x^2 + 5x - 3}{3x^{\frac{1}{2}}}$ **l** $\dfrac{2x^2 - x + 1}{5x^{\frac{3}{2}}}$

4 Find $\dfrac{d^2y}{dx^2}$ in these cases.

a $y = 6x^3 + 3x^2 - 4x$ **b** $\dfrac{dy}{dx} = 5x^{\frac{1}{3}}$ **c** $y = 10x - 7$ **d** $\dfrac{dy}{dx} = \dfrac{4x - 3}{\sqrt{x}}$

5 Find $\dfrac{ds}{dt}$ and $\dfrac{d^2s}{dt^2}$ when

a $s = 5t - 10t^2$ **b** $s = 3t^3 - 4t^2 + 7t$

c $s = 3t - \dfrac{5}{t^2}$ **d** $s = ut + \frac{1}{2}at^2$ (a and u constant)

6 Find the y-coordinate, and the gradient, at the points on these curves corresponding to the given values of x.

a $y = x^2 - 2x + 1$, $x = 2$ **b** $y = x^2 + x + 1$, $x = 0$

c $y = x^2 - 2x$, $x = -1$ **d** $y = (x+2)(x-4)$, $x = 3$

e $y = \sqrt{x}(3 + x^2)$, $x = 1$ **f** $y = (4x - 5)^2$, $x = \frac{1}{2}$

g $y = x + \dfrac{1}{x}$, $x = 1$ **h** $y = \sqrt{x} + x^2$, $x = 4$

7 Find the coordinates of the points on these curves at which the gradient has the given values.

a $y = x^2$, gradient $= 8$ **b** $y = x^3$, gradient $= 12$

c $y = x(2 - x)$, gradient $= 2$ **d** $y = x^2 - 3x + 1$, gradient $= 0$

e $y = x^3 - 2x + 7$, gradient $= 1$ **f** $y = x^{\frac{1}{3}}$, gradient $= \frac{1}{12}$

g $y = x^2 - x^3$, gradient $= -1$ **h** $y = x(x - 3)^2$, gradient $= 0$

i $y = x - \dfrac{1}{x}$, gradient $= 5$ **j** $y = \frac{1}{3}x^{\frac{3}{2}} - x^{\frac{1}{2}}$, gradient $= \frac{3}{4}$

8 The curve $y = ax^2 + bx$ passes through the point $(2, 4)$ with gradient 8. Find a and b.

9 The curve $y = cx + \dfrac{d}{x}$ has gradient 6 at the point $(\frac{1}{2}, 1)$. Find c and d.

10 Given that $f(x) = 2x^3 - x + \dfrac{1}{x}$, find the value of

a $f'(1)$ **b** $f''(1)$ **c** $\dfrac{1}{f(1)}$

11 Show that if $y = 2x - x^2$ then

$$y\frac{d^2y}{dx^2} - 2\frac{dy}{dx} + 2y = 4(x - 1)$$

12 If $y = x^3 + x^{\frac{5}{2}}$ find the value of y' and of y'', when $x = 4$.

13 A child's height h cm at age a years can be modelled by the equation

$$h = -\frac{a^4}{500} + (a - \tfrac{1}{2})^2 + 55$$

for ages $11 \leqslant a \leqslant 16$. Find the child's annual growth rate at age 12, and at age 15.

14 The temperature $\theta°C$ measured at a distance x cm from a candle flame is given by

$$\theta = 16 + \frac{450}{x^{\frac{3}{2}}}$$

for distances $x \geqslant 1$.

  a Find the temperature gradient (i.e. the rate at which temperature decreases with distance) 10 cm from the flame.

  b Find the rate of change of temperature gradient with respect to distance, 5 cm from the flame.

15 Some sugar is put into a cup of coffee, which is then stirred. The sugar concentration, c, measured in grams/litre, is given by

$$c = \frac{t^2}{200}(200 - t^2)$$

where t is the time in seconds after the sugar is added, for $0 \leqslant t \leqslant 10$.

  a Find the rate at which the concentration is increasing after 8 seconds.

  b Show that, after 10 seconds, the rate of change of concentration is zero.

9.3   Tangents and normals

The **tangent** to a curve at a given point is the straight line which touches the curve at that point.

The **normal** to a curve at a given point is a straight line through the point perpendicular to the tangent at the point.

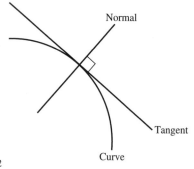

Remember   From Chapter 6, if two lines have gradients m_1 and m_2 then $m_1 m_2 = -1 \Leftrightarrow$ the lines are perpendicular.

Example 9   Find the equations of the tangent and the normal to $y = 2x^3 - 3x$ at the point where $x = 1$.

$$y = 2x^3 - 3x$$

$$\frac{dy}{dx} = 6x^2 - 3$$

When $x = 1$ $y = 2 \times 1^3 - 3 \times 1$ and $\dfrac{dy}{dx} = 6 \times 1^2 - 3$

$$= 2 - 3 = -1 \qquad\qquad\qquad = 6 - 3 = 3$$

So the tangent is the line through $(1, -1)$ with gradient 3.

Using $y - y_1 = m(x - x_1)$

$$y - (-1) = 3(x - 1)$$
$$y + 1 = 3x - 3$$
$$y - 3x = -4$$

This is the equation of the tangent.

Gradient of the tangent is 3.

So gradient of the normal is $-\frac{1}{3}$.

So the normal is the line through $(1, -1)$ with gradient $-\frac{1}{3}$.

| $m_1 m_2 = -1$ |
| $3 \times -\frac{1}{3} = -1$ |

$$y + 1 = -\tfrac{1}{3}(x - 1)$$

Multiply both sides by 3.

$$3y + 3 = -x + 1$$
$$3y + x = -2$$

So the tangent is $y - 3x = -4$ and the normal is $3y + x = -2$.

The equation of the lines can be given in different forms. For good style, use the same form for both answers.

Note   A sketch is not necessary but can be helpful. On a graphic calculator, the tangent and normal will *not* look perpendicular if the scales on the axes are different.

Exercise 9C   **1**   Find the equations of the tangents to these curves at the points corresponding to the given values of x.

     a $y = x^2$, $x = 2$ **b** $y = 3x^2 + 2$, $x = 4$

     c $y = \dfrac{2}{x^2}$, $x = 2$ **d** $y = 3x^2 - x + 1$, $x = 0$

     e $y = \dfrac{x + 3}{\sqrt{x}}$, $x = 4$ **f** $y = 2x + \dfrac{1}{x}$, $x = \frac{1}{2}$

   2   Find the equations of the normals to the curves in Question 1 at the points corresponding to the given values of x.

3 The tangent to $y = ax^2 + bx$ has gradient 3 at the point (2, 10). Find a and b.

4 **a** Find the values of x for which the gradient of the curve
$y = 2x^3 + 3x^2 - 12x + 3$ is zero.

b Hence find the equations of the tangents to the curve which are parallel to the x-axis.

5 **a** Find the gradient of the curve $y = 9x - x^3$ at the point where $x = 1$.

b Find the equation of the tangent to the curve at this point.

c Find the coordinates of the point where this tangent meets the line $y = x$.

6 **a** Find the coordinates of O, A and B, the points of intersection of the curve
$y = x^3 - 3x^2 + 2x$ with the x-axis.

b Find the equation of the tangents to the curve at O, A and B.

7 **a** Find the points of intersection of $y = 4 - x^2$ and $y = 3x$.

b Find the equations of the tangents to $y = 4 - x^2$ at the points of intersection.

c The tangents intersect at P. Find the coordinates of P.

8 **a** Show that there is only one point on the curve $y = 6x^3 + 6x^2 + 2x - 1$ where the gradient is zero. Find the coordinates of this point.

b State the equations of the tangent and the normal to the curve at the point.

9 Find the equation of the tangent to $y = 3x^3 - 4x^2 + 2x - 10$ at the point where the curve meets the y-axis.

10 The normal to the curve $y = x^{\frac{1}{2}} + x^{\frac{1}{3}}$ at the point (1, 2) meets the axes at $(h, 0)$ and $(0, k)$. Find h and k.

11 Tangents and normals are drawn to the curve $y = \dfrac{1}{x^2}$ at the points A(−1, 1) and B(1, 1).

Given that the tangents and normals meet the y-axis at the points C and D, find

a the length of CD

b the area of the quadrilateral ACBD

12 The normals to the curve $y = \sqrt{x}$ at the points where $x = 1$ and $x = 4$ meet at P. Find the coordinates of P.

13 The tangent to the curve $y = ax^2 + bx$ at the point where $x = 1$ has gradient 1 and passes through the point (4, 5). Find a and b.

14 The normal to the curve $y = ax^{\frac{1}{2}} + bx$ at the point where $x = 1$ has gradient 1 and intercepts the y-axis at (0, −4). Find a and b.

15 a Find the coordinates of the points on the curve $8y = 4 - x^2$ at which the gradients are $\frac{1}{2}$ and $-\frac{1}{2}$.

b Find the equations of the tangents to the curve at these points.

c Show that the tangents intersect at the point $(0, 1)$.

16 a Find the equations of the normals to the parabola $4y = x^2$ at the points $(-2, 1)$ and $(-4, 4)$.

b Show that the point of intersection of these two normals lies on the parabola.

Exercise 9D (Review)

1 Differentiate the following with respect to x.

a $10x^5$ **b** $\dfrac{3}{4}x^4$ **c** $\dfrac{5}{x}$ **d** $\dfrac{1}{2x}$ **e** $\dfrac{2}{5x}$

f $x^{\frac{1}{2}}$ **g** $-4x^{\frac{3}{2}}$ **h** $\dfrac{3}{x^2}$ **i** $\dfrac{1}{2\sqrt{x}}$ **j** $-\dfrac{3}{2x^{\frac{2}{3}}}$

2 Find $f'(x)$.

a $f(x) = 3x^4 + 12x^3 - 2x^2 + 6$ **b** $f(x) = \dfrac{3}{x} - \dfrac{1}{3x^2} + \sqrt{x}$

3 Find $\dfrac{dy}{dx}$ and $\dfrac{d^2y}{dx^2}$.

a $y = \dfrac{5}{2}x^4 - \dfrac{2}{3}x^3 + x - 1 - \dfrac{1}{x}$ **b** $y = 4\sqrt{x} + x\sqrt{x}$

4 Given that $y = 3x^3 + 2 - \dfrac{4}{x}$ find

a $\dfrac{dy}{dx}$ **b** the gradient when $x = -1$.

5 Find the coordinates of the points on these curves where the gradient is 12.

a $y = x^3$ **b** $y = x - \dfrac{11}{x}$ **c** $y = 48\sqrt{x}$ **d** $y = 2x^3 - 3x^2 + 4$

6 Find the equations of the tangent and the normal to the curve $y = x^3 + x^2$ at the point $(1, 2)$.

7 a Find the equations of the tangent and normal to the curve $xy = 4$ at the point where $x = 2$.

b Show that the tangent does not meet the curve again.

c Show that the normal does intersect the curve again and find the coordinates of the point of intersection.

8 Given that $y = ax^2 + bx - a^2$ has gradient -4 at $(-2, -13)$, find possible values for a and b.

9 The radius, r cm, of a circle at time t s is given by $r = \dfrac{t^3}{3} - 2t$, $t \geqslant 0$.

Find the rate at which the radius is changing when $t = 1$ and when $t = 2$.

10 Find the equations of the normals to the curve $xy = 3$ which are parallel to the line $3x - y - 2 = 0$.

A 'Test yourself' exercise can be found in the CD-ROM (Ty9).

➤ # Key points

Differentiation

$\dfrac{\mathrm{d}y}{\mathrm{d}x}$ is the **rate of change** of y with respect to x.

$$\mathrm{f}'(x) = \lim_{h \to 0} \frac{\mathrm{f}(x+h) - \mathrm{f}(x)}{h}$$

The **first derivative** (the gradient) can be written $\dfrac{\mathrm{d}y}{\mathrm{d}x}$, y' or $\mathrm{f}'(x)$.

The **second derivative** (the gradient of the gradient) can be written $\dfrac{\mathrm{d}^2 y}{\mathrm{d}x^2}$, y'' or $\mathrm{f}''(x)$.

$$y = c \;\; \Rightarrow \frac{\mathrm{d}y}{\mathrm{d}x} = 0$$

$$y = kx \;\; \Rightarrow \frac{\mathrm{d}y}{\mathrm{d}x} = k$$

$$y = kx^n \Rightarrow \frac{\mathrm{d}y}{\mathrm{d}x} = nkx^{n-1}$$

Rewrite expressions as sums of terms with powers of x before differentiating, i.e. write

$$y = (x+1)(3x-1) \quad \text{as} \quad y = 3x^2 + 2x - 1 \quad \text{or} \quad y = \frac{x^5 - x^2}{\sqrt{x}} \quad \text{as} \quad y = x^{\frac{9}{2}} - x^{\frac{3}{2}}$$

Integration was first introduced into mathematics in connection with calculating area. Integration may be thought of as the inverse process of differentiation; each process 'undoes' the other. In this chapter, to show how the powerful concept arises, we shall begin by introducing a new sort of equation, a differential equation.

After working through this chapter you should

■ *understand integration as the reverse of differentiation*

■ *be able to integrate x^n*

■ *given $f'(x)$ and a point on the curve, be able to find the equation of the curve.*

10.1 The reverse of differentiation

In Chapter 9, on differentiation, we found rates of change of functions and gradients of curves. This chapter, on integration, reverses the process: given a rate of change or a gradient function, can we find the original function?

For example, if you suspect that '*the rate at which bacteria increase is proportional to the number of bacteria*', can an expression be found for the number of bacteria at any given time? Using n for the number of bacteria at time t, the statement can be written as

$$\frac{dn}{dt} = kn$$

> $\frac{dn}{dt}$ is the rate of change of n with respect to t.

where k is called the constant of proportionality.

This is an equation containing a differential coefficient, $\frac{dn}{dt}$, and is called a

differential equation. Solving such differential equations, i.e. in this case, expressing n in terms of t, is an important branch of calculus with applications in physics, chemistry, biology, medicine, economics and many other fields.

Note   Any equation with terms in $\frac{dy}{dx}$ or $\frac{d^2y}{dx^2}$ etc. is a differential equation.

Starting with the example $\frac{dy}{dx} = 2x$, can an expression for y in terms of x be found?

$\frac{dy}{dx} = 2x$ is the result of differentiating, with respect to x, equations such as $y = x^2$ or $y = x^2 + 7$ or $y = x^2 - 2$. In fact $y = x^2 + c$, where c is any constant will, when differentiated, give $\frac{dy}{dx} = 2x$.

So $\dfrac{\mathrm{d}y}{\mathrm{d}x} = 2x \Leftrightarrow y = x^2 + c$ where c is a constant.

$y = x^2 + c$ is the **general solution** of the differential equation.

The process of finding, for example, $y = x^2 + c$ given $\dfrac{\mathrm{d}y}{\mathrm{d}x} = 2x$ is called **integration**.

$\dfrac{\mathrm{d}y}{\mathrm{d}x} = 2x$ and $y = x^2 + c$ can be illustrated graphically by the family of parabolas.

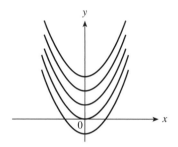

Finding the constant c

If a curve is known to pass through a given point, then the constant c can be found.

Example 1   Given that $\dfrac{\mathrm{d}y}{\mathrm{d}x} = 2x$, find the equation of the curve that passes through the point $(1, 2)$.

$\dfrac{\mathrm{d}y}{\mathrm{d}x} = 2x \Leftrightarrow y = x^2 + c$

> This is the general solution of the differential equation.

When $x = 1 \qquad y = 2$

$2 = 1 + c$

$\therefore \qquad\qquad c = 1$

> Substituting $x = 1$ and $y = 2$ in $y = x^2 + c$ gives an equation which can be solved to find c.

So the required curve is $y = x^2 + 1$.

> This is a particular solution of the differential equation.

Note $y = x^2 + c$ is the **general solution** of the differential equation $\dfrac{\mathrm{d}y}{\mathrm{d}x} = 2x$.

$y = x^2 + 1$ is a **particular solution** of the differential equation $\dfrac{\mathrm{d}y}{\mathrm{d}x} = 2x$

Using the integral sign

Integration was originally discovered, not as the reverse of differentiation, but as a process of summation (adding). This is covered later in the A-level course. The notation for integration evolved from the summation approach. The sign for integration, the integral sign $\int$, is an elongated S.

$\dfrac{\mathrm{d}y}{\mathrm{d}x} = 2x \Rightarrow y = x^2 + c$ can be expressed using the integral sign.

$$\int 2x \,\mathrm{d}x = x^2 + c$$

$\int 2x \,\mathrm{d}x$ is read as 'the integral of $2x \,\mathrm{d}x$' or 'the integral of $2x$ with respect to x'.

$\mathrm{d}x$ denotes 'integration with respect to x', i.e. x is the variable.

c is called the **constant of integration** or the **arbitrary constant**.

Before finding a rule for integrating x^n look at Example 2. Check the results by differentiating.

Example 2

a $\dfrac{\mathrm{d}y}{\mathrm{d}x} = x^2 \Leftrightarrow y = \dfrac{x^3}{3} + c$ i.e. $\int x^2 \,\mathrm{d}x = \dfrac{x^3}{3} + c$

b $\dfrac{\mathrm{d}y}{\mathrm{d}x} = 3x + \dfrac{1}{x^2} \Leftrightarrow y = \dfrac{3x^2}{2} - \dfrac{1}{x} + c$ i.e. $\int \left(3x + \dfrac{1}{x^2}\right) \mathrm{d}x = \dfrac{3x^2}{2} - \dfrac{1}{x} + c$

c $\dfrac{\mathrm{d}y}{\mathrm{d}x} = k \Leftrightarrow y = kx + c$ i.e. $\int k \,\mathrm{d}x = kx + c \ (k \neq 0)$

Rules for differentiating and integrating x^n

$$y = x^n \Rightarrow \dfrac{\mathrm{d}y}{\mathrm{d}x} = nx^{n-1}$$

The process for differentiating x^n is to multiply by the power and reduce the power by one.

The reverse process, for integrating, is to add one to the power and divide by the new power.

➤ $$\int x^n \,\mathrm{d}x = \dfrac{x^{n+1}}{n+1} + c \qquad \text{for } n \neq -1$$

Division by zero is not defined, so $n = -1$ is excluded.

Note Since any constant will disappear on differentiation, a constant must always be added on integration.

Example 3   Find $\int x^{-3}\,dx$.

$$\int x^{-3}\,dx = \frac{x^{-3+1}}{-3+1} + c$$

$$= -\frac{x^{-2}}{2} + c$$

> To integrate x^n, add 1 to the power and divide by the new power. Remember the constant.

Note   All powers of x can be integrated using this rule except x^{-1}.

> *Imagine, historically, the frustration of finding one value of n which was excluded from the rule. When $n = -1$ the integral is $\int \frac{1}{x}\,dx$. A new function had to be discovered (or invented) to integrate $\frac{1}{x}$. This function is introduced later in the A-level course.*

Integration of a polynomial

For a polynomial, the derived function is the sum of the derived functions of each term. Similarly, the integral of a polynomial is the sum of the integrals of each term.

Example 4 $\int \left(3x^4 + \frac{x}{2}\right)dx = \int 3x^4\,dx + \int \frac{x}{2}\,dx = \frac{3x^5}{5} + \frac{x^2}{4} + c$

> The arbitary constants from each integral can be combined in one constant.

Two general results for integrating follow from the results for differentiating on page 156.

➤
$$\int \{f(x) \pm g(x)\}\,dx = \int f(x)\,dx \pm \int g(x)\,dx$$

$$\int k f(x)\,dx = k \int f(x)\,dx$$

Example 5   Find these integrals.

a $\int \left(\frac{5}{x^2} + 3\sqrt{x} - \frac{2}{7x^3}\right)dx$

> Use index notation for each power of x.

$$= \int \left(5x^{-2} + 3x^{\frac{1}{2}} - \frac{2}{7}x^{-3}\right)dx$$

> Add 1 to each power and divide by the new power. *Note*: dividing by $\frac{3}{2}$ is the same as multiplying by $\frac{2}{3}$.

$$= \frac{5x^{-1}}{-1} + 3x^{\frac{3}{2}} \times \frac{2}{3} - \frac{2}{7} \times \frac{x^{-2}}{-2} + c$$

> Remember the constant.

$$= -\frac{5}{x} + 2x^{\frac{3}{2}} + \frac{1}{7x^2} + c$$

b $\displaystyle\int\left(t-\frac{3}{t}\right)^2 dt$

> Multiply out the brackets before integrating.

$$=\int\left(t^2-6+\frac{9}{t^2}\right)dt$$

$$=\int(t^2-6+9t^{-2})\,dt$$

> Integrating 6, a constant, gives $6t$.

$$=\frac{t^3}{3}-6t-9t^{-1}+c$$

> $\displaystyle\int 9t^{-2}\,dt=\frac{9t^{-2+1}}{-2+1}+c=-9t^{-1}+c$

$$=\frac{t^3}{3}-6t-\frac{9}{t}+c$$

c $\displaystyle\int\frac{x^3+\sqrt{x}}{x}\,dx$

> Divide both terms in the numerator by the denominator, x, before integrating.

$$=\int\left(\frac{x^3}{x}+\frac{x^{\frac{1}{2}}}{x}\right)dx$$

$$=\int(x^2+x^{-\frac{1}{2}})\,dx$$

$$=\frac{x^3}{3}+2x^{\frac{1}{2}}+c$$

> $-\frac{1}{2}+1=\frac{1}{2}$. Dividing by $\frac{1}{2}$ is the same as multiplying by 2.

$$=\frac{x^3}{3}+2\sqrt{x}+c$$

d $\displaystyle\int dx=x+c$

> $\displaystyle\int dx=\int 1\times dx$ and $\displaystyle\int 1\,dx=x+c$. If $\dfrac{dy}{dx}=1$, $y=\int 1dx=\int dx$.

e $\displaystyle\int kt^2 dt$

> k is a constant.

$$=\frac{kt^3}{3}+c$$

Exercise 10A

1 Find these integrals.

a $\displaystyle\int 3x^2\,dx$ **b** $\displaystyle\int 3x\,dx$ **c** $\displaystyle\int 3x^4\,dx$

d $\displaystyle\int(3+2x)\,dx$ **e** $\displaystyle\int(x-x^2)\,dx$ **f** $\displaystyle\int 2\,dx$

g $\displaystyle\int(ax+b)\,dx$ **h** $\displaystyle\int m\,dx$ **i** $\displaystyle\int x\,dx$

2 Using the integral sign, as in Question 1, integrate these expressions with respect to x.

  a x^{-3} b $\dfrac{1}{x^2}$ c $2x^{-4}$ d $\dfrac{3}{x^2}$

  e $-\dfrac{1}{4x^3}$ f $\dfrac{3}{2x^4}$ g $x^{\frac{1}{2}}$ h $x^{\frac{1}{3}}$

  i $x^{-\frac{1}{3}}$ j $-\dfrac{1}{\sqrt{x}}$ k $\dfrac{2}{3x^{\frac{2}{3}}}$ l $-\dfrac{x^{-\frac{3}{4}}}{4}$

3 Find y in terms of x.

  a $\dfrac{dy}{dx} = 4x^3$ b $\dfrac{dy}{dx} = 6 + x$ c $\dfrac{dy}{dx} = (2x + 3)^2$

  d $\dfrac{dy}{dx} = \left(x + \dfrac{1}{x}\right)^2$ e $\dfrac{dy}{dx} = \dfrac{x^2 + 3x}{\sqrt{x}}$ f $\dfrac{dy}{dx} = x(x + 1)^2$

4 Find $\displaystyle\int y\,dx$.

  a $y = 3 + 5x$ b $y = 6x^2$ c $y = \dfrac{1}{x^2}$ d $y = (\sqrt{x} + 1)^2$

  e $y = \dfrac{x^5 + x^6}{x^4}$ f $y = \dfrac{5 + x}{x^{\frac{1}{4}}}$ g $y = \dfrac{3}{4\sqrt{x}}$ h $y = (x^2 + 2)^2$

5 Find A in terms of x.

  a $\dfrac{dA}{dx} = 4x^5$ b $\dfrac{dA}{dx} = \dfrac{x^{-5}}{4}$ c $\dfrac{dA}{dx} = 4$ d $\dfrac{dA}{dx} = \dfrac{6 + \sqrt{x}}{\sqrt{x}}$

6 Find $\displaystyle\int f(t)\,dt$ for these functions.

  a $f(t) = at$, a is a constant b $f(t) = \frac{1}{3}t^3$

  c $f(t) = (t + 1)(t + 3)$ d $f(t) = \dfrac{1}{t^{n+1}}$

  e $f(t) = \dfrac{1}{t^2} + \dfrac{3 + t}{t^4}$ f $f(t) = \sqrt{t}(t + 1)$

7 Find

  a $\displaystyle\int ax\,dx$, a is a constant b $\displaystyle\int ay^2\,dy$, a is a constant

  c $\displaystyle\int \dfrac{k}{x^2}\,dx$ d $\displaystyle\int \dfrac{(y^2 + 2)(y^2 - 3)}{y^2}\,dy$

  e $\displaystyle\int x^2(\sqrt{x} + 1)\,dx$ f $\displaystyle\int \sqrt[5]{y}\,dy$

  g $\displaystyle\int \left(\dfrac{3}{\sqrt[4]{x}} - \dfrac{2}{\sqrt[3]{x}}\right)dx$ h $\displaystyle\int x^{\frac{1}{3}}\left(x^{\frac{1}{2}} + x^{-\frac{1}{2}}\right)dx$

  i $\displaystyle\int \dfrac{3x^3 + x - 2\sqrt{x}}{x}\,dx$

10.2 Applying integration

Example 6 Find the equation of the curve whose gradient at (x, y) is given by $x^3 + \dfrac{3}{x^3}$ and which passes through the point $(1, -1)$.

$$\frac{dy}{dx} = x^3 + 3x^{-3}$$

> $\dfrac{dy}{dx}$ is the gradient. Write $\dfrac{3}{x^3}$ as $3x^{-3}$.

$$\Rightarrow \quad y = \frac{x^4}{4} - \frac{3x^{-2}}{2} + c$$

> Integrate term by term. Remember the constant.

When $x = 1$, $y = -1$

> $(1, -1)$ lies on the curve so $x = 1$, $y = -1$ satisfies the equation.

So $\quad -1 = \frac{1}{4} - \frac{3}{2} + c$

> Hence the value of c can be found.

$\therefore \quad c = \frac{1}{4}$

So $\quad y = \dfrac{x^4}{4} - \dfrac{3}{2x^2} + \dfrac{1}{4}.$

> This is the equation of the curve.

Example 7 Given that $f''(x) = 7$, $f'(2) = 8$ and $f(2) = -1$, find $f(x)$.

$$f''(x) = 7$$

> Integrate $f''(x)$, the second derivative, to find $f'(x)$.

$$\Rightarrow \quad f'(x) = 7x + c$$

But $f'(2) = 8$.

> Use $f'(2) = 8$ to find c.

So $\quad 7 \times 2 + c = 8$

$\therefore \quad c = -6$

So $\quad f'(x) = 7x - 6$

> Integrate $f'(x)$ to find $f(x)$.

and $\therefore \quad f(x) = \dfrac{7x^2}{2} - 6x + k$

But $f(2) = -1$.

> Use $f(2) = -1$ to find k.

So $\quad \dfrac{7 \times 2^2}{2} - 6 \times 2 + k = -1$

$\qquad 14 - 12 + k = -1$

$\therefore \qquad\qquad k = -3$

So $f(x) = \dfrac{7x^2}{2} - 6x - 3.$

> State $f(x)$.

Exercise 10B

1 Find the equation of a curve whose gradient, $\dfrac{dy}{dx}$, is given by $2x + 5$ and which passes through the point $(3, -1)$.

2 A curve passes through the point $(2, 0)$ and its gradient is given by $3x^2 - \dfrac{1}{x^2}$. Find the equation of the curve.

3 If $f'(x) = 2x - \dfrac{1}{x^2}$ and $f(1) = 1$, find $f(x)$.

4 If $f'(x) = 3x^2 - \sqrt{x}$ and $f(0) = 4$, find $f(x)$.

5 The gradient of a curve at the point (x, y) is $3x^2 - 8x + 3$. Find

  a   the equation of the curve, given that it passes through the origin

  b   the other points of intersection of the curve with the x-axis.

6 The gradient of a curve is given by $\dfrac{dy}{dx} = 8x - 3x^2$.

  Given that the curve passes through the origin, find

  a   the equation of the curve

  b   where the curve cuts the x-axis

  c   the equation of the tangent to the curve parallel to the x-axis.

7 **a**   Find y in terms of x if $\dfrac{dy}{dx} = 6x^2 - \dfrac{1}{x^3}$, given that $y = 3$ when $x = 1$.

  b   Find s in terms of t if $\dfrac{ds}{dt} = 3t - \dfrac{8}{t^2}$, given that $s = 1\frac{1}{2}$ when $t = 1$.

8 **a**   Find A in terms of x if $\dfrac{dA}{dx} = (2x + 1)(x^2 - 1)$, given that $A = 0$, when $x = 1$.

  b   Find the value of A when $x = -1$.

9 Find the general solution, and, for the conditions given, a particular solution of each of these differential equations.

  a $\dfrac{dy}{dx} = 3x$, $y = 3$ when $x = 0$

  b $\dfrac{dy}{dx} = \dfrac{4}{x^2} + 5$, $y = -10$ when $x = -4$

  c $\dfrac{dv}{dt} = 6t + 3t^2$, $v = 12$ when $t = 2$

  d $\dfrac{dy}{dx} = \left(1 + \dfrac{1}{x}\right)\left(1 - \dfrac{1}{x}\right)$, $y = 0$ when $x = -1$

10 Given that $\dfrac{d^2y}{dx^2} = 6x - 1$ and that when $x = 2$, $\dfrac{dy}{dx} = 4$ and $y = 0$, find y in terms of x.

11 Express s as a function of t given that $\dfrac{d^2s}{dt^2} = a$ where a is a constant, and that when $t = 0$, $s = 0$ and $\dfrac{ds}{dt} = u$.

Exercise 10C (Review)

1 Find

a $\int (x^4 + 3x^2 - 5)dx$

b $\int (2x+1)(3x+1)dx$

c $\int (x^3 + x^2 + x + 1 + x^{-2} + x^{-3})dx$

d $\int \left(2x^3 + \dfrac{x}{3} - \dfrac{x^{2.5}}{2}\right)dx$

e $\int \dfrac{x^{\frac{3}{2}} - x^{\frac{1}{2}}}{x^2}dx$

f $\int (x^{\frac{1}{2}} + 1)(x^{\frac{3}{2}} - 1)dx$

g $\int \dfrac{x^3 + x}{\sqrt{x}}dx$

h $\int \left(\dfrac{3}{2\sqrt{x}} + x^{\frac{2}{3}} + 4\sqrt{x}\right)dx$

i $\int \left(\dfrac{2}{x^3} - \dfrac{1}{2x^2} + \dfrac{4}{x^{\frac{3}{2}}}\right)dx$

2 Given that $\dfrac{dy}{dx} = x^2 + 2$ and that $y = 6$ when $x = 3$, find

a y in terms of x

b the value of y when $x = 1$.

3 Given that $f'(x) = (3x+1)(x-2)$ and $f(-1) = \frac{1}{2}$, find $f(x)$.

4 Find the equation of a curve whose gradient, $\dfrac{dy}{dx}$, is given by $(4x - 3)$, and which passes through the point $(-2, 6)$.

5 If $f'(x) = \dfrac{2}{x^2} - x$ and $f(2) = 3$, find $f(x)$.

6 The gradient of a curve is given by $\dfrac{dy}{dx} = ax + b$

Given that the curve passes through $(0, 0)$, $(1, 2)$ and $(-1, 4)$, find

a the values of a and b

b the equation of the curve.

A 'Test yourself' exercise can be found on the CD-ROM (Ty10).

Key points

Integration

$\int x^n\, dx = \dfrac{x^{n+1}}{n+1} + c \quad n \neq -1 \quad c$ is the constant of integration.

To integrate x^n add one to the power and divide by the new power.

$\int \{f(x) \pm g(x)\}\, dx = \int f(x)\, dx \pm \int g(x)\, dx$

$\int kf(x)\, dx = k \int f(x)\, dx$

11 Proof

Mathematical proof is the process of starting with an assumption, or a statement which is given, and, by logical argument, arriving at a conclusion.

If you have worked through the first ten chapters of this book you will have seen many proofs in the worked examples and you will have proved many results in the exercises.

The techniques of proof can take various forms. Throughout this book problems are solved and results proved by logical deduction. An example of another method of proof, known as 'reductio ad absurdum', is given on the CD-ROM (E1.2). Other forms of proof are considered later in the A-level course.

There is an extension chapter on the CD (E11.1) which gives practice in the terms 'implies', 'is implied by', 'necessary and sufficient' and the notation $\Rightarrow$, $\Leftarrow$ and $\Leftrightarrow$. A shortened version using only the implication signs is given here.

11.1 Mathematical proof

Any question which is worded:

'Prove that ...' or 'Given ..., prove...' or 'Prove ..., given ...'

> The word 'show' often replaces 'prove'.

requires you to form a logical argument. In each case you must start either with what is given or with standard results. Each step of the proof must be deduced logically from the previous step until you arrive at what you are trying to prove. At any stage you may use standard results such as Pythagoras' theorem or trigonometric identities.

11.2 Implication signs

Every step of a proof is a mathematical statement. Examples of statements are

- $\triangle ABC$ is isosceles
- $\sin \theta = \frac{3}{5}$
- the gradient of $y = mx + c$ is m

The relationship between statements can be expressed in different ways. One way is using implication signs.

$\Rightarrow$ implies
$\nRightarrow$ does not imply
$\Leftarrow$ is implied by
$\Leftrightarrow$ implies and is implied by

Consider three statements about a triangle ABC

- △ABC is isosceles

- AB = AC

- $\angle B = \angle C$

AB = AC ⇒ △ABC is isosceles
Read 'AB = AC implies triangle ABC is isosceles'.

However △ABC is isosceles ⇏ AB = AC
Read 'Triangle ABC is isosceles does not imply AB = AC'.

It could be that △ABC is isosceles with AB ≠ AC. The two equal sides, which are necessary since △*ABC* is isosceles, could be *AB* and *BC*.

Also △ABC is isosceles ⇐ AB = AC
Read 'Triangle ABC is isosceles is implied by AB = AC'.

Consider now the statements
AB = AC and $\angle B = \angle C$

$$AB = AC \Rightarrow \angle B = \angle C$$

and $\quad \angle B = \angle C \Rightarrow AB = AC$

so $\quad AB = AC \Leftrightarrow \angle B = \angle C$
Read 'AB = AC implies and is implied by $\angle B = \angle C$'.

Example 1   Link the statements $a = 0$ and $ab = 0$ using implication signs.

$a = 0 \Rightarrow ab = 0$
a being zero is enough to ensure that *ab* is zero.

The statements cannot be linked using ⇔ since $ab = 0 \nRightarrow a = 0$; b could be zero instead.

However $ab = 0 \Rightarrow$ Either $a = 0$ or $b = 0$ or both.

Example 2   Link the statements $x^3 > 0$ and $x > 0$ using implication signs.

$$x^3 > 0 \Rightarrow x > 0$$

and $\quad x > 0 \Rightarrow x^3 > 0$
The cube of a +ve number is +ve and the cube root of a +ve number is +ve, so $x^3 > 0$ implies $x > 0$ and vice versa.

so $\quad x^3 > 0 \Leftrightarrow x > 0$

Exercise 11A   **1** For these pairs of statements replace the … with ⇒, ⇐ or ⇔. Assume N is a positive integer.

    a $x^2 = 36$ … $x = -6$ **b** $x(x + 1) = 0$ … $x = 0$

    c $x(x + 1) = 0$ … $x = 0$ or $x = -1$ **d** $x^2 < x$ … $x < 1$

    e The last digit of N is 1. … The last digit of N^2 is 1.

    f The triangle is equilateral. … The three sides of the triangle are equal.

    g $b^2 - 4ac < 0$ … $ax^2 + bx + c = 0$ has no real roots.

    h N is a multiple of 5. … Last digit of N is zero.

    i $y = 3x + 1$ … $x = \dfrac{y - 1}{3}$

    j The nth term of a sequence is $3n + 2$. … The terms of the sequence differ by 3.

k $f(x) = 3x^2 + 4x + 1 \dots f'(x) = 6x + 4$

l In the triangle ABC $a^2 + b^2 = c^2$. ... The triangle ABC is right angled.

2 Prove that the sum of the first n even numbers is $n(n + 1)$.

3 a Prove that the sum of n terms of an arithmetic series, with first term a and last term l, is $\dfrac{n}{2}(a + l)$

   b Prove that the sum of n terms of an arithmetic series, with first term a and common difference d, is $\dfrac{n}{2}\left(2a + (n - 1)d\right)$.

4 Prove that, for a geometric series with first team a and common ratio r

   a the sum of n terms is $\dfrac{a(r^n - 1)}{r - 1}$

   b for $|r| < 1$, the sum of infinity is $\dfrac{a}{1 - r}$

5 Given that $ax^2 + bx + c = 0$, prove, by completing the square, that

$$x = \frac{-b \pm \sqrt{(b^2 - 4ac)}}{2a}.$$

6 Prove that when the polynomial $p(x)$ is divided by $(ax - b)$ the remainder is $p(\frac{b}{a})$.

7 Prove that there is no value of k for which $x^2 + (k - 2)x + k^2 + 2 = 0$ has real roots.

8 Given that A is $(-8, 8)$, B is $(6, 4)$ and C is $(3, 2)$, prove that $\triangle$ABC is right angled.

9 Prove that the line $y = 3x + 4$ is a tangent to the curve $x^2 + xy + 1 = 0$.

10 Prove that $\dfrac{3}{3 - \sqrt{2}} - \dfrac{4}{\sqrt{2} + 2} = \dfrac{17\sqrt{2} - 19}{7}$

11 Prove that the perpendicular bisector of AB where A is $(-4, 6)$ and B is $(4, -2)$ passes through the point $(5, 7)$.

12 Given that the lines $3y + ax = 6$ and $2y + bx = -4$ are perpendicular, prove that $ab = -6$.

13 Prove that $\displaystyle\sum_{1}^{8} (3r - 5) = 68$

14 Prove that the normal to the curve $y = 2x^2 - 3x^{-\frac{1}{2}}$ at the point $(1, -1)$ passes through the point $(12, 3)$.

15 a By completing the square, prove that the roots of $x^2 + px + q = 0$ are

$$x = \frac{-p \pm \sqrt{p^2 - 4q}}{2}$$

   b Hence prove that $x^2 + px + q = 0$ has real roots if and only if $p^2 - 4q > 0$.

16 Prove that $x^2 + 1 \geqslant 2x$ for all $x > 0$.

17 Prove that the equation $4 = x(2 - x)$ has no real roots.

1 The line L passes through the points $A(1, 3)$ and $B(-19, -19)$.

    a Calculate the distance between A and B.

    b Find an equation of L in the form $ax + by + c = 0$, where a, b and c are integers.

London June 1997

2 The points $A(-2, 4)$, $B(6, -2)$ and $C(5, 5)$ are the vertices of $\triangle ABC$ and D is the mid-point of AB.

    a Find an equation of the line passing through A and B in the form $ax + by + c = 0$, where a, b, and c are integers to be found.

    b Show that CD is perpendicular to AB.

Edexcel Specimen paper

3 **a** Find an equation of the line l which passes through the points $A(1, 0)$ and $B(5, 6)$.

    The line m with equation $2x + 3y = 15$ meets l at the point C.

    b Determine the coordinates of C.

    The point P lies on m and has x-coordinate -3.

    c Show, by calculation, that $PA = PB$.

London June 1996

4 The straight line l_1 with equation $y = \frac{3}{2}x - 2$ crosses the y-axis at the point P. The point Q has coordinates $(5, -3)$.

    a Calculate the coordinates of the mid-point of PQ.

    The straight line l_2 is perpendicular to l_1 and passes through Q.

    b Find an equation for l_2 in the form $ax + by = c$, where, a, b and c are integer constants.

    The lines l_1 and l_2 intersect at the point R.

    c Calculate the exact coordinates of R.

Edexcel Jan 2003

5
$$\frac{\mathrm{d}y}{\mathrm{d}x} = 5 + \frac{1}{x^2}.$$

    a Use integration to find y in terms of x.

    b Given that $y = 7$ when $x = 1$, find the value of y at $x = 2$.

Edexcel June 2003

6 **a** Find the sum of the integers which are divisible by 3 and lie between 1 and 400.

    b Hence, or otherwise, find the sum of the integers, from 1 to 400 inclusive, which are *not* divisible by 3.

London Jan 1997

7 Find

   a $\int (\frac{2}{3}x^3 + 5x^{-\frac{1}{3}})\mathrm{d}x$ **b** $\int \frac{(2x+1)^2}{x^{\frac{1}{2}}}\mathrm{d}x$

8 **a** Find the sum of all the integers between 1 and 1000 which are divisible by 7.

   b Hence, or otherwise, evaluate $\sum_{r=1}^{142}(7r+2)$.

Edexcel P1 May 2002

9 Differentiate with respect to x

$$2x^3 + \sqrt{x} + \frac{x^2 + 2x}{x^2}.$$

Edexcel Nov 2002 (part)

10 **a** Find an equation of the straight line passing through the points with coordinates $(-1, 5)$ and $(4, -2)$, giving your answer in the form $ax + by + c = 0$, where a, b and c are integers.

The line crosses the x-axis at the point A and the y-axis at the point B, and O is the origin.

   b Find the area of $\triangle OAB$.

London Jan 1997

11 $y = 7 + 10x^{\frac{3}{2}}$

   a Find $\frac{\mathrm{d}y}{\mathrm{d}x}$.   **b** Find $\int y\,\mathrm{d}x$.

Edexcel Jan 2003

12 $y = \frac{x^3 + \sqrt{x}}{x}$

   a Find $\frac{\mathrm{d}y}{\mathrm{d}x}$.   **b** Find $\frac{\mathrm{d}^2 y}{\mathrm{d}x^2}$.   **c** Find $\int y\,\mathrm{d}x$.

13 **a** An arithmetic series has first term a and common difference d. Prove that the sum of the first n terms of this series is

$$\frac{1}{2}n[2a + (n-1)d]$$

The first three terms of an arithmetic series are k, 7.5 and $k + 7$ respectively.

   b Find the value of k.

   c Find the sum of the first 31 terms of this series.

Edexcel Mock paper

14 The points A and B have coordinates $(4, 6)$ and $(12, 2)$ respectively.
The straight line l_1 passes through A and B.

   a Find an equation for l_1 in the form $ax + by = c$, where a, b and c are integers.

The straight line l_2 passes through the origin and has gradient -4.

   b Write down an equation for l_2.

The lines l_1 and l_2 intercept at the point C.

   c Find the exact coordinates of the mid-point of AC.

Edexcel June 2003

15 $f(x) = 9 - (x - 2)^2$

    a Write down the maximum value of $f(x)$.

    b Sketch the graph of $y = f(x)$, showing the coordinates of the points at which the graph meets the coordinate axes.

The points A and B on the graph of $y = f(x)$ have coordinates $(-2, -7)$ and $(3, 8)$ respectively.

    c Find, in the form $y = mx + c$, an equation of the straight line through A and B.

    d Find the coordinates of the point at which the line AB crosses the x-axis.

The mid-point of AB lies on the line with equation $y = kx$, where k is a constant.

    e Find the value of k.

Edexcel Nov 2002

16 In the first month after opening, a mobile phone shop sold 280 phones. A model for future trading assumes that sales will increase by x phones per month for the next 35 months, so that $(280 + x)$ phones will be sold in the second month, $(280 + 2x)$ in the third month, and so on.

Using this model with $x = 5$, calculate

    a   **i** the number of phones sold in the 36th month,
         ii the total number of phones sold over the 36 months.

The shop sets a sales target of 17 000 phones to be sold over the 36 months.

Using the same model,

    b find the least value of x required to achieve this target.

Edexcel June 2003

17 The figure shows a sketch of the curve with equation $y = f(x)$, $-1 \leqslant x \leqslant 4$. The curve cuts the y-axis at A(0, 1), touches the x-axis at B(1, 0), passes through a turning point C(2, $\frac{5}{2}$) and cuts the x-axis at D(3, 0).

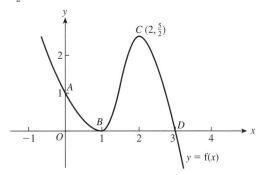

In separate diagrams show a sketch of the curve with equation

    a $f(x + 1)$ **b** $f(-x)$

marking on each sketch the coordinates of points at which the curve

    i meets the axes,

    ii has a turning point.

18 A curve C has equation $y = x^3 - 5x^2 + 5x + 2$.

a Find $\dfrac{dy}{dx}$ in terms of x.

The points P and Q lie on C. The gradient of C at both P and Q is 2. The x-coordinate of P is 3.

b Find the x-coordinate of Q.

c Find an equation for the tangent to C at P, giving your answer in the form $y = mx + c$, where m and c are constants.

This tangent intersects the coordinate axes at the points R and S.

d Find the length of RS, giving your answer as a surd.

Edexcel Jan 2002

19 The curve C has equation $y = f(x)$. Given that

$$\frac{dy}{dx} = 3x^2 - 20x + 29$$

and that C passes through the point $P(2, 6)$,

a find y in terms of x.

b Verify that C passes through the point $(4, 0)$.

c Find an equation of the tangent to C at P.

The tangent to C at the point Q is parallel to the tangent at P.

d Calculate the exact x-coordinate of Q.

Edexcel Nov 2002

20 $f'(x) = \dfrac{x^3 + 1}{x^2}$

a Use integration to find $f(x)$ in terms of x.

b Find the equation of the curve in the form $y = f(x)$, given that $y = 5$ when $x = 2$.

21 Each year, for 40 years, Anne will pay money into a savings scheme. In the first year she pays in £500. Her payments then increase by £50 each year, so that she pays in £550 in the second year, £600 in the third year, and so on.

a Find the amount that Anne will pay in the 40th year.

b Find the total amount that Anne will pay in over the 40 years.

Over the same 40 years, Brian will also pay money into the savings scheme. In the first year he pays in £890 and his payments then increase by £d each year.

Given that Brian and Anne will pay in exactly the same amount over the 40 years,

c find the value of d.

Edexcel June 2001

22 The straight line l_1 has equation $4y + x = 0$.

The straight line l_2 has equation $y = 2x - 3$.

a On the same axes, sketch the graphs of l_1 and l_2. Show clearly the coordinates of all points at which the graphs meet the coordinate axes.

The lines l_1 and l_2 intersect at the point A.

b Calculate, as exact fractions, the coordinates of A.

c Find an equation of the line through A which is perpendicular to l_1. Give your answer in the form $ax + by + c = 0$, where a, b and c are integers.

Edexcel Jan 2002

23 The points $A(-1, -2)$, $B(7, 2)$ and $C(k, 4)$, where k is a constant, are the vertices of $\triangle ABC$. Angle ABC is a right angle.

a Find the gradient of AB.

b Calculate the value of k.

c Show that the length of AB may be written in the form $p\sqrt{5}$, where p is an integer to be found.

d Find the exact value of the area of $\triangle ABC$.

e Find an equation for the straight line l passing through B and C. Give your answer in the form $ax + by + c = 0$, where a, b and c are integers.

The line l crosses the x-axis at D and the y-axis at E.

f Calculate the coordinates of the mid-point of DE.

Edexcel June 2001

24

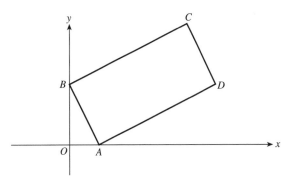

The points A (3, 0) and B (0, 4) are two vertices of the rectangle $ABCD$, as shown in the diagram.

a Write down the gradient of AB and hence the gradient of BC.

The point C has coordinates $(8, k)$, where k is a positive constant.

b Find the length of BC in terms of k.

Given that the length of BC is 10 and using your answer to part **b**,

c find the value of k,

d find the coordinates of D.

Edexcel Jan 2001

25 a Find an equation of the line p which passes through the point $(-3, 2)$ and which is parallel to the line q with equation $7x - 2y - 14 = 0$.

The lines p and q meet the y-axis at the points A and B respectively.

b Find the distance between AB.

Edexcel Mock paper 1

26

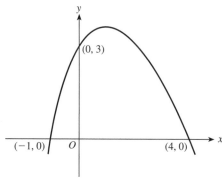

The curve with equation $y = f(x)$ meets the coordinate axes at the points $(-1, 0)$, $(4, 0)$ and $(0, 3)$, as shown in the figure. Using a separate diagram for each, sketch the curve with equation

a $y = f(x - 1)$,

b $2y = -f(x)$.

On each sketch, write in the coordinates of the points at which the curve meets the coordinate axes.

London June 1997

27 The curve C has equation $y = 3x^2 - \sqrt{x}$, $x > 0$.
The point A$(\frac{1}{4}, k)$ lies on the curve.

a Find the value of k.

b Find the equation of the tangent to the curve C at the point A in the form $ax + by + c = 0$.

The tangent meets the y-axis at the point M.

c Find the coordinates of M.

d Find the equation of the normal to the curve C at the point A in the form $ax + by + c = 0$.

The normal meets the y-axis at the point N.

e Find the coordinates of N.

f Find the area of triangle AMN.

28 The diagram shows part of the curve with equation $y = \mathrm{f}(x)$. The curve cuts the x-axis at the points A(1, 0) and B(4, 0). The point C(2, 3) is a maximum point.

Using a separate diagram for each, sketch the curve with equation

a $y = \mathrm{f}(-x)$ **b** $y = \mathrm{f}(2x)$ **c** $y = 3\mathrm{f}(x)$ **d** $y = \mathrm{f}(\frac{1}{2}x)$

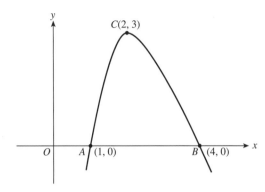

On each sketch mark the coordinates of the positions of A, B and C after the transformations.

29 Find the set of values of x for which $(x - 1)(x - 4) < 2(x - 4)$.

London Jan 1999

30 A sequence of terms $\{u_n\}$ is defined, for $n \geqslant 1$, by the recurrence relation

$$u_{n+2} = 2ku_{n+1} + 15u_n,$$

where k is a constant. Given that $u_1 = 1$ and $u_2 = -2$,

a find an expression, in terms of k, for u_3.

b Hence find an expression, in terms of k, for u_4.

Given also that $u_4 = -38$,

c find the possible values of k.

London Jan 1999

31 The curve C has equation $y = 4x - x^2$.

a Sketch the curve C.

The curve cuts the x-axis at the origin and at a point A(k, 0).

b Find the value of k.

c Find the equation of the normal to the curve C at the point A.

The normal at A meets the curve again at a point P.

d Find the coordinates of P.

- The paper is 1 hour 30 minutes long.

- No calculators may be used.

1. Express $\dfrac{1 + \sqrt{2}}{3 - \sqrt{2}}$ in the form $a + b\sqrt{2}$.

(4)

---

2. Find $\displaystyle\int (4x + \sqrt[3]{x})\,\mathrm{d}x$

(4)

---

3. The sum S of an arithmetic series is given by

$$S = \sum_{r=1}^{20} (5r + 3).$$

(a) Write down the first three terms of the series.

(2)

(b) Find the common difference of the series.

(1)

(c) Calculate the value of S.

(2)

---

4. (a) Express 16^x in the form 2^{ax} where a is an integer to be found.

(2)

(b) Hence solve the equation

$$2^{x^2} = 16^x.$$

(3)

---

5. The points $A(-4, -2)$, $B(6, 4)$ and $C(4, -4)$ are the vertices of triangle ABC.

(a) Find an equation for the line passing through A and B in the form $ax + by + c = 0$, where a, b and c are integers.

(4)

Point D is the mid-point of AB.

(b) Show that CD is perpendicular to AB.

(4)

6.

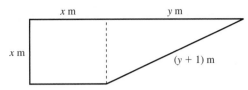

The diagram represents the floor plan of an attic room which consists of a square of side x m and a triangle of sides x m, y m and $(y + 1)$ m.

The perimeter of the room is 10 m.

(a) Show that $3x + 2y = 9$.

(2)

The area of the floor is $5.5\,\text{m}^2$.

(b) By forming another equation in x and y, find the value of x and the value of y.

(9)

---

7. $x_{n+1} = \dfrac{a - 3x_n}{x_n}, \qquad n > 0, \quad a > 3, \quad x_1 = 1.$

(a) Find, in terms of a, expressions for

    (i) x_2,   (ii) x_3

(3)

Given that $x_3 > 2$,

(b) find the range of values of a.

(4)

---

8. (a) Find, in terms of k, the roots of the equation

$$x^2 + 2kx - 5 = 0.$$

(3)

(b) Prove that, for all real values of k, the roots of $x^2 + 2kx - 5 = 0$ are real and different.

(2)

(c) Given that $k = \sqrt{3}$, find the roots of the equation in the form $p\sqrt{3} + q\sqrt{2}$ where p and q are integers.

(3)

---

9. The curve C has equation $y = 2x^3 - x^2$. The points A and B both lie on the curve C having coordinates $(1, 1)$ and $(2, 12)$ respectively.

(a) Show that the gradient of C at B is 5 times the gradient of C at A.

(5)

(b) Find an equation for the tangent to C at B.

(3)

10.

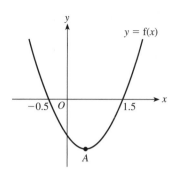

The figure shows a sketch of the curve with equation $f(x) = 4x^2 - 4x - 3$.
Given that $4x^2 - 4x - 3 = a(x+b)^2 + c$, for all x, where a, b and c are constants,

(a) calculate the value of a, b and c.

(4)

(b) Hence, or otherwise, find the coordinates of the minimum point A.

(2)

On separate diagrams sketch the curve with equation

(c) $y = f(x+3)$,

(3)

(d) $y = 3f(x)$.

(2)

On each diagram, show clearly the coordinates of the minimum point, and of each point at which the curve crosses the x-axis.

---

11. The curve C has equation $y = f(x)$ and the point $P(2, 10.25)$ lies on C.

Given that $f'(x) = (x^{\frac{3}{2}} - 2x^{-\frac{3}{2}})^2 + 2$,

(a) show that $f'(x)$ can be written as $Ax^3 + \dfrac{B}{x^3} + D$ where A, B and D are constants.

(4)

(b) Find $f(x)$.

(4)

(c) Verify that the point $Q(1, 7)$ lies on C.

(2)

(d) Find an equation of the normal to C at Q.

(4)

END

12 More algebra

In this chapter more advanced tools of algebra, particularly concerned with division, are introduced.

After working through this chapter you should be

- *familiar with the terms 'quotient' and 'remainder' when they are used in dividing polynomials*
- *able to carry out simple algebraic division, division by $(x + a)$ or $(x - a)$*
- *able to use the Factor Theorem and the Remainder Theorem*
- *able to factorise cubic expressions such as $x^3 + 3x^2 - 4$*
- *able to find unknowns in an identity.*

12.1 Identities

In Chapter 3 the idea of an identity was introduced.

Remember An identity is an equation in which the two expressions are equal for all values of the variable.

To prove an identity, one has to show that the expressions on both sides of the equals sign are identical. One method of doing this is to start with one side and by algebraic manipulation show that it is equal to the other side.

Example 1 Prove that $x^2 + 2x - 2 \equiv (x + 1)(x - 1) + 2(x + 1) - 3$.

$\text{RHS} \equiv (x + 1)(x - 1) + 2(x + 1) - 3$ Start with the RHS as it can be simplified.

$\phantom{\text{RHS}} \equiv x^2 - 1 + 2x + 2 - 3$

$\phantom{\text{RHS}} \equiv x^2 + 2x - 2$ Show, by simplifying, that it equals the LHS.

$\phantom{\text{RHS}} \equiv \text{LHS}$

In the following examples the identities contain unknowns, A, B, etc. which have to be found. One method of finding unknowns is by **substitution**. Since the expressions are equal for *all* values of x *any* value can be substituted. Choose values, where possible, that eliminate one of the unknowns.

Example 2 Find A and B given that $6x + 9 \equiv A(x + 1) + B(x - 2)$.   Use substitution.

$6x + 9 \equiv A(x + 1) + B(x - 2)$ Since this is an identity and thus true for all values of x, any value can be substituted.

Putting $x = -1$

$-6 + 9 = A \times 0 - 3B$ Putting $x = -1$ eliminates one of the terms on the RHS since it makes the bracket $(x + 1)$ zero.

$3 = -3B$

$B = -1$

Putting $x = 2$

$12 + 9 = 3A + B \times 0$

$21 = 3A$

$A = 7$

So $A = 7$ and $B = -1$, and $6x + 9 \equiv 7(x + 1) - (x - 2)$.

> Putting $x = 2$ eliminates the other term on the RHS since it makes the bracket $(x - 2)$ zero.

> This can easily be checked by multiplying out the RHS.

In Example 3 it is not possible to choose values of x which eliminate terms. Two possible methods are shown.

Example 3   Find A and B given that $4x^2 - 3x + 10 \equiv A(x^2 + 1) + B(2x^2 - x + 4)$

Method 1

$4x^2 - 3x + 10 \equiv A(x^2 + 1) + B(2x^2 - x + 4)$

Putting $x = 0$

$10 = A \times 1 + B \times 4$

$10 = A + 4B$ ①

Putting $x = 1$

$4 - 3 + 10 = A(1 + 1) + B(2 - 1 + 4)$

$11 = 2A + 5B$ ②

Subtracting ② from $2 \times$ ①

$9 = 3B$

$B = 3$

Substituting in ①

$10 = A + 4 \times 3$

$A = -2$

So $A = -2$ and $B = 3$

> There are no real values of x which make $(x^2 + 1)$ or $(2x^2 - x + 4)$ zero. However, since this is an identity, any value of x can be substituted. Equations connecting A and B can be found and then solved.

> Remember to check the solution.

Another technique for solving identities is by **equating coefficients**.
Given that, for example, $ax^3 + bx^2 + cx + d \equiv Ax^3 + Bx^2 + Cx + D$, it can be shown that $a = A$, $b = B$, $c = C$ and $d = D$, i.e. the coefficients of the x^3, x^2 and x terms must be equal. Similarly the constant terms must be equal.

Method 2

$4x^2 - 3x + 10 \equiv A(x^2 + 1) + B(2x^2 - x + 4)$

$\equiv Ax^2 + A + 2Bx^2 - Bx + 4B$

$\equiv (A + 2B)x^2 - Bx + (A + 4B)$

> Multiplying out brackets and collecting like terms make it easier to compare coefficients. With practice, the coefficients can be worked out mentally.

Equating coefficients of x terms:

$-3 = -B$

$B = 3$

> The coefficient of x on the LHS is -3 and on the RHS is $-B$. These must be equal.

Equating constant terms:

$10 = A + 4B$ ①

But $B = 3$

Substituting $B = 3$ in ①

> The value of B has been found, so it can be substituted here to find A.

$10 = A + 4 \times 3$

$\quad = A + 12$

$A = -2$

So $A = -2$ and $B = 3$

Many identities can be solved by using a combination of both methods: substitution and equating coefficients.

Example 4 Find A, B and C given that $3x^2 - 12x + 16 \equiv A(x + B)^2 + C$.

Hence state the vertex of the parabola $y = 3x^2 - 12x + 16$.

$3x^2 - 12x + 16 \equiv A(x + B)^2 + C$

$\qquad\qquad\qquad \equiv Ax^2 + 2ABx + AB^2 + C$

> Multiplying out brackets and collecting like terms make it easier to compare coefficients. With practice, the coefficients can be worked out mentally.

Equating coefficients of x^2 terms:

$3 = A$

$A = 3$

> The coefficient of x^2 on the LHS is 3 and on the RHS is A. These must be equal.

Equating coefficients of x terms:

$-12 = 2AB$

But $A = 3$

$\therefore -12 = 2 \times 3 \times B$

$B = -2$

> The value of A has been found, so it can be substituted here to find B.

Equating constant terms:

$16 = AB^2 + C$ ①

Substituting $A = 3$ and $B = -2$ in ①

$16 = 3 \times (-2)^2 + C$

$\quad = 12 + C$

So $C = 4$

> The values of A and B have been found, so they can be substituted here to find C.

Hence $A = 3$, $B = -2$ and $C = 4$

So $3x^2 - 12x + 16 \equiv 3(x - 2)^2 + 4$

> Check by substituting for A, B and C. This gives $3(x - 2)^2 + 4$. Multiply out the brackets to check that $3x^2 - 12x + 16$ is obtained.

The vertex of the parabola

$y = 3x^2 - 12x + 16$ is at $(2, 4)$

> **Recall** The vertex of $y = a(x + p)^2 + q$ is $(-p, q)$.

Compare this method of completing the square with the method on page 50.

1 Find the values of A and B in these identities.

a $5x - 14 \equiv A(x - 1) + B(x - 4)$ b $2x + 6 \equiv A(x + 2) + B(x + 4)$

c $-x \equiv A(x - 3) - B(x - 2)$ d $5x + 17 \equiv A(x + 2) - B(x - 5)$

e $2x - 4 \equiv A(3 + x) + B(7 - x)$ f $5x - 7 \equiv A(x - 1) + B(2x - 3)$

g $8x + 1 \equiv A(3x - 1) + B(2x + 3)$

2 Find the values of A, B and C in these identities.

a $6x^2 - 25x + 23 \equiv A(x - 1)(x - 2) + B(x - 2)(x - 3) + C(x - 3)(x - 1)$

b $4x^2 - 2x + 4 \equiv Ax(x + 2) + B(x + 2)(x - 2) + Cx(x - 2)$

c $4x^2 + 4x - 26 \equiv A(x + 2)(x - 4) + B(x - 4)(x - 1) + C(x - 1)(x + 2)$

d $2x^2 - 22x + 53 \equiv A(x - 5)(x - 3) + B(x - 3)(x + 2) + C(x + 2)(x - 5)$

e $2 \equiv A(x - 1)(x + 1) + Bx(x + 1) + Cx(x - 1)$

f $2x^2 + 9x - 10 \equiv A(x - 3)(x + 4) + B(x + 2)(x + 4) + C(x + 2)(x - 3)$

3 Prove that

$(2x + 3)(x - 7) - 2(x + 8)(x - 2) \equiv 11 - 23x$

4 Find the values of A, B and C in these identities.

a $8 - x \equiv A(x - 2)^2 + B(x - 2)(x + 1) + C(x + 1)$

b $2x^3 - 15x^2 - 10 \equiv A(x - 2)(x + 1) + B(x + 1)(2x^2 + 1) + C(2x^2 + 1)(x - 2)$

c $22 - 4x - 2x^2 \equiv A(x - 1)^2 + B(x - 1)(x + 3) + C(x + 3)$

d $3x^2 + 6x - 4 \equiv A(x + 1)^2 + B$

e $5x^2 + 7x + 9 \equiv A(x + 1)^2 + B(x + 1) + C$

f $x^2 + 1 \equiv A(x - 2)^2 + B(x - 2) + C$

5 By equating coefficients, find A, B and C.

a $2x^2 - 12x + 23 \equiv A(x - B)^2 + C$ b $5 + 4x - 2x^2 \equiv A + B(x - C)^2$

c $4x^2 - 40x + 50 \equiv A(x - B)^2 + C$ d $-4x^2 + 3x + 1 \equiv A(x - B)^2 + C$

6 By finding p, q and r, express these quadratic expressions in completed square form and hence state the coordinates of the vertex of their graphs.

a $3x^2 - 6x - 4 \equiv p(x - q)^2 + r$ b $5x^2 + 20x - 2 \equiv p(x - q)^2 + r$

c $3x^2 + 16x + 3 \equiv p(x + q)^2 + r$ d $-2x^2 - x + 3 \equiv p(x + q)^2 + r$

7 Find R in these identities.

a $2x^2 + 3x - 4 \equiv (x + 2)(Px + Q) + R$

b $6x^4 - 2x^3 + 5x^2 - 3x + 6 \equiv (x - 1)(Ax^3 + Bx^2 + Cx + D) + R$

8 Find the values of A, B and C in these identities.

a $A(x^2 + 1) + B(2x^2 + 3) + Cx(x + 1) \equiv 3x^2 + 2x + 2$

b $(Ax + B)(x^2 + 2) + Cx^2(x - 1) \equiv 2x^3 + 5x^2 + 8x + 6$

c $A(x^2 + x + 1) + (Bx + C)(3x - 2) \equiv 11x^2 - 7x + 4$

12.2 Division of polynomials

Before tackling division in algebra, it is useful to think about dividing integers, and, in particular, how long division is carried out.

Think about dividing 36 by 5. This can be written as $36 \div 5$ or $\dfrac{36}{5}$.

$\dfrac{36}{5} = 7\frac{1}{5}$ or $\dfrac{36}{5} = 7$ remainder 1.

This can be expressed as

$$36 \ = \ 5 \ \times \ 7 \ + \ 1$$

Dividend   Divisor   Quotient   Remainder

Now think of dividing $(x^2 + x - 3)$ by $(x - 1)$. This can be written as

$(x^2 + x - 3) \div (x - 1)$ or $\dfrac{x^2 + x - 3}{x - 1}$.

As you will see below

$$x^2 + x - 3 \text{ can be written as } (x - 1)(x + 2) - 1$$

Dividend   Divisor   Quotient   Remainder

A method is needed to find the quotient and remainder in all situations.

The method of long division in algebra has much in common with long division of integers.

The result of this division can be written as

$$7846 = 23 \times 341 + 3$$

or $\dfrac{7846}{23} = 341$ remainder 3

or $\dfrac{7846}{23} = 341\frac{3}{23}$

Consider dividing $(x^2 + x - 3)$ by $(x - 1)$.

The divisor and dividend should always be arranged in descending powers of the variable.

$$\begin{array}{r} x+2 \\ x-1\overline{\smash{\big)}\,x^2+x-3} \\ \underline{x^2-x} \\ 2x-3 \\ \underline{2x-2} \\ -1 \end{array}$$

■ Divide the term on the left of the dividend (x^2) by the term on the left of the divisor (x). This gives

$$\frac{x^2}{x} = x$$

This is the first term of the quotient.

■ Multiply the whole divisor ($x-1$) by this term of the quotient (x). The result is $x^2 - x$. Write this below the dividend, lining up like terms. Then subtract it from the dividend.

■ Bring down from the dividend the next term (-3). (Sometimes more than one term needs to be brought down.) This forms the new dividend.

■ Repeat the process. Divide the term on the left of the new dividend ($2x$) by the term on the left of the divisor (x). This gives

$$\frac{2x}{x} = 2$$

This is the second term of the quotient.

■ Multiply the whole divisor ($x-1$) by this term of the quotient (2). This result is $2x - 2$. Write this below the dividend and subtract it from the new dividend ($2x - 3$). The result is -1.

■ The process cannot be repeated as x cannot be divided into -1. So -1 is the remainder. So

$$x^2 + x - 3 \equiv (x-1)(x+2) - 1$$

■ Finally, check the division by multiplying.

Note When to stop! When the divisor is linear the remainder is a constant, so when a constant is obtained (such as -1 above) the division stops.

In general, the division stops when the remainder is a polynomial of lower degree than the divisor.

> The algorithm (set of instructions) for division of polynomials
>
> ■ **Arrange the divisor and dividend in descending powers of the variable, lining up like terms, and leaving gaps in the dividend for 'missing' terms.**
> ■ **Divide the term on the left of the dividend by the term on the left of the divisor. The result is a term of the quotient.**
> ■ **Multiply the whole divisor by this term of the quotient and subtract the product from the dividend.**
> ■ **Bring down as many terms as necessary to form a new dividend.**
> ■ **Repeat the instructions until all terms of the dividend have been used.**

Example 5   Divide $x^3 - 3x^2 + 6x + 5$ by $x - 2$.

$$
\begin{array}{r}
x^2 - x + 4 \\
x - 2 \overline{\smash{)}\, x^3 - 3x^2 + 6x + 5} \\
\underline{x^3 - 2x^2} \\
-x^2 + 6x \\
\underline{-x^2 + 2x} \\
4x + 5 \\
\underline{4x - 8} \\
+ 13
\end{array}
$$

So $x^3 - 3x^2 + 6x + 5 \equiv (x - 2)(x^2 - x + 4) + 13$.

This can also be written as

$$\frac{x^3 - 3x^2 + 6x + 5}{x - 2} \equiv x^2 - x + 4 + \frac{13}{x - 2}, \quad x \neq 2$$

Example 6   Divide $3x^4 - 2x^2 - 1$ by $x + 1$.

$$
\begin{array}{r}
3x^3 - 3x^2 + x - 1 \\
x + 1 \overline{\smash{)}\, 3x^4 - 2x^2 - 1} \\
\underline{3x^4 + 3x^3} \\
-3x^3 - 2x^2 \\
\underline{-3x^3 - 3x^2} \\
x^2 \\
\underline{x^2 + x} \\
-x - 1 \\
\underline{-x - 1} \\
0
\end{array}
$$

> There are no terms in x^3 or x in the dividend, so gaps must be left.

> There is no remainder so $(x + 1)$ is a factor of $3x^4 - 2x^2 - 1$.

So $3x^4 - 2x^2 - 1 \equiv (x + 1)(3x^3 - 3x^2 + x - 1)$.

Example 7
Extension   Divide $x^3 - y^3$ by $x - y$.

$$
\begin{array}{r}
x^2 + xy + y^2 \\
x - y \overline{\smash{)}\, x^3 - y^3} \\
\underline{x^3 - x^2 y} \\
x^2 y \\
\underline{x^2 y - xy^2} \\
xy^2 - y^3 \\
\underline{xy^2 - y^3} \\
0
\end{array}
$$

> There are no terms with x^2 or x in the dividend, so gaps must be left.

> No remainder.

So $x^3 - y^3 \equiv (x - y)(x^2 + xy + y^2)$

or $\dfrac{x^3 - y^3}{x - y} \equiv x^2 + xy + y^2, \qquad x \neq y$

Alternative method of division

With practice, long division in algebra can be avoided. Instead, a method of building up the quotient can be used.

Consider dividing $(x^3 + x^2 - 9x + 6)$ by $(x - 2)$.

$$x^3 + x^2 - 9x + 6 = (x - 2)(?\quad ?\quad ?)$$

Once mastered, this method is easier than it looks!

The quotient bracket can be built up, term by term.

To obtain x^3 (the first term of the dividend), the quotient must start with x^2.

$$x^3 + x^2 - 9x + 6 = (\overbrace{x - 2)(x^2}\quad ?\quad ?)$$

$\quad\quad\quad$ Dividend $\quad\quad$ Divisor $\quad\quad$ Quotient

Consider now how the term, x^2, of the dividend will be obtained. It will come from two products:

$$-2 \times x^2 = -2x^2 \quad \text{and} \quad x \times x \text{ term}$$

$\quad$ In divisor $\quad$ In quotient $\quad\quad$ In divisor $\quad$ In quotient

$-2x^2$ will need $3x^2$ added to it to give the x^2 required.

To obtain $3x^2$, the x term of the quotient will have to be $3x$.

$$x^3 + x^2 - 9x + 6 = (\overbrace{x - 2)(x^2 + 3x}\quad ?)$$

Check that this gives $x^3 + x^2$.

Consider now how the x term, $-9x$, of the dividend will be obtained. It will come from two products:

$$-2 \times 3x = -6x \quad \text{and} \quad x \times \text{constant term}$$

$\quad$ In divisor $\quad$ In quotient $\quad\quad$ In divisor $\quad$ In quotient

$-6x$ will need $-3x$ more to give the $-9x$ required.

To obtain $-3x$, the constant term of the quotient will have to be -3.

So $x^3 + x^2 - 9x + 6 = (x - 2)(x^2 + 3x - 3)$.

Check the division by multiplying the divisor by the quotient.

There is no remainder because $-2 \times -3 = 6$.

For complicated examples later in the course such as $(x^5 - x^4 - 2x^2 - 9x + 3)$ divided by $(x^2 + 3)$ the method of building up the quotient is to be recommended.

For a simpler method of division go to the CD-ROM (E12.2).

Exercise 12B

1   Find the quotient when each of these polynomials is divided by the expression in brackets.

   a $x^3 + 4x^2 + x - 6$ $(x - 1)$

   b $x^3 + 4x^2 - 9x - 36$ $(x + 3)$

   c $x^3 + 3x^2 + 3x + 2$ $(x + 2)$

   d $x^3 - 3x^2 - 10x + 24$ $(x - 4)$

   e $2x^3 + x^2 - 13x + 6$ $(x - 2)$

   f $x^3 - 3x - 2$ $(x - 2)$

   g $x^3 - x^2 - 4$ $(x - 2)$

   h $2x^3 - x - 1$ $(x - 1)$

   i $x^4 - x^3 + 2x^2 + x - 3$ $(x + 1)$

   j $2x^4 - 9x^3 + 13x^2 - 15x + 9$ $(x - 3)$

   k $x^4 + x^3 - 14x^2 + 4x + 6$ $(x - 3)$

   l $x^4 + 5x^3 - x - 5$ $(x + 5)$

2   Find the quotient and the remainder when these polynomials are divided by the expression in brackets.

   a $x^3 + 3x^2 + 6x + 1$ $(x + 1)$ b $x^3 + 5x^2 - 6x - 2$ $(x + 1)$

   c $x^3 - 6x^2 + 11x - 6$ $(x - 1)$ d $x^3 - 3x^2 - x - 1$ $(x + 2)$

   e $2x^3 - x + 51$ $(x + 3)$ f $2x^3 - x^2 - 4x + 12$ $(x + 2)$

   g $x^3 - 2x^2 - 16x + 16$ $(x - 5)$ h $2x^3 + 5x^2 - 7x + 4$ $(x + 4)$

3   Find the quotient and the remainder when these polynomials are divided by the expression in brackets.

   a $8x^3 - 24x + 9$ $(2x - 3)$

   b $2x^3 + 5x^2 - 14x + 3$ $(2x - 3)$

   c $4x^4 - x^3 + 17x^2 + 11x + 4$ $(4x + 3)$

   d $3x^3 + 8x^2 - 6x + 12$ $(3x - 1)$

   e $2x^3 - 5x^2 + 5x + 4$ $(2x + 1)$

   f $12x^4 - 7x^2 + 1$ $(2x + 1)$

   g $-x^3 - x^2 - 2x + 1$ $(-x + 1)$

   h $6x^3 + 29x^2 - 40x - 42$ $(6x + 5)$

12.3 Remainder and factor theorems

When $f(x) = x^2 + x - 3$ is divided by $(x - 1)$ there is a quotient and a remainder.

$$f(x) \equiv x^2 + x - 3 \equiv (x - 1) \times \text{Quotient} + \text{Remainder}$$

Substituting $x = 1$ gives

> $x = 1$ is chosen to make the bracket $(x - 1)$ zero.

$$f(1) = 1^2 + 1 - 3 = 0 \times \text{Quotient} + \text{Remainder}$$

$$f(1) = -1 = \text{Remainder}$$

So the remainder is $f(1)$ and it has been found without carrying out the division.

The quotient has not been found.

> Check: $x^2 + x - 3 \equiv (x - 1)(x + 2) - 1$

The remainder theorem gives a quick way of finding remainders, without the need to carry out long division.

- The remainder on division by $x - 1$ is $f(1)$.
- The remainder on division by $x + 2$ is $f(-2)$.
- The remainder on division by $2x - 1$ is $f\left(\frac{1}{2}\right)$.

> $2x - 1 = 0 \Leftrightarrow x = \frac{1}{2}$

Example 8

In this example, to illustrate the method, more steps are shown than necessary.

Find the remainder when $2x^3 + x - 7$ is divided by $x + 2$.

Let $f(x) = 2x^3 + x - 7$

> Define the expression as $f(x)$ so that function notation can be used.

$$f(x) \equiv 2x^3 + x - 7 \equiv (x + 2) \times \text{Quotient} + \text{Remainder}$$

Putting $x = -2$

> $x = -2$ is chosen to make the bracket $(x + 2)$ zero.

$$f(-2) = 2 \times (-2)^3 + (-2) - 7 = \text{Remainder}$$

$\therefore$ the remainder is -25.

Example 9

Prove that if $f(x)$ is divided by $(x - a)$ the remainder is $f(a)$.

Assume that when $f(x)$ is divided by $(x - a)$ the quotient is a polynomial $g(x)$ and the remainder is R.

Then $f(x) \equiv (x - a)g(x) + R$

Putting $x = a$ in this identity gives

$$f(a) = 0 \times g(a) + R$$

Hence $R = f(a)$.

So, if polynomial $f(x)$ is divided by $(x - a)$ the remainder is $f(a)$.
The remainder theorem states this in a more general form.

> **Remainder Theorem**
>
> If f(x) is divided by $(x - a)$ the remainder is f(a).
>
> If f(x) is divided by $(ax - b)$ the remainder is $f\left(\dfrac{b}{a}\right)$.

Note: $x = \dfrac{b}{a} \Leftrightarrow ax - b = 0$

If there is zero remainder on division, then the divisor is a **factor**.

So, if f$(a) = 0$ then $(x - a)$ is a factor of f(x).

The factor theorem (a corollary of the remainder theorem) states this in a more general form.

> **Factor Theorem**
>
> $f(a) = 0 \Leftrightarrow (x - a)$ **is a factor of** $f(x)$.
>
> $f\left(\dfrac{b}{a}\right) = 0 \Leftrightarrow (ax - b)$ **is a factor of** $f(x)$.

Example 10   Find the remainder when $2x^3 - 3x^2 + x + 5$ is divided by $2x - 1$.

Let $f(x) = 2x^3 - 3x^2 + x + 5$

> Define the expression as $f(x)$ so that function notation can be used.

The remainder on division by $2x - 1$ is $f\left(\frac{1}{2}\right)$.

> $2x - 1 = 0 \Leftrightarrow x = \frac{1}{2}$

$$f\left(\tfrac{1}{2}\right) = 2 \times \left(\tfrac{1}{2}\right)^3 - 3 \times \left(\tfrac{1}{2}\right)^2 + \tfrac{1}{2} + 5$$
$$= \tfrac{1}{4} - \tfrac{3}{4} + \tfrac{1}{2} + 5$$
$$= 5$$

So the remainder is 5.

Example 11   When the expression $x^3 + ax^2 + 2x + 1$ is divided by $x - 2$, the remainder is three times as great as when the expression is divided by $x - 1$. Find the value of a.

Let $f(x) = x^3 + ax^2 + 2x + 1$

$f(2) = 8 + 4a + 4 + 1 = 13 + 4a$

> $f(2)$ is the remainder on division by $x - 2$.

$f(1) = 1 + a + 2 + 1 = 4 + a$

> $f(1)$ is the remainder on division by $x - 1$.

But $f(2) = 3f(1)$

So $13 + 4a = 3(4 + a)$
$$= 12 + 3a$$
$\therefore \qquad\quad a = -1$

Factorising polynomials

> **Steps for factorising a polynomial in x**
> - Put the expression equal to $f(x)$ so that function notation can be used.
> - Find one linear factor. $(x - a)$ will be a factor if $f(a) = 0$.
>   Try $f(1)$ first. If $f(1) = 0$ then $(x - 1)$ is a factor.
>   If $f(1) \neq 0$, try $f(-1)$. If $f(-1) = 0$ then $(x + 1)$ is a factor.
>   If $f(-1) \neq 0$, list (at least mentally) other possible linear factors and test until one factor is found.
> - When one factor is found, divide the expression by that factor.
> - Repeat the process until the polynomial is fully factorised.

Example 12 Show that $(x + 1)$ is a factor of $x^3 + 2x^2 - 5x - 6$ and hence factorise the expression fully.

Let $f(x) = x^3 + 2x^2 - 5x - 6$

$f(-1) = (-1)^3 + 2(-1)^2 - 5 \times (-1) - 6$

$\quad = -1 + 2 + 5 - 6 = 0$

$\therefore \quad (x + 1)$ is a factor.

$x^3 + 2x^2 - 5x - 6 \equiv (x + 1)(x^2 + x - 6)$

$\quad \equiv (x + 1)(x + 3)(x - 2)$

> Define the expression as $f(x)$ so that function notation can be used.

> $(x + 1)$ is a factor of $f(x)$ if, and only if, $f(-1) = 0$.

> The function can now be divided by $(x + 1)$ giving $x^2 + x - 6$.

> The expression is fully factorised when no further factors can be found.

Example 13 Factorise $x^3 - 3x^2 - 4x + 12$.

Possible factors are:

$(x \pm 1)$, $(x \pm 2)$, $(x \pm 3)$, $(x \pm 4)$, $(x \pm 6)$, $(x \pm 12)$

Try $(x \pm 1)$ first.

Let $\qquad f(x) = x^3 - 3x^2 - 4x + 12$

$\qquad f(1) = 1 - 3 - 4 + 12 \neq 0$

$\therefore \quad (x - 1)$ is *not* a factor.

$\qquad f(-1) = -1 - 3 + 4 + 12 \neq 0$

$\therefore \quad (x + 1)$ is *not* a factor.

$\qquad f(2) = 8 - 12 - 8 + 12 = 0$

$\therefore \quad (x - 2)$ is a factor.

$x^3 - 3x^2 - 4x + 12 = (x - 2)(x^2 - x - 6)$

$\qquad = (x - 2)(x - 3)(x + 2)$

> If this cubic expression has a linear factor, then the bracket would start with x and end with a factor of 12.

> Use function notation.

> Try possible values, a, until some value gives $f(a) = 0$.

> So $(x - a)$ is factor.

> Algebraic division can be used if necessary.

Example 14 List the possible linear factors of $2x^3 + 3x^2 - 5$.

The possible factors of $2x^3 + 3x^2 - 5$ are

$(x \pm 1), (2x \pm 1), (x \pm 5), (2x \pm 5)$

> To obtain $2x^3$ the linear factor must contain only x or $2x$. Similarly, to obtain 5, it must contain 1 or 5.

Example 15 Factorise fully $f(x) = x^3 + x^2 + x - 3$. Hence solve $f(x) = 0$.

$f(x) = x^3 + x^2 + x - 3$

$f(1) = 1 + 1 + 1 - 3 = 0$

$\Rightarrow \quad (x - 1)$ is a factor

So $x^3 + x^2 + x - 3 = (x - 1)(x^2 + 2x + 3)$.

> $x^2 + 2x + 3$ cannot be factorised.

$f(x) = 0 \Rightarrow (x - 1)(x^2 + 2x + 3) = 0$

$\Rightarrow \quad x = 1 \ \text{ or } \ x^2 + 2x + 3 = 0$

The discriminant of $x^2 + 2x + 3 = 0$ is $b^2 - 4ac = 4 - 12 = -8 < 0$.

$\therefore \quad x^2 + 2x + 3 = 0$ has no real roots so $f(x) = 0 \Rightarrow x = 1$.

Example 16 Extension Factorise $x^3 + y^3$.

Substituting $x = -y$ makes the expression zero.

$\therefore \quad (x + y)$ is a factor

$x^3 + y^3 = (x + y)(x^2 - xy + y^2)$

> Let $f(x) = x^3 + y^3$.
> Since $f(-y) = (-y)^3 + y^3 = 0$, $x + y$ is a factor.

Exercise 12C Factor theorem

1 Find the values of $f(0)$, $f(1)$, $f(-1)$, $f(2)$ and $f(-2)$ and state one factor of each expression.

  a $f(x) = x^3 + 3x^2 - 4x - 12$ **b** $f(x) = 3x^3 - 2x - 1$

  c $f(x) = x^5 + 2x^4 + 3x^3$ **d** $f(x) = x^4 - 4x^2 + 3$

2 Show that

  a $x - 1$ is a factor of $f(x) = x^5 - 3x^2 + 2$

  b $x + 2$ is a factor of $f(x) = 2x^3 - 3x^2 - 12x + 4$

  c $2x - 1$ is a factor of $f(x) = 2x^3 + x^2 - 3x + 1$

3 Find the values of a in these expressions when

  a $x^3 + x^2 + ax + 8$ is divisible by $x - 1$

  b $x^4 - 3x^2 + 2x + a$ is divisible by $x + 1$

  c $x^5 + 4x^4 - 6x^2 + ax + 2$ is divisible by $x + 2$

4 Show that $x - 1$ is a factor of $x^3 - 7x + 6$, and find the other factors of the expression.

  Hence solve $x^3 - 7x + 6 = 0$.

5 Find a factor of $f(x)$ using the factor theorem and hence factorise $f(x)$.

  a $f(x) = x^3 - x^2 - 4x + 4$ **b** $f(x) = x^3 + 2x^2 - 5x - 6$

  c $f(x) = x^3 - 2x^2 - 5x + 6$ **d** $f(x) = 2x^3 + x^2 - 8x - 4$

6 Solve $f(x) = 0$ for the functions in Question 5.

7 Show that $2x^3 + x^2 - 13x + 6$ is divisible by $x - 2$, and find the other factors of the expression.
Hence solve $2x^3 + x^2 - 13x + 6 = 0$.

8 Find the values of a and b so that $x^3 + ax^2 + bx - 6$ is divisible by $x - 1$ and $x + 2$.
Factorise the expression.

9 Find the values of a and b so that $2x^4 + 3x^3 + ax^2 - x + b$ is divisible by $x + 1$ and $2x - 1$.
Factorise the expression.

10 a Factorise $x^2 - 3x - 4$.

   b Show that $x^2 - 3x - 4$ is a factor of $f(x)$ where
$f(x) = x^4 - 3x^3 - 12x^2 + 24x + 32$.

   c Solve $f(x) = 0$.

1 Find the remainders when

  a $x^3 + 3x^2 - 4x + 2$ is divided by $x - 1$

  b $x^3 - 2x^2 + 5x + 8$ is divided by $x - 2$

  c $x^5 + x - 9$ is divided by $x + 1$

  d $x^3 + 3x^2 + 3x + 1$ is divided by $x + 2$

  e $4x^3 - 5x + 4$ is divided by $2x - 1$

  f $4x^3 + 6x^2 + 3x + 2$ is divided by $2x + 3$

2 Find the values of a in these expressions when

  a $x^3 + ax^2 + 3x - 5$ has remainder -3 when divided by $x - 2$

  b $x^3 + x^2 - 2ax + a^2$ has remainder 8 when divided by $x - 2$

  c $x^3 - 3x^2 + ax + 5$ has remainder 17 when divided by $x - 3$

  d $x^5 + 4x^4 - 6x^2 + ax + 2$ has remainder 6 when divided by $x + 2$

3 Find the values of a and b if $x^4 + ax^3 + bx^2 - 2x + 8$ has remainder 6 when divided by $x - 1$ and remainder -24 when divided by $x + 2$.

4 The remainder when $2x^3 + ax^2 + 4$ is divided by $x - 2$ is five more than when it is divided by $x - 1$.
Find a.

5 The remainder when $ax^4 + 3x^2 - x + 2$ is divided by $x + 2$ is eight times the remainder when it is divided by $x - 1$.
Find a.

6   The expression $ax^2 + bx + c$ is divisible by $x - 1$, has remainder 2 when divided by $x + 1$, and has remainder 8 when divided by $x - 2$.
    Find the values of a, b and c.

7   $x - 1$ and $x + 1$ are factors of the expression $x^3 + ax^2 + bx + c$, and it leaves a remainder of 12 when divided by $x - 2$.
    Find the values of a, b and c.

Exercise 12E (Review)

1   Divide $(2x^4 + 6x^3 + x^2 - 5x + 2)$ by $(x + 2)$.

2   Divide $(x^3 + 3x^2 - 22x - 90)$ by $(x - 5)$.

3   Divide $(x^4 - x^3 - x^2 + 4x - 3)$ by $(x - 1)$.

4   Divide $(x^4 - 5x^2 + 13x + 3)$ by $(x + 3)$.

5   Find A and B in this identity.
    $5x + 31 \equiv A(x + 2) + B(x - 1)$

6   Find A, B and C in the identity
    $5x + 31 \equiv A(x + 2)(x - 1) + B(x - 1)(x - 5) + C(x - 5)(x + 2)$.

7   Divide $x^3 - 9x + 10$ by $x - 2$.
    Hence find the solutions to $x^3 - 9x + 10 = 0$ in the set

    a $\mathbb{Z}$ (integers)   b $\mathbb{Q}$ (rational numbers)   c $\mathbb{R}$ (real numbers)

8   Given that $x^3 - 2x^2 + x - 12 \equiv (x - 3)(x^2 + ax + b)$, find the values of a and b.

9   Given that $x + 2$ is a factor of $x^3 - 2x^2 + ax + 6$, find a.

10   Given that $x - 1$ and $x + 3$ are factors of $x^3 - x^2 + ax + b$, find a and b.

11   Factorise fully $2x^4 - x^3 - 5x^2 - 2x$.
     Hence solve $2x^4 - x^3 - 5x^2 - 2x = 0$.

12   Given that $f(x) = x^3 - 10x + 12$
     a   show that $f(2) = 0$
     b   solve $f(x) = 0$

13   The remainder when $2x^3 + ax^2 - x - 4$ is divided by $x - 2$ is 6.
     Find a.

14   The remainders when $3x^9 - ax + b$ is divided by $x - 1$ and by $x + 1$ are equal.
     a   Find a.
     b   Show that b can take any value.

15   When the polynomial $f(x) = 2x^3 + ax^2 + bx + c$ is divided by x, $x - 1$ and $2x + 1$ the remainders are 5, 10 and 1 respectively.
     Find a, b and c.

A 'Test yourself' exercise (Ty12) and an 'Extension exercise' (Ext12) can be found on the CD-ROM.

Key points

Solving identities

An identity is an equation which is true for all values of the variable.

- To prove the truth of an identity, one method is to start with one side and by algebraic manipulation show that it is equal to the other side.
- To find unknowns, either substitute values or equate (compare) coefficients.

Remainder theorem

If $f(x)$ is divided by $(x - a)$ then the remainder is $f(a)$.

If $f(x)$ is divided by $(ax - b)$ then the remainder is $f\left(\dfrac{b}{a}\right)$.

Factor theorem

$f(a) = 0 \Leftrightarrow (x - a)$ is a factor of $f(x)$.

$f\left(\dfrac{b}{a}\right) = 0 \Leftrightarrow (ax - b)$ is a factor of $f(x)$.

Factorising polynomials

- Put the expression equal to $f(x)$ so that function notation can be used.
- Find one linear factor. $(x - a)$ will be a factor if $f(a) = 0$.
  Try $f(1)$ first. If $f(1) = 0$ then $(x - 1)$ is a factor.
  If $f(1) \neq 0$, try $f(-1)$. If $f(-1) = 0$ then $(x + 1)$ is a factor.
  If $f(-1) \neq 0$, list (at least mentally) other possible linear factors and test until one factor is found.
- When one factor is found, divide the expression by that factor.
- Repeat the process until the polynomial is fully factorised.

From wheels to jet engines, from oil wells to domes, the properties of circles are central to life as we know it. In order to use circles and their properties, in design, for example, we need to define and describe them. In this chapter, circles are expressed as equations in coordinates x and y.

After working through this chapter you should

■ *be familiar with the following circle properties:*

   i the angle in a semicircle is a right angle

   ii the perpendicular from the centre to a chord bisects the chord

   iii the perpendicularity of radius and tangent

■ *know that the equation of a circle, centre the origin, radius r is $x^2 + y^2 = r^2$*

■ *know that the equation of a circle, centre (a, b), radius r is $(x - a)^2 + (y - b)^2 = r^2$*

■ *be able to find the radius and the coordinates of the centre of the circle given the equation of a circle.*

13.1   Properties of a circle

A circle is defined as the set of points in a plane which are at a fixed distance, the **radius**, from a fixed point, the **centre**.

The circle is the most symmetrical of all plane figures, having an infinite number of axes of symmetry. Every diameter is an axis of symmetry.

Circles have many geometric properties. Some properties are:

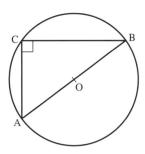

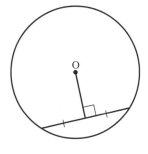

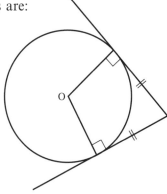

The angle in a semicircle is a right angle.

The perpendicular from the centre to a chord bisects the chord and, conversely, the perpendicular bisector of a chord passes through the centre.

A tangent is perpendicular to the radius through the the point of contact.

The lengths of the tangents from an external point to the points of contact with the circle are equal.

The properties and their converses can be expressed in different ways. For example:

If the ends of a diameter of a circle AB are joined to any point C on the circle then $\angle ACB = 90°$.

For any right-angled triangle ACB with hypotenuse AB, a circle drawn with AB as diameter will pass through C.

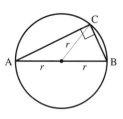

 If you are not familiar with these properties you may find the practice questions on the CD-ROM (A15.3) useful.

13.2 The equation of a circle

A circle can be defined by

a   its centre and radius

b   any three points on it.

Consider a circle, centre at the origin O, with radius r.

Let P(x, y) be any point on the circle and R be the foot of the perpendicular from P to the x-axis, so that OR = x and PR = y.

Then, by Pythagoras' theorem

$$OP^2 = OR^2 + PR^2$$
$$r^2 = x^2 + y^2$$

> The equation of a circle, centre the origin and of radius r is
> $$x^2 + y^2 = r^2$$

Consider now a circle of radius r whose centre is at the point C(a, b).

Let P(x, y) be any point on the circle and complete the triangle CPR as shown, so that CR is parallel to the x-axis and PR is parallel to the y-axis.

Now, CR = $x - a$ and PR = $y - b$.

Then, by Pythagoras' theorem
$$CP^2 = CR^2 + PR^2$$
$$r^2 = (x - a)^2 + (y - b)^2$$

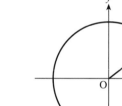

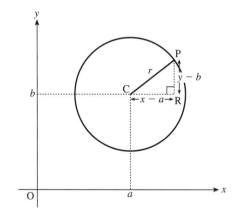

204

The equation of a circle centre (a, b) and radius r is

$$(x - a)^2 + (y - b)^2 = r^2$$

> The circle $x^2 + y^2 = r^2$ has been translated a units to the right and b up, a translation of $\begin{pmatrix} a \\ b \end{pmatrix}$.

This equation for a circle can be expanded to give

$$x^2 - 2ax + a^2 + y^2 - 2by + b^2 - r^2 = 0$$
$$x^2 + y^2 - 2ax - 2by + a^2 + b^2 - r^2 = 0$$

Notice that in the equation of a circle the coefficients of x^2 and y^2 are the same, and there is no term in xy. The form $(x - a)^2 + (y - b)^2 = r^2$ is the more useful because the centre and radius of the circle can be seen immediately.

Example 1   Find the centre and radius of the circles

a $(x - 7)^2 + (y + 2)^2 = 64$

b $x^2 + 2x + y^2 - 8y + 8 = 0$

Solution   **a**   The general equation of a circle centre (a, b), radius r is $(x - a)^2 + (y - b)^2 = r^2$

> $(x - 7)^2 + (y + 2)^2 = 64$
> $\Rightarrow (x - 7)^2 + (y - (-2)^2 = 8^2$

Comparing $(x - 7)^2 + (y + 2)^2 = 64$ with the general equation, the circle $(x - 7)^2 + (y + 2)^2 = 64$ has the centre $(7, -2)$, radius 8.

b

$$x^2 + 2x + y^2 - 8y + 8 = 0$$
$$\Rightarrow (x + 1)^2 - 1 + (y - 4)^2 - 16 + 8 = 0$$
$$\Rightarrow \qquad (x + 1)^2 + (y - 4)^2 = 9$$
$$\Rightarrow \qquad (x + 1)^2 + (y - 4)^2 = 3^2$$

> Completing the square for the x terms and for the y terms:
> $x^2 + 2x = (x + 1)^2 - 1$
> $y^2 - 8y = (y - 4)^2 - 16$

So the circle has centre $(-1, 4)$ and radius 3.

 To find the radius and centre of a circle whose equation is given in the form $x^2 + y^2 + 2gx + 2fy + c = 0$, complete the squares for the x terms and for the y terms before comparing with $(x - a)^2 + (y - b)^2 = r^2$.

Example 2   Find the Cartesian equations of these circles.

> A Cartesian equation is a relationship connecting x and y.

    a   Centre $(4, -3)$ and radius 7

    b   Centre $(2, 5)$, passing through the origin

    c   Centre $(-2, 4)$ and radius $\frac{3}{2}$

Solution   **a**   The equation of the circle is

$$(x-4)^2 + (y-(-3))^2 = 7^2$$
$$(x-4)^2 + (y+3)^2 = 49$$

> Use $a = 4$, $b = -3$, $r = 7$.

> This can also be written as $x^2 + y^2 - 8x + 6y - 24 = 0$

    b

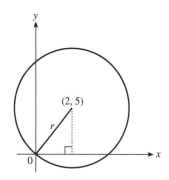

> A sketch helps to understand the set-up of the question.

By Pythagoras' theorem

$$r^2 = 2^2 + 5^2$$
$$= 29$$

> There is no need to find r. Only r^2 is required for the equation of the circle.

So the equation of the circle is

$$(x-2)^2 + (y-5)^2 = 29$$

> Use $a = 2$, $b = 5$ and $r^2 = 29$.

    c   The equation of the circle is

> Use $a = -2$, $b = 4$ and $r = \frac{3}{2}$.

$$(x-(-2))^2 + (y-4)^2 = \left(\tfrac{3}{2}\right)^2$$
$$(x+2)^2 + (y-4)^2 = \tfrac{9}{4}$$
$$x^2 + 4x + 4 + y^2 - 8y + 16 = \tfrac{9}{4}$$
$$4x^2 + 4y^2 + 16x - 32y + 71 = 0$$

> The equation can be left in this form; or multiply through by 4 to remove fractions.

Example 3   Decide which of these equations represent circles. Find the coordinates of the centre and the radius of each circle.

a $2x^2 + 3y^2 - 2x + 4y - 12 = 0$

b $4x^2 + 4y^2 - 20x + 8y + 9 = 0$

c $x^2 + y^2 - 4x + 6y + 18 = 0$

Solution   **a** $2x^2 + 3y^2 - 2x + 4y - 12 = 0$

The coefficients of x^2 and y^2 are different and so this cannot be the equation of a circle.

In both **b** and **c** the coefficients of x^2 and y^2 are the same and there is no xy term, so these could represent circles.

b
$$4x^2 + 4y^2 - 20x + 8y + 9 = 0$$
$$x^2 - 5x + y^2 + 2y + \tfrac{9}{4} = 0$$
$$\left(x - \tfrac{5}{2}\right)^2 + (y + 1)^2 - \tfrac{25}{4} - 1 + \tfrac{9}{4} = 0$$
$$\left(x - \tfrac{5}{2}\right)^2 + (y + 1)^2 = 5$$

> Divide through by 4 to make the coefficients of x^2 and y^2 one, and collect the x and y terms together.

> Complete the square for x and y.
> Note: $\tfrac{25}{4} + 1 - \tfrac{9}{4} = \tfrac{25 + 4 - 9}{4} = \tfrac{20}{4} = 5$

This is the equation of a circle, centre $\left(\tfrac{5}{2}, -1\right)$, radius $\sqrt{5}$.

c
$$x^2 + y^2 - 4x + 6y + 18 = 0$$
$$x^2 - 4x + y^2 + 6y + 18 = 0$$
$$(x - 2)^2 + (y + 3)^2 - 4 - 9 + 18 = 0$$
$$(x - 2)^2 + (y + 3)^2 + 5 = 0$$

> Collect x and y terms together.

> Complete the squares.

> This gives $r^2 = -5$ which does not give a real value for r.

This cannot be the equation of a circle, because r^2 is negative.

Given any three points not on the same straight line, one, and only one, circle can be drawn to pass through the three points. In Example 4 below, three points are given and two methods are used to find the equation of the circle.

Example 4   Find the equation of the circle which passes through the points A(6, 2), B(8, −2) and C(0, 2).

Solution   *Method 1*

> The first method uses the fact that the perpendicular bisectors of the chords AB and AC will intersect at the centre of the circle.

> To find the equations of the perpendicular bisectors, find the mid-points and gradients of the chords AB and AC.

A is (6, 2) and C is (0, 2).
So AC is parallel to the x-axis.
The mid-point of AC is (3, 2) and the perpendicular bisector of AC has equation $x = 3$.

Notice that the y coordinates are the same.

If two points on the circle have the same x coordinates (or y coordinates) the perpendicular bisector will be parallel to the y (or x) axis and simple to find.

The centre of the circle lies on the perpendicular bisector and therefore the centre of the circle will have x coordinate 3.

The mid-point of AB is (7, 0) and the gradient of AB is

Use mid-point $= \left(\dfrac{x_1 + x_2}{2}, \dfrac{y_1 + y_2}{2} \right)$

$$\frac{2 - (-2)}{6 - 8} = \frac{4}{-2} = -2$$

Use gradient $= \dfrac{y_2 - y_1}{x_2 - x_1}$

So the gradient of the perpendicular bisector of AB is $\frac{1}{2}$ and its equation is

If two perpendicular lines have gradients m_1 and m_2 then $m_1 m_2 = -1$

$$y - 0 = \tfrac{1}{2}(x - 7)$$
$$y = \tfrac{1}{2}x - \tfrac{7}{2}$$

Use $y - y_1 = m(x - x_1)$.

D, the centre of the circle, lies on $y = \tfrac{1}{2}x - \tfrac{7}{2}$ and has x coordinate 3. Substituting $x = 3$ gives

$$y = \tfrac{1}{2} \times 3 - \tfrac{7}{2}$$
$$= -2$$

Hence the centre of the circle is at D(3, −2).

If the radius of the circle is r, then

$$r^2 = AD^2$$
$$= (6 - 3)^2 + (2 - (-2))^2$$
$$= 3^2 + 4^2$$
$$= 25$$

To find the radius of the circle, find the distance of D from any of A, B or C. There is no need to find r, because only r^2 is required for the equation of the circle.

So, the equation of the circle is

$$(x - 3)^2 + (y - (-2))^2 = 25$$
$$(x - 3)^2 + (y + 2)^2 = 25$$

Use $a = 3$, $b = -2$ and $r^2 = 25$.

This equation should now be checked using the coordinates of A, B and C.

Method 2   The second method uses the general equation for a circle.

Suppose A, B and C lie on the circle with equation

$$x^2 + 2gx + y^2 + 2fy + c = 0$$

then

Substituting the coordinates of A, B and C into this equation produces three simultaneous equations.

at A(6, 2) $6^2 + 2g \times 6 + 2^2 + 2f \times 2 + c = 0$

at B(8, −2) $8^2 + 2g \times 8 + (-2)^2 + 2f \times (-2) + c = 0$

at C(0, 2) $2^2 + 2f \times 2 + c = 0$

$$12g + 4f + c = -40 \qquad ①$$

$$16g - 4f + c = -68 \qquad ②$$

$$4f + c = -4 \qquad ③$$

These equations can now be solved simultaneously to find f, g and c.

The solution is $g = -3$, $f = 2$, $c = -12$ and the equation of the circle is

$$x^2 - 6x + y^2 + 4y - 12 = 0$$

This is equivalent to the equation obtained by method 1.

Exercise 13A

1 Find the centre and radius of each of these circles.

   a $(x - 5)^2 + (y + 1)^2 = 4^2$ **b** $(x + 6)^2 + (y + 2)^2 = 100$

   c $(x - 4)^2 + (y - 7)^2 - 1 = 0$ **d** $(x + m)^2 + (y - n)^2 = c$

2 By completing the square for the x terms and for the y terms, find the centre and radius of each of these circles.

   a $x^2 + 6x + y^2 - 2y = 15$ **b** $x^2 - 4x + y^2 - 8y - 44 = 0$

   c $x^2 + 2x + y^2 + 2y = 2$ **d** $x^2 + y^2 = 12x - 10y - 12$

3 Find the Cartesian equations of the circles with these centres and radii.

   a Centre (3, 5), radius 2   **b** Centre (4, 7), radius 3

   c Centre (5, −2), radius 5   **d** Centre (−4, 3), radius $\frac{1}{2}$

   e Centre (0, 6), radius $\sqrt{3}$ **f** Centre (4, −3), radius $\frac{1}{2}\sqrt{5}$

   g Centre (−1, −7), radius 10   **h** Centre $\left(\frac{1}{4}, -\frac{1}{2}\right)$, radius $\frac{3}{2}$

4 Find the centre and radius of each of these circles.

   a $x^2 + y^2 - 4x - 6y - 3 = 0$ **b** $x^2 + y^2 - 4x - 2y - 4 = 0$

   c $x^2 + y^2 + 10x - 14y + 10 = 0$ **d** $x^2 + y^2 - 4x + 2y - 1 = 0$

   e $x^2 + y^2 + 8x - 4y = 0$ **f** $2x^2 + 2y^2 + 2x - 6y - 9 = 0$

   g $100x^2 + 100y^2 - 120x + 100y - 39 = 0$

   h $x^2 + y^2 - 2ax - 2ay = 0$

5 Which of these equations represent a circle?

a $x^2 + y^2 - 7 = 0$ b $x^2 + y^2 + 3 = 0$

c $x^2 + y^2 + 8x - 3y + 20 = 0$ d $2x^2 + 2y^2 - 3x + 5y - 40 = 0$

e $3x^2 + 4y^2 - 6x + 5y - 30 = 0$ f $x^2 + y^2 - 4x + 6y + 13 = 0$

g $ax^2 + ay^2 = 7$ h $x^2 + y^2 + c = 0$

6 Find the equation of the circle with centre (2, 5) which touches the x-axis.

7 Find the equation of the circle whose centre is (3, −5) and which passes through the point (6, −7).

8 The points (10, −4) and (2, 2) are the ends of a diameter of a circle. Find the equation of the circle.

9 Find the equations of the circles passing through these sets of points:

a (5, 0), (3, 4) and (−5, 0)   b (6, 3), (−5, 2) and (7, 2)

c (2, 0), (3, 1) and (−6, 10)   d (2, 2), (1, 1) and (7, −3)

10 A circle has radius 7 and its centre lies in the first quadrant. Find the equation of the circle, given that it touches both the x-axis and the y-axis.

11 A circle passes through the points (0, −3) and (0, 3) and has radius 5. Draw a sketch to show the two possible positions of the circles and find their equations.

12 A circle has its centre at the origin and cuts the x-axis at the points A(−c, 0) and B(c, 0). Write down the equation of the circle.

The point P(h, k) lies on the circle. Find expressions for the gradients of AP and BP. By showing that the product of the gradients of AP and BP is −1, prove that $\angle APB = 90°$.

(This question illustrates the circle property that the angle in a semicircle is a right angle.)

13 Write down the equation of the circle, centre O, radius 13. Show that A(12, 5) and B(5, 12) lie on the circle. Find the mid-point, C, of the chord AB. Show that AB is perpendicular to OC.

(This question illustrates the circle property that the line joining the mid-point of a chord to the centre of the circle is perpendicular to the chord.)

13.3 Tangents to a circle

To find the equation of the tangent to a circle at any point on its circumference, use the fact that the tangent at a point is perpendicular to the radius of the circle through that point.

Example 5   Verify that the point (1, 2) lies on the circle $x^2 + y^2 - 6x + 4y - 7 = 0$ and find the equations of the tangent and normal at this point.

$$x^2 + y^2 - 6x + 4y - 7 = 0$$

When $x = 1$, $y = 2$

$$\text{LHS} = x^2 + y^2 - 6x + 4y - 7$$

Remember: to verify that a point does lie on a curve, start with one side of the equation (LHS in this case) and show that it equals the other side.

$$= 1^2 + 2^2 - 6 + 8 - 7$$

$$= 0$$

$$= \text{RHS}$$

∴ (1, 2) does lie on the circle $x^2 + y^2 - 6x + 4y - 7 = 0$.

$$x^2 + y^2 - 6x + 4y - 7 = 0$$

$$(x - 3)^2 + (y + 2)^2 - 9 - 4 - 7 = 0$$

$$(x - 3)^2 + (y + 2)^2 = 20$$

To find the gradient of the radius through (1, 2), first find the centre of the circle.

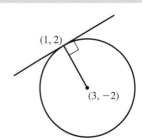

Therefore the centre of the circle is at (3, −2).

Hence, the gradient of the radius through (1, 2) is

$$\frac{2 - (-2)}{1 - 3} = \frac{4}{-2} = -2$$

and the gradient of the tangent at (1, 2) is $\frac{1}{2}$.

Remember that for perpendicular lines, the product of their gradients is −1.

Therefore the equation of this tangent is

Use $y - y_1 = m(x - x_1)$

$$y - 2 = \tfrac{1}{2}(x - 1)$$

Multiply through by 2 to remove fractions.

$$2y - 4 = x - 1$$

$$x - 2y + 3 = 0$$

The normal is perpendicular to the tangent, hence the radius through (1, 2) is normal to the circle at (1, 2).

The normal has gradient -2 and passes through $(1, 2)$.
Therefore its equation is

$$y - 2 = -2(x - 1)$$

Use $y - y_1 = m(x - x_1)$

$$y - 2 = -2x + 2$$

$$y = -2x + 4$$

➤ | To find the equation of the tangent to a circle at a point on its circumference, use the fact that the tangent is perpendicular to the radius.

To find the equation of the normal to a circle at a point on its circumference use the fact that the normal is perpendicular to the tangent.

Example 6 Find the length of the tangents from the point $(8, 5)$ to the circle $(x - 2)^2 + (y + 1)^2 = 16$.

$(x - 2)^2 + (y + 1)^2 = 16$ is the equation of a circle with centre $(2, -1)$ and radius 4.

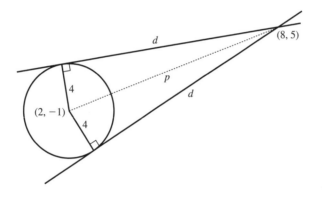

Draw a sketch.
Note: there is no need to show the axes on the diagram.

By symmetry, the two tangents from the point $(8, 5)$ are of equal length.

To find d using Pythagoras' theorem the length p, or p^2, is needed.

$$p^2 = (8 - 2)^2 + (5 - (-1))^2$$

$$= 6^2 + 6^2$$

$$= 72$$

To calculate p^2 use

$$p^2 = (x_2 - x_1)^2 + (y_2 - y_1)^2$$

for the points $(2, -1)$ and $(8, 5)$.

The radius is perpendicular to the tangent therefore, by Pythagoras' theorem,

$$p^2 = d^2 + 4^2$$

$$d^2 = p^2 - 4^2$$

$$= 72 - 16$$

$$= 56$$

$$d = \sqrt{56}$$

$$= 2\sqrt{14}$$

d must be positive, so only the positive root is taken.

∴ The length of the tangents from the point $(8, 5)$ to the circle is $2\sqrt{14}$.

Example 7   Find the points of intersection, if any, of the circle $x^2 + y^2 = 80$ with the lines

a $y = -2x$ **b** $y = 30 - 2x$ **c** $y = 20 - 2x$

Sketch the circle and lines on the same diagram.

a   The line $y = -2x$ meets the circle

$x^2 + y^2 = 80$ where

$x^2 + (-2x)^2 = 80$

$\Rightarrow x^2 + 4x^2 = 80$

$\Rightarrow \quad 5x^2 = 80$

$\Rightarrow \quad x^2 = 16$

$\Rightarrow \quad x = \pm 4$

> To find the points of intersection, solve the equations
> $y = -2x$ and $x^2 + y^2 = 80$ simultaneously.

When $x = 4$, $y = -8$ and when $x = -4$, $y = 8$.

So $y = -2x$ intersects the circle at $(4, -8)$ and $(-4, 8)$.

> To find the values of y when $x = \pm 4$, substitute $x = \pm 4$ in the equation of the line, not the circle. Substituting $x = 4$, for example, in the equation of the circle would give two values for y, only one of which satisfies $y = -2x$.

b   The line $y = 30 - 2x$ meets the circle

$x^2 + y^2 = 80$ where

$x^2 + (30 - 2x)^2 = 80$

$\Rightarrow x^2 + 900 - 120x + 4x^2 = 80$

$\Rightarrow \quad 5x^2 - 120x + 820 = 0$

$\Rightarrow \quad x^2 - 24x + 164 = 0$

> To find the points of intersection, solve the equations
> $y = 30 - 2x$ and $x^2 + y^2 = 80$ simultaneously.

> Divide through by 5.

The discriminant $b^2 - 4ac = 24^2 - 4 \times 1 \times 164$

$= 576 - 656$

< 0

> A negative discriminant $\Rightarrow$ no real roots $\Rightarrow$ the line and circle do not intersect.

So $x^2 - 24x + 164 = 0$ has no real roots and hence the line $y = 30 - 2x$ does not intersect the circle.

c The line $y = 20 - 2x$ meets the circle

$x^2 + y^2 = 80$ where

$$x^2 + (20 - 2x)^2 = 80$$

$\Rightarrow x^2 + 400 - 80x + 4x^2 = 80$

$\Rightarrow \quad 5x^2 - 80x + 320 = 0$

$\Rightarrow \quad x^2 - 16x + 64 = 0$

$\Rightarrow \quad (x - 8)^2 = 0$

$\Rightarrow \quad x = 8 \text{ (repeated root)}$

> To find the points of intersection, solve the equations
> $y = 30 - 2x$ and $x^2 + y^2 = 80$ simultaneously.

> Discriminant $= 0 \Rightarrow$ equal roots $\Rightarrow$ the line and circle touch.

Since the equation has a repeated root (two equal roots) the line $y = 20 - 2x$ is a tangent to the circle.
When $x = 8$, $y = 4$ so the line $y = 20 - 2x$ touches the circle at $(8, 4)$.

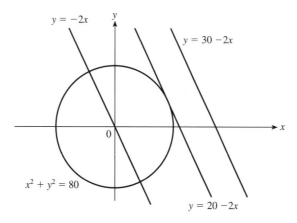

Exercise 13B

1 Find the coordinates of the centre, C, of the circle, whose equation is
$x^2 - 4x + y^2 - 6y = 4$.
Show that the point T(6, 4) lies on the circle.
Find the gradient of CT and hence deduce the gradient of the tangent at T.
Find the equations of the tangent and of the normal to the circle at T.

2 Find the equations of the tangents to these circles at the points given.

   a $(x - 1)^2 + (y + 2)^2 = 13;$ (3, 1)   **b** $(x + 3)^2 + (y - 5)^2 = 34;$ (0, 0)

   c $x^2 + y^2 - 6x + 4y = 0;$ (6, −4)   **d** $2x^2 + 2y^2 - 8x - 5y - 1 = 0;$ (1, −1)

3 Find the equations of the normals to these circles at the points given.

   a $x^2 + y^2 + 2x + 4y - 12 = 0;$ (3, −1)

   b $x^2 + y^2 + 2x - 2y - 8 = 0;$ (2, 2)

4 Draw a sketch of the circle $(x + 4)^2 + y^2 = 25$ and state its radius and the coordinates of its centre, C.

On the same diagram, mark the point P(9, 0) and draw one of the tangents to the circle from P. Mark the point of contact as T.

Draw CT and give the reason why angle CTP is a right angle.

Use Pythagoras' theorem in triangle CPT to calculate the length of the tangent from P to the circle.

5 Find the lengths of the tangents from the given points to these circles.

a $x^2 + y^2 + 4x - 6y + 10 = 0$; (0, 0)

b $x^2 + y^2 - 4x - 8y - 5 = 0$; (8, 2)

c $x^2 + y^2 + 6x + 10y - 2 = 0$; (−2, 3)

d $x^2 + y^2 - 10x + 8y + 5 = 0$; (5, 4)

6 The tangent to the circle $x^2 + y^2 - 2x - 6y + 5 = 0$ at the point (3, 4) meets the x-axis at M. Find the distance of M from the centre of the circle.

7 Find the equations of the tangents to the circle $x^2 + y^2 - 6x + 4y + 5 = 0$ at the points where it meets the x-axis.

8 Prove that the line $y = 2x$ is a tangent to the circle $x^2 + y^2 - 8x - y + 5 = 0$ and find the coordinates of the point of contact.

9 Prove that the line $x - 2y + 12 = 0$ touches the circle $x^2 + y^2 - x - 31 = 0$ and find the coordinates of the point of contact.

10 Find the coordinates of the centre, C, of the circle $x^2 + y^2 - 8x + 4y = 30$.

By solving the equations $x^2 + y^2 - 8x + 4y = 30$ and $y = x + 4$, show that $y = x + 4$ is a tangent to the circle and find the coordinates of T, the point of contact of the tangent and the circle.

Find the gradient of CT and hence show that CT is perpendicular to the tangent.

(This question illustrates the circle property that a tangent is perpendicular to the radius through the point of contact.)

11 A line, l, has equation $y = mx$ and a circle, C, has equation $x^2 + y^2 - 6x - 4y + 9 = 0$

a Given that l is a tangent to C, find the possible values of m.

b Find the range of values of m, given that l intersects C in two distinct points.

c Find the range of values of m, given that l and C do not intersect.

12 Given that the line with equation $y = x + k$ is a tangent to the circle $x^2 + y^2 = 25$, find the two possible values of k.

13 Prove that, for any value of c, the circle whose equation is $(x + c)^2 + (y - c)^2 = c^2$ touches both the x-axis and the y-axis.

14 Show that a circle with centre (a, a) and radius a touches both the x-axis and the y-axis.

Hence, or otherwise, find the centres of the two circles which pass through $(1, 2)$ and touch both the axes.

Exercise 13C (Review)

1 Find the centre and radius of each of these circles.

a $x^2 + 4x + y^2 + 6y - 17 = 0$ **b** $x^2 + y^2 - 6x - 8y = 0$

2 The points $(2, -3)$ and $(8, 7)$ are the ends of a diameter of a circle.
Find the coordinates of the centre of the circle and the length of the diameter.
What is the equation of the circle?

3 Find the equations of the circles which pass through each of these sets of points.

a $(1, 0), (7, -6), (7, 0)$ **b** $(6, 4), (5, 9), (-10, 6)$ **c** $(0, 0), (3, 7), (10, 0)$

4 Show that $y - 3x - 5 = 0$ is a tangent to the circle
$x^2 + y^2 - 2x + 4y - 5 = 0$ and find the coordinates of the point of contact.

5 Find the length of the tangents from the point $(7, -2)$ to the circle
$$x^2 + (y - 5)^2 = 49$$

6 Find the equation of the tangent to the circle whose equation is
$(x + 5)^2 + (y - 1)^2 = 65$ at the point $(3, 2)$.

7 A chord of the circle $(x + 2)^2 + (y - 9)^2 = 26$ has equation $y = x + 5$.

Find the coordinates of the end points of the chord and hence find the equation of the perpendicular bisector of the chord.

Verify that the perpendicular bisector passes through the centre of the circle.

8 The points A$(1, 5)$ and B$(7, 9)$ are the ends of a diameter of a circle.

Show that P$(2, 4)$ lies on the circle by showing that AP and BP are perpendicular and using the property that the angle in a semicircle is a right angle.

A 'Test yourself' exercise (Ty13) and an 'Extension exercise' (Ext13) can be found on the CD-ROM.

➤ # Key points

Circles

- The equation of a circle centre the origin, radius r is
  $$x^2 + y^2 = r^2$$

- The equation of a circle centre (a, b), radius r is
  $$(x - a)^2 + (y - b)^2 = r^2$$

- To find the radius and centre of a circle whose equation is given in the form $x^2 + y^2 + 2gx + 2fy + c = 0$, complete the squares for the x terms and for the y terms before comparing with $(x - a)^2 + (y - b)^2 = r^2$

- To find the equation of the tangent to a circle at a point on its circumference, use the fact that the tangent is perpendicular to the radius.

- To find the equation of the normal to a circle at a point on its circumference, use the fact that the normal is perpendicular to the tangent.

The binomial expansion is a method of raising a binomial expression (i.e. one with two terms)* to any power. You will frequently come across expressions such as $(a + b)^5$ and it may be necessary to expand the brackets. Although it is easy enough to work out $(a + b)^2$ or $(a + b)^3$, methods are needed to help expand a binomial expression raised to higher powers. The binomial expansion is a useful tool and is widely used in probability as well as in other areas.

After working through this chapter you should be

■ *able to use Pascal's triangle to expand $(a + b)^n$ for small positive integers n*

■ *familiar with notations $n!$ and $\dbinom{n}{r}$*

■ *able to use the binomial expansion to expand $(a + b)^n$ for all positive integers n*

■ *able to expand $(1 + x)^n$ for all positive integers n.*

14.1   Binomial expansion using Pascal's triangle

Consider

$$(a + b)^0 = \qquad\qquad \mathbf{1}$$
$$(a + b)^1 = \qquad\quad \mathbf{1}a + \mathbf{1}b$$
$$(a + b)^2 = \quad \mathbf{1}a^2 + \mathbf{2}ab + \mathbf{1}b^2$$
$$(a + b)^3 = \mathbf{1}a^3 + \mathbf{3}a^2b + \mathbf{3}ab^2 + \mathbf{1}b^3$$

The table of coefficients looks like this

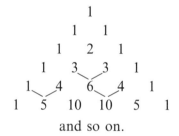

and so on.

Each entry in the table is the sum of the two above, so the next row would be 1, 6, 15, 20, 15, 6, 1.

This triangle is known as the Chinese triangle – Jia Xian, a Chinese mathematician, was using it in the mid-eleventh century – or **Pascal's triangle** *after the mathematician and philosopher, Blaise Pascal (1623–1662).*

Notice that in the expansion of $(a + b)^3$ the first term is a^3 (or a^3b^0). In the second term, $3a^2b$, the power of a is *reduced* by one, and the power of b *increased* by one, and so on, until the last term, which is b^3 (or a^0b^3).

The coefficients in each row of Pascal's triangle are symmetrical.

---

*A *monomial* expression has one term. A *trinomial* expression has three terms. A *polynomial* expression has many terms.

To expand $(a + b)^4$ look at the row of Pascal's triangle which starts 1, **4**, ...
The letters in the terms are $a^4 (= a^4b^0)$, a^3b, a^2b^2, ab^3, $b^4 (= a^0b^4)$.

So $(a + b)^4 = a^4 + 4a^3b + 6a^2b^2 + 4ab^3 + b^4$

Writing the expansion as

$(a + b)^4 = a^4b^0 + 4a^3b + 6a^2b^2 + 4ab^3 + a^0b^4$

illustrates that, when a binomial expression is raised to the **4th** power,

- there are $\mathbf{4} + 1 = 5$ terms
- the coefficients of the first two terms are 1 and **4**
- the sum of the powers of a and b is **4**.

In general, in the expansion of $(a + b)^n$

- there are $n + 1$ terms
- the coefficients of the first two terms are 1 and n
- the sum of the powers of a and b in each term is n.

Example 1   Expand, using Pascal's triangle:

a $(a + b)^6$
$(a + b)^6 = a^6 + 6a^5b + 15a^4b^2 + 20a^3b^3$
$\qquad\qquad + 15a^2b^4 + 6ab^5 + b^6$

> The row of the triangle which starts 1, 6, ... gives the coefficients: 1, 6, 15, 20, 15, 6, 1. The letters in the terms are a^6, a^5b, a^4b^2, ...

b $(1 + x)^5$
$(1 + x)^5 = 1 + 5x + 10x^2 + 10x^3 + 5x^4 + x^5$

> The coefficients are 1, 5, 10, 10, 5, 1. $a = 1, b = x$

c $(2x - 3)^4$
$(2x - 3)^4 = (2x + (-3))^4$
$\qquad = (2x)^4 + 4(2x)^3(-3) + 6(2x)^2(-3)^2$
$\qquad\quad + 4(2x)(-3)^3 + (-3)^4$
$\qquad = 16x^4 - 96x^3 + 216x^2 - 216x + 81$

> Here $a = 2x$, $b = -3$. Note how, with b −ve, the signs of the terms alternate.

Note   Putting, for example, $x = 1$ in both LHS and RHS of parts **b** and **c** will give some check that the expansion is correct.

Exercise 14A   **1**   Expand

a $(1 + x)^4$ **b** $(1 - x)^5$ **c** $(1 + x)^6$ **d** $(1 - x)^6$

2   Expand

a $(1 + 2x)^4$ **b** $(1 - 3y)^5$ **c** $(1 + 4z)^3$ **d** $\left(1 - \dfrac{x}{2}\right)^3$

e $(a + b)^7$ **f** $(a^2 - b^2)^5$ **g** $(a - b)^3(a + b)^3$

3   Expand

a $(x + y)^4$ **b** $(x + y)^5$ **c** $(3 + x)^4$ **d** $(x - 2)^3$

e $\left(2 - \dfrac{x}{2}\right)^3$ **f** $\left(\dfrac{1}{x} + x^2\right)^3$ **g** $(2 - x^3)^4$

4 **a** Expand

    i $(a+b)^3$ **ii** $(a-b)^3$

  b Hence, or otherwise, simplify, leaving surds in the answer where appropriate

    i $(1+\sqrt{2})^3 + (1-\sqrt{2})^3$ **ii** $(1+\sqrt{2})^3 - (1-\sqrt{2})^3$

    iii $(\sqrt{6}+\sqrt{2})^3 - (\sqrt{6}-\sqrt{2})^3$

5 Simplify, leaving surds in the answer where appropriate

  a $(2+\sqrt{3})^4 + (2-\sqrt{3})^4$ **b** $(2+\sqrt{6})^4 - (2-\sqrt{6})^4$

  c $(\sqrt{2}+\sqrt{3})^4 + (\sqrt{2}-\sqrt{3})^4$

14.2   Notation $n!$ and $\binom{n}{r}$

Pascal's triangle is a convenient way of expanding $(a+b)^n$ when n is small, but it is inconvenient for large values of n.

An alternative approach to finding the coefficients depends on the **theory of combinations**. In this alternative approach, the formula for the number of ways of choosing r objects from n different objects is used. This approach will be discussed in **14.3** but, before that, new notation is needed, namely $n!$ (say 'n factorial') and $\binom{n}{r}$ or ^nC_r, (say 'n choose r').

$n! = n(n-1)(n-2)(n-3) \times \ldots \times 2 \times 1$, so, for example,
$4! = 4 \times 3 \times 2 \times 1 = 24$.
By definition, $0! = 1$.

$$\binom{n}{r} = {}^nC_r = \frac{n!}{r!(n-r)!}, \text{ so, for example,}$$

$$\binom{5}{2} = {}^5C_2 = \frac{5!}{2!3!} = \frac{5 \times \overset{2}{\cancel{4}} \times \overset{1}{\cancel{3}} \times \overset{1}{\cancel{2}} \times 1}{(\underset{1}{\cancel{2}} \times 1) \times (\underset{1}{\cancel{3}} \times \underset{1}{\cancel{2}} \times 1)} = 10$$

The value of $\binom{n}{r}$ gives the number of ways of choosing r objects from n different objects. So, for example, the number of ways of choosing two objects from five different objects is $\binom{5}{2} = 10$.

> If the five different objects are A, B, C, D and E then the ten possible choices are AB, AC, AD, AE, BC, BD, BE, CD, CE and DE.

$$\binom{n}{r} = {}^nC_r = \frac{n!}{r!(n-r)!}$$

where $n! = n(n-1)(n-2)(n-3) \times \ldots \times 2 \times 1$ and, by definition, $0! = 1$

Many calculators will evaluate ^nC_r and n! but this example shows how to evaluate ^nC_r or $\binom{n}{r}$ without a calculator. The majority of the factors in the numerator and denominator cancel, so the calculation of $\binom{n}{r}$ is simple.

Example 2 Evaluate $\binom{n}{r}$ without a calculator.

a $\binom{4}{1} = \dfrac{4!}{1!3!} = \dfrac{4 \times \cancel{3} \times \cancel{2} \times 1}{\cancel{3} \times \cancel{2} \times 1} = 4$

Note: When $r = 1$, the factors in the numerator cancel *all* the terms in the denominator.

b $\binom{4}{0} = \dfrac{4!}{0!4!} = 1$

Note: $0! = 1$

c $\binom{10}{3} = \dfrac{10!}{3!7!}$

$= \dfrac{10 \times 9 \times 8 \times \cancel{7} \times \cancel{6} \times \cancel{5} \times \cancel{4} \times \cancel{3} \times \cancel{2} \times 1}{(3 \times 2 \times 1) \times (\cancel{7} \times \cancel{6} \times \cancel{5} \times \cancel{4} \times \cancel{3} \times \cancel{2} \times 1)}$

$= \dfrac{10 \times \overset{3}{\cancel{9}} \times \overset{4}{\cancel{8}}}{\cancel{3} \times \cancel{2} \times 1} = 120$

d $\binom{n}{2} = \dfrac{n!}{2!(n-2)!} = \dfrac{n(n-1)\cancel{(n-2)(n-3)} \times \cdots \times \cancel{2} \times 1}{(2 \times 1) \times \cancel{(n-2)(n-3)} \times \cdots \times \cancel{2} \times 1)} = \dfrac{n(n-1)}{2!}$

e $\binom{n}{3} = \dfrac{n!}{3!(n-3)!} = \dfrac{n(n-1)(n-2)}{3!}$

f For $n, r \in \mathbb{Z}^+$, $r > 0$ and $n \geqslant r$

$\binom{n}{r} = \dfrac{n(n-1)(n-2) \times \cdots \times (n-r+1)}{r!}$

After cancelling $(n - r)!$ there will be $r!$ in the denominator and r factors left in the numerator, counting down from n.

Example 3 Evaluate on a calculator **a** 10!   **b** $\binom{18}{6}$

a $10! = 3\,628\,800$

Press 10! ENTER

b $\binom{18}{6} = 18\,564$

On some calculators, pressing 18 ^nC_r 6 ENTER will give the answer. Find out how to obtain the correct answer on your calculator.

Exercise 14B

1 Evaluate, *without* using a calculator

a $3!$

b $4!$

c $5!$

d $\dfrac{10!}{8!}$

e $\dfrac{7!}{4!}$

f $\dfrac{12!}{9!}$

g $\dfrac{11!}{7!4!}$

h $\dfrac{6!2!}{8!}$

i $\dbinom{6}{3}$

j $\dbinom{5}{4}$

k $\dbinom{8}{2}$

l $\dbinom{21}{2}$

2 Evaluate, with a calculator

a $12!$

b $16!$

c $\dbinom{16}{4}$

d $\dbinom{8}{6}$

e $\dbinom{31}{7}$

f $\dbinom{6}{0}$

g $\dbinom{42}{38}$

h $\dbinom{42}{4}$

3 Evaluate $\dbinom{4}{0}$, $\dbinom{4}{1}$, $\dbinom{4}{2}$, $\dbinom{4}{3}$ and $\dbinom{4}{4}$ on a calculator and compare the results to the line of Pascal's triangle which starts 1, 4, …

4 Explain why $\dbinom{42}{38}$, the number of ways of choosing 38 objects from 42 different objects, is the same as $\dbinom{42}{4}$, the number of ways of choosing 4 objects from 42 different objects.

14.3 Formula for binomial expansion

Consider $(a+b)^4 = (a+b)(a+b)(a+b)(a+b)$.

When multiplying the brackets, either an a or a b is chosen from each bracket. Choosing an a from *each* bracket can be done in only one way. This gives the term a^4.

Choosing an a from three brackets and a b from one bracket can be done in four ways, since the b can be chosen from any one of the 4 brackets, i.e. there are $\binom{4}{1}$ choices.

Similarly, an a from two brackets and a b from two brackets can be chosen in $\binom{4}{2}$ ways.

So $(a+b)^4 = \dbinom{4}{0}a^4 + \dbinom{4}{1}a^3b + \dbinom{4}{2}a^2b^2 + \dbinom{4}{3}ab^3 + \dbinom{4}{4}b^4$

$\qquad\qquad = a^4 + 4a^3b + 6a^2b^2 + 4ab^3 + b^4$

Note $\binom{n}{0} = 1$ and $\binom{n}{1} = n$ for all values of n, so the first two coefficients in the expansion of $(a+b)^n$ are always 1 and n.

$$(a+b)^n$$

$$= \binom{n}{0}a^n + \binom{n}{1}a^{n-1}b + \binom{n}{2}a^{n-2}b^2 + \cdots + \binom{n}{r}a^{n-r}b^r + \cdots + \binom{n}{n}b^n$$

$$= a^n + na^{n-1}b + \frac{n(n-1)}{2!}a^{n-2}b^2 + \cdots + \frac{n(n-1)\cdots(n-r+1)}{r!}a^{n-r}b^r + \cdots + b^n$$

Putting $a = 1$ and $b = x$ in the formula gives

$$(1+x)^n = \binom{n}{0} + \binom{n}{1}x + \binom{n}{2}x^2 + \binom{n}{3}x^3 + \cdots + \binom{n}{r}x^r + \cdots + \binom{n}{n}x^n$$

$$= 1 + nx + \frac{n(n-1)}{2!}x^2 + \frac{n(n-1)(n-2)}{3!}x^3 + \cdots + x^n$$

The values of $\binom{n}{r}$ (or ^nC_r) correspond to the entries in Pascal's triangle.

$$\binom{0}{0} = 1 \qquad \text{Row 0}$$

$$\binom{1}{0} = 1 \quad \binom{1}{1} = 1 \qquad \text{Row 1}$$

$$\binom{2}{0} = 1 \quad \binom{2}{1} = 2 \quad \binom{2}{2} = 1 \qquad \text{Row 2}$$

$$\binom{3}{0} = 1 \quad \binom{3}{1} = 3 \quad \binom{3}{2} = 3 \quad \binom{3}{3} = 1 \qquad \text{Row 3}$$

$$\binom{4}{0} = 1 \quad \binom{4}{1} = 4 \quad \binom{4}{2} = 6 \quad \binom{4}{3} = 4 \quad \binom{4}{4} = 1 \qquad \text{Row 4}$$

$\cdots$ $\cdots$

$$\binom{n}{0} = 1 \quad \binom{n}{1} = n \quad \binom{n}{2} = \cdots \qquad \text{Row } n$$

Entry 0   Entry 1   Entry 2

If the rows and the entries in each row are counted, starting from zero, then $\binom{n}{r}$ (or ^nC_r) is equal to the rth entry in the nth row of the triangle.

For example, $\binom{4}{2}$ is entry number 2 in row 4.

Example 4   **a**   Expand $(1 + x)^7$ as far as the term in x^4.

b   By putting $x = 0.01$, find the value of 1.01^7 correct to 5 decimal places, *without* using a calculator.

c   By replacing x by $(-3x)$ in the solution to **a**, find the expansion of $(1 - 3x)^7$ in ascending powers of x as far as the term in x^4.

Remember   'Ascending powers of x' means the terms are in order of the powers of x with the smallest power first.

'Descending powers of x' means the terms are in order of the powers of x with the largest power first.

Solution   **a** $(1 + x)^7 = 1 + 7x + \dfrac{7 \times \cancel{6}^{3}}{\cancel{2} \times 1}x^2 + \dfrac{7 \times \cancel{6} \times 5}{\cancel{3} \times \cancel{2} \times 1}x^3 + \dfrac{7 \times \cancel{6} \times 5 \times \cancel{4}}{\cancel{4} \times \cancel{3} \times \cancel{2} \times 1}x^4 + \cdots$

As far as the term in x^4

$(1 + x)^7 = 1 + 7x + 21x^2 + 35x^3 + 35x^4$

> Using
> $(1 + x)^n$
> $= 1 + nx + \dfrac{n(n-1)}{2!}x^2 + \dfrac{n(n-1)(n-2)}{3!}x^3 + \cdots$

b   Putting $x = 0.01$

$(1 + 0.01)^7 \approx 1 + 7 \times 0.01 + 21 \times 0.01^2 + 35 \times 0.01^3$

> The x^4 and higher power terms will not affect the fifth decimal place.

$\approx 1 + 0.07 + 0.0021 + 0.000\,035$

$\approx 1.072\,135$

So $1.01^7 = 1.072\,14$ (to 5 d.p.)

c   From part **a**, as far as the term in x^4,

$(1 + x)^7 = 1 + 7x + 21x^2 + 35x^3 + 35x^4$

Replacing x by $(-3x)$ gives

$(1 - 3x)^7 = 1 + 7 \times (-3x) + 21 \times (-3x)^2 + 35 \times (-3x)^3 + 35 \times (-3x)^4$

$= 1 - 21x + 189x^2 - 945x^3 + 2835x^4$

Example 4b illustrates how a binomial expansion can be used for an approximation. In the expansion, the variable x was raised to higher and higher powers. When a small value was substituted for x, the terms in x^4 and x^5 and higher powers became small enough to be neglected. The number of terms needed for an approximation depends on the size of x and on the accuracy required.

Example 5   Expand $(2 - x)^{10}$, in ascending powers of x, as far as the term in x^3.

$(2 - x)^{10} = 2^{10} + 10 \times 2^9(-x) + \dfrac{\cancel{10}^{5} \times 9}{\cancel{2} \times 1} \times 2^8(-x)^2 + \dfrac{10 \times \cancel{9}^{3} \times \cancel{8}^{4}}{\cancel{3} \times \cancel{2} \times 1} \times 2^7(-x)^3 + \cdots$

$= 1024 - 5120x + 11\,520x^2 - 15\,360x^3$ (as far as the term in x^3).

Example 6   Using a calculator, find the first four terms of the expansion of $(3 + x)^{11}$.

$$(3 + x)^{11} = 3^{11} + 11 \times 3^{10}x + {}^{11}C_2 \times 3^9x^2 + {}^{11}C_3 \times 3^8x^3 + \cdots$$

> Use a calculator to evaluate coefficients.

$$= 177\,147 + 649\,539x + 1\,082\,565x^2 + 1\,082\,565x^3 + \cdots$$

Example 7   Express $(2 - \sqrt{5})^4$ in the form $m + n\sqrt{5}$.

$$(2 - \sqrt{5})^4 = 2^4 + \binom{4}{1} \times 2^3 \times (-\sqrt{5}) + \binom{4}{2} \times 2^2 \times (-\sqrt{5})^2$$

> Or use coefficients 1, 4, 6, 4, 1, from Pascal's triangle.

$$+ \binom{4}{3} \times 2 \times (-\sqrt{5})^3 + (-\sqrt{5})^4$$

> Take care with the negative terms.

$$= 16 - 4 \times 8\sqrt{5} + 6 \times 4 \times 5 - 4 \times 2 \times 5\sqrt{5} + 25$$

> *Note*: $(-\sqrt{5})^2 = 5$

$$= 16 - 32\sqrt{5} + 120 - 40\sqrt{5} + 25$$

> Collect terms.

$$= 161 - 72\sqrt{5}$$

Example 8   Given that the first three terms in the binomial expansion of $(1 + ax)^b$ are $1 - 12x + 54x^2$, find the values of a and b.

$$(1 + ax)^b = 1 + b(ax) + {}^bC_2(ax)^2 + \cdots$$
$$= 1 - 12x + 54x^2 + \cdots$$

> Equate coefficients of x and x^2.

$$ab = -12 \qquad \text{①}$$

$${}^bC_2a^2 = \frac{b(b-1)}{2 \times 1} \times a^2 = 54 \qquad \text{②}$$

> Solve the simultaneous equations ① and ②.

Rearranging ①

$$a = -\frac{12}{b}$$

Substituting in ②

$$\frac{b(b-1)}{2} \times \frac{12^2}{b^2} = 54$$

> Cancel $2b$ on LHS.

$$\frac{(b-1) \times 12 \times 6}{b} = 54$$

> Multiply both sides by b.

$$72(b-1) = 54b$$

> Divide both sides by 18.

$$4(b-1) = 3b$$

> Multiply out the bracket and rearrange.

$$4b - 4 = 3b$$

So $b = 4$

From ① $a = -3$

So $a = -3$, $b = 4$.

Exercise 14C

1 a Expand $(1 + y)^3$. By replacing y by $(-3y)$ in the expansion of $(1 + y)^3$ find the expansion of $(1 - 3y)^3$ in ascending powers of y.

b Expand $(z + 1)^4$. Hence find the expansion of $(2z + 1)^4$ in descending powers of z.

2 Expand

a $(1 + x)^4$ **b** $(1 - x)^5$ **c** $(1 + x)^6$ **d** $(1 - x)^7$

3 Expand, in ascending powers of x, as far as the term in x^2

a $(1 + x)^8$ **b** $(1 - x)^9$ **c** $(1 + x)^{10}$ **d** $(1 - x)^{11}$

4 Expand

a $(1 + 3x)^3$ **b** $(1 - 2y)^5$ **c** $(1 + 4z)^4$ **d** $\left(1 - \dfrac{x}{3}\right)^3$

e $\left(1 + \dfrac{3x}{2}\right)^4$ **f** $(x - 1)^5$ **g** $(2x + 1)^3$ **h** $\left(\dfrac{x}{2} - 1\right)^4$

5 Expand

a $(x + y)^4$ **b** $(a + b)^6$ **c** $(x^2 - y^2)^4$ **d** $(2 - x)^4$

e $\left(3 - \dfrac{x}{2}\right)^3$ **f** $(3 - x^3)^4$ **g** $(3a + 4b)^4$ **h** $\left(x - \dfrac{1}{x}\right)^3$

i $(1 + x)^3 (1 - x)^3$ **j** $\left(1 + \dfrac{2}{x}\right)^4$

6 Expand, in ascending powers of x, as far as the term in x^2

a $(1 + 3x)^7$ **b** $(2 - x)^8$ **c** $\left(1 + \dfrac{x}{3}\right)^{10}$ **d** $\left(1 - \dfrac{x}{2}\right)^{12}$

7 Express in the form $a + b\sqrt{2}$

a $(3 + \sqrt{2})^3$ **b** $(\sqrt{2} - 1)^4$ **c** $(1 - 3\sqrt{2})^5$

8 Expand $(a + b)^3$ and $(a - b)^3$. Hence, or otherwise, simplify these expressions, leaving surds in the answer where appropriate.

a $(1 + \sqrt{3})^3 + (1 - \sqrt{3})^3$ **b** $(1 + \sqrt{3})^3 - (1 - \sqrt{3})^3$

c $(2 + \sqrt{3})^3 - (2 - \sqrt{3})^3$ **d** $(\sqrt{6} + \sqrt{3})^3 + (\sqrt{6} - \sqrt{3})^3$

9 Simplify, leaving surds in the answer where appropriate

a $(1 + \sqrt{3})^4 + (1 - \sqrt{3})^4$ **b** $(2 + \sqrt{5})^4 - (2 - \sqrt{5})^4$

c $(2 + \sqrt{3})^4 + (2 - \sqrt{3})^4$ **d** $(\sqrt{6} + \sqrt{3})^4 - (\sqrt{6} - \sqrt{3})^4$

10 a Given that $x = 0.01$ find, *without* a calculator, x^2, x^3 and x^4.

b Repeat part **a** for $x = 0.001$.

11 a Expand $(1 - x)^7$, in ascending powers of x, up to and including the term in x^4.

b Putting $x = 0.01$, find the value of 0.99^7 correct to 5 decimal places, *without* using a calculator.

12 a Expand $\left(1 + \dfrac{x}{4}\right)^4$ up to, and including, the term in x^2.

b Substituting $x = 0.1$ in the expansion find, *without* using a calculator, the value of 1.025^4, correct to 3 decimal places.

13 a Find the first three terms of the expansion of $(2 + x)^5$ in ascending powers of x.

b Putting $x = 0.001$, find, *without* using a calculator, the value of 2.001^5, correct to 5 decimal places.

14 Given that $(1 + cx)^n = 1 + 24x + 264x^2 + \ldots$, find c and n.

15 Given that $(1 + ax)^n = 1 - 20x + 150x^2$ up to, and including, the term in x^2, find a and n.

16 Given that $(2 + kx)^6 = 64 - 576x + cx^2$, find k and c.

17 The coefficient of x^3 is seven times the coefficient of x in the expansion of $(1 + x)^n$, $n \in \mathbb{Z}^+$. Find n.

For more practice on the binomial expansion go to the CD-ROM (see E14.3).

**Exercise 14D
(Review)**

Numerical answers should be calculated without a calculator. A calculator may, however, be used for checking purposes.

1 Find the first three terms in these expansions in ascending powers of x.

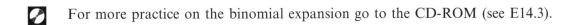

a $(1 + x)^4$ **b** $(1 - 2x)^3$ **c** $\left(1 + \dfrac{x}{3}\right)^5$ **d** $(2 - x)^4$ **e** $\left(x + \dfrac{1}{x}\right)^6$

2 a Expand $(1 + x)^4$ in ascending powers of x.

b Hence, or otherwise, show that $(1 + x)^4 - (1 - x)^4 = 8x(1 + x^2)$.

c Hence solve $(1 + x)^4 - (1 - x)^4 = 0$.

3 Expand $(a + b)^{11}$ in descending powers of a as far as the fourth term.

4 Write down the expansion of $(a - b)^5$ and, by putting $a = 10$ and $b = \frac{1}{2}$, use the result to find the value of $\left(9\frac{1}{2}\right)^5$ to the nearest 100.

5 Simplify $\left(\sqrt{2} + \sqrt{3}\right)^4 - \left(\sqrt{2} - \sqrt{3}\right)^4$.

6 When $(1 + ax)^n$ is expanded in ascending powers of x, the first three terms are $1 - 28x + 294x^2$. Find a and n.

7 a Expand $(1 - 2x)^9$ in ascending powers of x up to and including the term in x^3.

b Use the expansion to find an approximation to $(0.98)^9$, correct to 4 decimal places.

8 In the expansion of $\left(1 + \dfrac{x}{2}\right)^n$ in ascending powers of x the coefficient of x^2 is 30.

   a Find n.

   b Find the first four terms of the expansion.

9 **a** Find the first four terms in the expansion of $\left(2 + \frac{1}{4}x\right)^{10}$, in ascending powers of x.

   b Hence find the value of 2.025^{10}, correct to the nearest whole number.

10 Expand

   a $\left(x + \dfrac{1}{x}\right)^3$ **b** $\left(x + \dfrac{2}{x}\right)^3$

A 'Test yourself' exercise (Ty14) and an 'Extension exercise' (Ext14) can be found on the CD-ROM.

➤ Key points

Binomial expansion

The coefficients of the terms of a binomial expansion $(a+b)^n$ can be obtained from Pascal's triangle (suitable for small positive integers n) or by the formula.

For $(a+b)^n$, use the row of the triangle that starts $1, n, \ldots$

$$
\begin{array}{ccccccccc}
& & & & 1 & & & & \\
& & & 1 & & 1 & & & \\
& & 1 & & 2 & & 1 & & \\
& 1 & & 3 & & 3 & & 1 & \\
1 & & 4 & & 6 & & 4 & & 1
\end{array}
$$

$$(a+b)^4 = \boxed{1}\,a^4 + \boxed{4}\,a^3b + \boxed{6}\,a^2b^2 + \boxed{4}\,ab^3 + \boxed{1}\,b^4$$

Notation

$$\binom{n}{r} = {}^nC_r = \frac{n!}{r!(n-r)!}$$

where $n! = n(n-1)(n-2)(n-3) \times \ldots \times 2 \times 1$ and, by definition $0! = 1$.

Formula for binomial expansion

$(a+b)^n$

$$= \binom{n}{0}a^n + \binom{n}{1}a^{n-1}b + \binom{n}{2}a^{n-2}b^2 + \cdots + \binom{n}{r}a^{n-r}b^r + \cdots + \binom{n}{n}b^n$$

$$= a^n + na^{n-1}b + \frac{n(n-1)}{2!}a^{n-2}b^2 + \ldots + \frac{n(n-1)\ldots(n-r+1)}{r!}a^{n-r}b^r + \ldots + b^n$$

To expand, for example, $(3x - 2y)^5$, replace a by $3x$ and b by $(-2y)$ in the formula. Take care of signs and coefficients, when $3x$ and $(-2y)$ are squared, cubed, etc.

$$(1+x)^n = \binom{n}{0} + \binom{n}{1}x + \binom{n}{2}x^2 + \binom{n}{3}x^3 + \ldots + \binom{n}{r}x^r + \ldots + \binom{n}{n}x^n$$

$$= 1 + nx + \frac{n(n-1)}{2!}x^2 + \frac{n(n-1)(n-2)}{3!}x^3 + \ldots + x^n$$

Imagine looking at a graph of temperature against time.

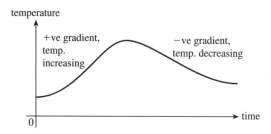

You could pick out the periods of time when the temperature was rising (the curve would have a positive gradient), and when the temperature was falling (the curve would have a negative gradient). Maximum temperatures would correspond to the peaks of the graph and minimum temperatures to the troughs; at these points the gradient would be zero.

Examining the gradient of a function, and, in particular, establishing the points where the gradient is zero, is a very useful approach in calculus. Most of this chapter deals with solving problems – maximum and minimum problems – by this technique.

After working through this chapter you should be able to

■ *find increasing and decreasing regions of functions*

■ *find stationary points (where $\dfrac{dy}{dx} = 0$) and identify their type*

■ *solve problems involving maxima and minima*

■ *sketch curves using various techniques.*

15.1 Increasing and decreasing functions

Before looking at the points of zero gradient, it is helpful to consider the range of values for which a function is increasing or decreasing.

Consider the graph of $y = x^2$.

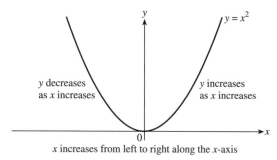

■ For $x < 0$, y decreases as x increases.
The gradient is negative.
For $x < 0$, y is a
decreasing function of x.

■ For $x > 0$, y increases as x increases.
The gradient is positive.
For $x > 0$, y is an
increasing function of x.

■ At $x = 0$, y is neither increasing nor decreasing.
At $x = 0$, y has a **stationary value**. The origin $(0, 0)$ is a **stationary point**.

> A function f which increases as x increases for a set of values is called an *increasing function* for that set of values.
>
> A function, f(x), is *increasing* for $a < x < b$ if f$'(x) > 0$ for $a < x < b$.
>
> A function f which decreases as x increases for a set of values is called a *decreasing function* for that set of values.
>
> A function, f(x), is *decreasing* for $a < x < b$ if f$'(x) < 0$, for $a < x < b$.

Remember $\dfrac{dy}{dx}$ can also be written as f$'(x)$ or y'.

Example 1   Given that $y = -7 + 9x^2 - 2x^3$, find the range of values of x for which y is an increasing function.

$$y = -7 + 9x^2 - 2x^3$$

So $\dfrac{dy}{dx} = 18x - 6x^2$ | Differentiate with respect to x.

y is an increasing function if $\dfrac{dy}{dx} > 0$ | *Note*: For a decreasing function, solve $\dfrac{dy}{dx} < 0$.

$\therefore$ y is an increasing function if $18x - 6x^2 > 0$ | Divide by 6.

$$3x - x^2 > 0$$
$$x(3 - x) > 0$$

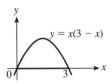

So $0 < x < 3$ | Solve the quadratic inequality by drawing a sketch or otherwise.

So y is an increasing function of x for $0 < x < 3$.

Exercise 15A

1   Find f$'(x)$ and the range of values of x for which f(x) is increasing.

   a   f$(x) = x^2 + 3x - 5$ **b**   f$(x) = 2 - 4x - x^2$

   c   f$(x) = 2x^3 - 9x^2 + 12x - 5$ **d**   f$(x) = 5x^2 - 2$

   e   f$(x) = x^3 - 48x$ **f**   f$(x) = 18x + x^3$

2   Find the range of values of x for which y is decreasing.

   a $y = 2x^2 - 2x - 1$ **b** $y = 4 + 3x - 7x^2$

   c $y = 2x^3 + 9x - 5$ **d** $y = x^3 - x^2$

   e $y = 3x^3 - 9x + 2$ **f** $y = 18x - 15x^2 - 4x^3$

15.2 Stationary points

➤ | **At a stationary point on a curve, the gradient is zero.** |
|---|

There are three types of stationary points: **maxima**, **minima** and **points of inflexion**.

Maximum and minimum points are called **turning points** because the graph turns at these points.

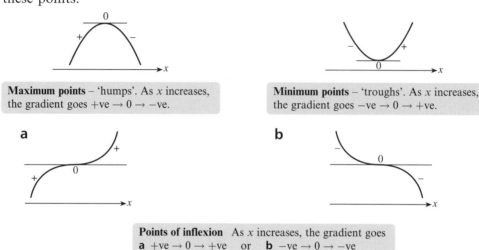

Maximum points – 'humps'. As x increases, the gradient goes $+$ve $\rightarrow 0 \rightarrow -$ve.	**Minimum points** – 'troughs'. As x increases, the gradient goes $-$ve $\rightarrow 0 \rightarrow +$ve.

a   **b**

Points of inflexion   As x increases, the gradient goes **a** $+$ve $\rightarrow 0 \rightarrow +$ve   or   **b** $-$ve $\rightarrow 0 \rightarrow -$ve

Consider the gradient of the parts of this curve.

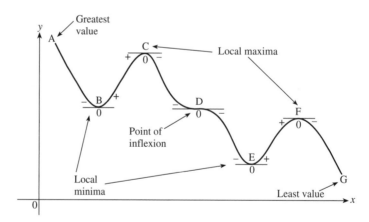

C and F are *maximum points* on the curve. The y-coordinates at C and F are the *maximum values* of the function in the region of those points, so these maximum values are called **local maxima**: they give the maximum value in the immediate vicinity of the point. Similarly B and E are **local minima**. D is also a stationary point: a *point of inflexion*.*

Notice that the value of the function at F (a local maximum) is less than the value at B (a local minimum).

The *greatest value* of the function (for the range of values shown) is its value at A and the *least value* is its value at G.

*Another type of point of inflexion (not a stationary point) is introduced later in the course.

Identifying the type of a stationary point

To investigate the stationary values of a function, first find the stationary points (where $\frac{\mathrm{d}y}{\mathrm{d}x} = 0$) and then use either of two methods.

Method 1: Work out the sign of the gradient on either side of the stationary point(s).

Note Be careful to take points in the neighbourhood of the stationary point. If too far away you may jump over another stationary point.

- For a maximum point, the gradient goes +ve → 0 → −ve, i.e. it decreases as x increases.

- For a minimum point, the gradient goes −ve → 0 → +ve, i.e. it increases as x increases.

Method 2: Find $\frac{\mathrm{d}^2 y}{\mathrm{d}x^2}$, the gradient of the gradient function.

- If $\frac{\mathrm{d}^2 y}{\mathrm{d}x^2} < 0$, i.e. the gradient is decreasing as x increases, the point is a maximum point.

- If $\frac{\mathrm{d}^2 y}{\mathrm{d}x^2} > 0$, i.e. the gradient is increasing as x increases, the point is a minimum point.

- If $\frac{\mathrm{d}^2 y}{\mathrm{d}x^2} = 0$ use method 1.*

> $\frac{\mathbf{d}y}{\mathbf{d}x} = \mathbf{0}$ **and** $\frac{\mathbf{d}^2 y}{\mathbf{d}x^2} < \mathbf{0} \Rightarrow$ **maximum point**
>
> $\frac{\mathbf{d}y}{\mathbf{d}x} = \mathbf{0}$ **and** $\frac{\mathbf{d}^2 y}{\mathbf{d}x^2} > \mathbf{0} \Rightarrow$ **minimum point**
>
> $\frac{\mathbf{d}y}{\mathbf{d}x} = \mathbf{0}$ **and** $\frac{\mathbf{d}^2 y}{\mathbf{d}x^2} = \mathbf{0} \Rightarrow$ **point of inflexion or maximum or minimum point.**
>
> **So the nature of a stationary point can be determined by the above or by considering the gradient of the curve on either side of the stationary point.**

*There are other methods available for dealing with this situation at a higher level.

Example 2   Find the coordinates of any stationary points of the curve $y = 3x - x^3$.

$$y = 3x - x^3$$

Differentiate y with respect to x.

$$\frac{dy}{dx} = 3 - 3x^2$$

At a stationary point, $\frac{dy}{dx} = 0$

so $3 - 3x^2 = 0$

Divide through by 3.

$$1 - x^2 = 0$$

$$x^2 = 1$$

$\therefore$ $x = \pm 1$

When $x = 1$, $y = 3 \times 1 - 1^3 = 2$ and
when $x = -1$, $y = 3 \times (-1) - (-1)^3 = -2$

Substitute for x in the equation of the curve.

So the stationary points are $(1, 2)$ and $(-1, -2)$

$$\frac{d^2y}{dx^2} = -6x$$

When $x = 1$, $\frac{d^2y}{dx^2} = -6 < 0$

Therefore $(1, 2)$ is a maximum point.

When $x = -1$, $\frac{d^2y}{dx^2} = 6 > 0$

The nature of the stationary points can usually be determined by finding the sign of $\frac{d^2y}{dx^2}$.

If $\frac{d^2y}{dx^2} = 0$ then Method 1 (page 233) should be used.

Note: If possible sketch the curve on a graphical calculator.

Therefore $(-1, -2)$ is a minimum point.

Example 3   Find the coordinates of any stationary points of the curve $y = 4x^3 - x^4$.
Sketch the curve, state the range of values for which the function is increasing and the number of distinct roots of $4x^3 - x^4 = 0$.

$$y = 4x^3 - x^4 \qquad ①$$

Differentiate y with respect to x.

$$\frac{dy}{dx} = 12x^2 - 4x^3$$

At a stationary point, $\frac{dy}{dx} = 0$

so $$12x^2 - 4x^3 = 0$$

Divide all terms by 4.

$$3x^2 - x^3 = 0$$

$$x^2(3 - x) = 0$$

$\therefore$ $x = 0$ or $x = 3$

When $x = 0$, $y = 0$, and when $x = 3$, $y = 4 \times 3^3 - 3^4 = 27$.

Substitute for x in ①

So the stationary points are $(0, 0)$ and $(3, 27)$.

To decide on the type of stationary points both methods will be used.

Using Method 2

$$\frac{d^2y}{dx^2} = 24x - 12x^2$$

When $x = 3$, $\frac{d^2y}{dx^2} = -36 < 0$

$\therefore$ at $x = 3$, there is a maximum point.

When $x = 0$, $\frac{d^2y}{dx^2} = 0$

Since $\frac{d^2y}{dx^2} = 0$, the point could be a maximum, a minimum or a point of inflexion.

So now use method 1: work out the sign of $\frac{dy}{dx}$ on either side of $x = 0$.

Using Method 1

Value of x	L e.g. $x = -1$	0	R e.g. $x = 1$	L e.g. $x = 2$	3	R e.g. $x = 4$
Sign of $\frac{dy}{dx}$ (i.e. sign of $12x^2 - 4x^3$)	+	0	+	+	0	−
		Point of inflexion			Maximum point	

$\frac{dy}{dx} = 4x^2(3 - x)$. So for $x \neq 0$, x^2 is always +ve and so only the sign of $3 - x$ need be considered.

The stationary points are $(0, 0)$ point of inflexion, and $(3, 27)$ maximum point.

The curve cuts the y-axis where $x = 0$.

When $x = 0$, $y = 0$.

$\therefore$ the curve cuts the y-axis at $(0, 0)$.

The curve cuts the x-axis where $y = 0$.

When $y = 0$ $4x^3 - x^4 = 0$

$$x^3(4 - x) = 0$$

So $x = 0$ or $x = 4$

$(0, 0)$ is the point of inflexion already found.

$(4, 0)$ is the other point where the curve cuts the x-axis.

The graph shows that the function $f(x) = 4x^3 - x^4$ is increasing for $x < 0$ and $0 < x < 3$ and that $4x^3 - x^4 = 0$ has two distinct real roots.

To sketch the curve, the intersections of the curve with the axes should be found.

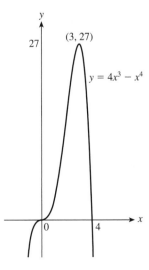

Note $x = 0$ is a triple root.*

*A triple root: curve cuts the x-axis with a point of inflexion. A double root: curve touches the x-axis.
 A single root: curve cuts the x-axis.

Exercise 15B

1 Find the coordinates of the points on these curves where the gradient is zero.

a $y = x^2 - 4x - 2$ **b** $y = x^2 + 6x + 7$ **c** $y = x(3x - 2)$

d $y = \dfrac{x^2 + 3x + 4}{x}$ **e** $y = (x + 8)(x - 3)$ **f** $y = x^4 - 2x^2 + 5$

g $y = 9 + 24x - 2x^3$ **h** $y = \sqrt{x} + \dfrac{1}{\sqrt{x}}$

2 $y = x^2 - 3x + 3$.

  a Show that $\dfrac{dy}{dx} = 0$ when $x = \frac{3}{2}$.

  b By considering the sign of $\dfrac{dy}{dx}$ for two values of x, one less than $\frac{3}{2}$ and one greater than $\frac{3}{2}$, or otherwise, show that $x = \frac{3}{2}$ gives a minimum value of y.

  c Find the coordinates of the minimum point on the curve $y = x^2 - 3x + 3$.

3 Find the coordinates of any stationary points on these curves, and state, giving reasons, whether each point is a maximum or a minimum.

  a $y = x^4$ **b** $y = 2x^3 - 15x^2 + 36x - 20$

  c $y = (x + 2)^2$ **d** $y = 5 - 9x + 6x^2 - x^3$

  e $y = x^4 + 3x^2$ **f** $y = 25x + \dfrac{4}{x}$

  g $y = 2x^{\frac{1}{2}} + 4x^{-\frac{1}{2}}$

4 Find the value(s) of x for which these *derived* functions are zero.

  a $f'(x) = (x - 3)(x + 7)$ **b** $f'(x) = x - \dfrac{1}{x}$ **c** $f'(x) = \sqrt{2x} - \dfrac{1}{\sqrt{2x}}$

  d $f'(x) = x + \dfrac{1}{x^2}$ **e** $f'(x) = \pi x - \dfrac{10}{x^2}$

5 Find the coordinates of any stationary points on these curves, and state, giving reasons, whether the points are maxima or minima or points of inflexion.

  a $y = x^3$ **b** $y = x^5 - 15x^3$

  c $y = x^4 + 3x^3$ **d** $y = 20 + 15x - x^2 - \dfrac{x^3}{3}$

  e $y = \dfrac{1 - 27x^2}{x^3}$ **f** $y = x^3 + 6x^2 + 12x - 4$

6 $f(x) = 3x^2 - 2x^3$.

  a Find $f'(x)$.

  b Show that the function $f(x)$ is stationary (i.e. $f'(x) = 0$) when $x = 0$ and $x = 1$ and find $f(0)$ and $f(1)$.

  c Show that $f(0)$ is a minimum value of the function and that $f(1)$ is a maximum value.

  d Solve $f'(x) > 0$ to find the range of values of x for which the function is increasing.

  e Solve $f'(x) < 0$ to find the range of values of x for which the function is decreasing.

   f Sketch $y = f(x)$.

   g Using the sketch, find the number of distinct real roots of $f(x) = 0$.

7 $f(x) = x^2 - x - 2$.

   a Find $f'(x)$.

   b Find the value(s) of x and $f(x)$ for which the function is stationary and the nature of the stationary value(s).

   c Find the range of values of x for which the function is increasing, and the range for which it is decreasing.

   d Sketch the curve $y = f(x)$.

   e Use the sketch to state the number of distinct real roots of the equation $f(x) = 0$.

8 Repeat Question 7 for these functions.

   i $f(x) = x^3 - 12x$ **ii** $f(x) = 3 - 2x - x^2$
   iii $f(x) = 3x^2 - x^3$ **iv** $f(x) = 3 - x^3$
   v $f(x) = x^3 - 5x^2 + 3x + 2$

9 Investigate the stationary points of these curves, stating the coordinates and the type of each stationary point.

   a $y = x + \dfrac{1}{x}$ **b** $y = \dfrac{3x^{\frac{1}{2}}}{4} - x^{\frac{3}{2}}$ **c** $y = 4x^2 + \dfrac{1}{x}$ **d** $y = x^2 + \dfrac{16}{x}$

10 Given that $y = 5x^2 + ax + b$ has a turning point at (b, a) where $a \neq 0$, find a and b.

11 Given that $y = ax^3 + bx^2 + 3x + 4$ and that $y = 5$ and $\dfrac{dy}{dx} = 2$ when $x = 1$

   a find a and b

   b show that the curve has no stationary point.

12 A ball is thrown in the air. At time t in seconds, its height h in metres above ground is given by $h = 2 + 9t - 5t^2$.

   a The formula given is valid only until the ball hits the ground.
       Find the time at which this happens.

   b Find the time at which the ball's height reaches a maximum and find this maximum height.

15.3 Maximum and minimum problems

As the examples in Exercises 15C and 15E will show, finding maximum and minimum points has many useful applications.

To tackle each problem follow these steps.

■ Draw a diagram, if relevant.

■ Choose letters to represent unknown quantities.

■ Express the quantity (e.g. y) to be maximised or minimised in terms of just *one* variable (e.g. x). (This may require some algebraic manipulation.)

■ Differentiate y with respect to x.

■ Solve $\dfrac{dy}{dx} = 0$ to find the value(s) of x at the maximum or minimum point(s).

■ Substitute the value(s) of x in the expression for y, checking that the value(s) give possible answers.

■ Check, if necessary, that the value gives a maximum or minimum.

Practice in applying algebra to problems can be found on the CD-ROM (A15.3).

Example 4   An open box is to be made from a rectangular sheet of card measuring 16 cm by 10 cm. Four equal squares are to be cut from each corner and the flaps folded up. Find the length of the side of a square which makes the volume of the box as large as possible. Find this largest volume.

As a practical exercise, take a sheet of paper 16 cm by 10 cm and cut off four equal squares from the corners. Fold up the flaps. Use intuition to cut off the corners which you think will give the largest volume box.

Let the length of the side of each square be x cm.   Draw a diagram.

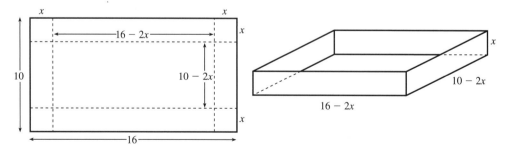

Let V cm^3 be the volume of the box.   Volume V is to be maximised.

238

Then $V = x(10 - 2x)(16 - 2x)$ | Express V in terms of x.

$\qquad = x(160 - 52x + 4x^2)$

$\qquad = 160x - 52x^2 + 4x^3$

$\dfrac{\mathrm{d}V}{\mathrm{d}x} = 160 - 104x + 12x^2$ | Differentiate V with respect to x.

At a maximum point, $\dfrac{\mathrm{d}V}{\mathrm{d}x} = 0$.

So $\qquad 160 - 104x + 12x^2 = 0$ | Divide all terms by 4.

$\qquad 3x^2 - 26x + 40 = 0$

$\qquad (3x - 20)(x - 2) = 0$ | Solve by the formula if the factors cannot be found.

$\qquad\qquad x = \frac{20}{3} \text{ or } x = 2$

$\qquad\qquad x = 6\frac{2}{3} \text{ or } x = 2$

The length of the side of the square must be less than half the side of the rectangle.
$\therefore\ x = 6\frac{2}{3}$ is *not* a possible solution.

So $x = 2$ is the solution.

When $x = 2 \qquad V = 2 \times (10 - 4) \times (16 - 4)$ | Substitute $x = 2$ in the expression for V.

$\qquad\qquad = 2 \times 6 \times 12$

$\qquad\qquad = 144$

To check that this is a maximum value,

$$\dfrac{\mathrm{d}^2 V}{\mathrm{d}x^2} = -104 + 24x$$

When $x = 2 \qquad \dfrac{\mathrm{d}^2 V}{\mathrm{d}x^2} = -104 + 48 < 0$

$\therefore\ x = 2$ gives a maximum value.

Alternatively, to check $x = 2$ gives a maximum value:

x	L $x = 1$	2	R $x = 3$
$\dfrac{\mathrm{d}V}{\mathrm{d}x}$	$+68$	0	-44
		Maximum point	

So the largest volume is $144\,\text{cm}^3$ and this is obtained by cutting off squares with length of each side $2\,\text{cm}$.

Compare your paper 'box' with the solution. Also try drawing $V = 160x - 52x^2 + 4x^3$ on a graphical calculator and interpreting the graph.

Example 5 A sports area is to be designed in the form of a rectangular field with a semicircular area at each end. The perimeter of the sports area is to be 1400 metres. What should be the dimensions of the sports area if the rectangular field is to have as large an area as possible? Give answers in metres correct to 1 decimal place.

Let the length be l m and radius of the semicircle be r m.

Draw a diagram.

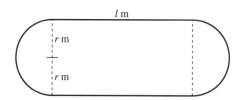

Let A m^2 be the area of the rectangular field.

A is the quantity to be maximised.

$$A = 2r \times l$$
$$= 2lr \qquad ①$$

A is expressed here in terms of two variables, r and l. A must be expressed in terms of one variable only, so another equation connecting the variables is needed.

Perimeter $= (2\pi r + 2l)\,$m

But perimeter is 1400 m

The perimeter is known; this gives the second equation.

so $\quad 2\pi r + 2l = 1400$

Divide all terms by 2.

$$\pi r + l = 700$$

Rearrange.

$$l = 700 - \pi r \qquad ②$$

Substituting in ①

$$A = 2(700 - \pi r)r$$

A is now expressed in terms of only one variable, r.

$$= 1400r - 2\pi r^2$$

Remember π is a constant.

so $\quad \dfrac{\mathrm{d}A}{\mathrm{d}r} = 1400 - 4\pi r$

Differentiate A with respect to r.

At a maximum point, $\dfrac{\mathrm{d}A}{\mathrm{d}r} = 0$.

So $\quad 1400 - 4\pi r = 0$

$$r = \frac{1400}{4\pi}$$

$$= \frac{350}{\pi}$$

$$= 111.4 \text{ (correct to 1 d.p.)}$$

The side of the rectangular area is $2r$.

$$2r = \frac{700}{\pi}$$

$$= 222.8 \text{ (correct to 1 d.p.)}$$

Substitute $r = \dfrac{350}{\pi}$ in ②.

$$l = 700 - \pi r$$

$$= 700 - \pi \times \frac{350}{\pi}$$

$$= 350$$

> Use the exact value for r. This makes the calculation easier as π cancels.

$A = 1400r - 2\pi r^2$ is a quadratic expression. The coefficient of r^2 is $-$ve.

> Alternatively show that $\dfrac{d^2 A}{dr^2} < 0$ when $r = \dfrac{350}{\pi}$ to prove that A has a maximum value.

$\therefore$ A has a maximum value.

So the dimensions of the rectangular field for maximum area are $350\,\mathrm{m}$ by $222.8\,\mathrm{m}$, giving semicircular ends of radius $111.4\,\mathrm{m}$.

Note To solve maximum and minimum problems, express the quantity (e.g. y) to be maximised or minimised in terms of one variable (e.g. x) and find where $\frac{dy}{dx} = 0$.

Exercise 15C

1 A farmer has $100\,\mathrm{m}$ of metal railing with which to form two adjacent sides of a rectangular enclosure, the other two sides being two existing walls meeting at right angles. Let the length of one side of the enclosure be $x\,\mathrm{m}$.

    a Draw a diagram and write down an expression for the length of the other side of the enclosure.

    b Obtain an expression for the area, $A\,\mathrm{m}^2$, of the enclosure in terms of x.

    c Find $\dfrac{dA}{dx}$ and solve $\dfrac{dA}{dx} = 0$.

    d Hence find the dimensions of the enclosure which give the maximum area and find the maximum area.
(Remember to include proof that the answer does give a *maximum* area.)

2 A rectangular sheep pen is to be made out of $1000\,\mathrm{m}$ of fencing, using an existing straight hedge for one of the sides. Find the maximum area possible, and the dimensions necessary to achieve this.

3 An aeroplane flying level at $250\,\mathrm{m}$ above the ground suddenly swoops down to drop supplies, and then regains its former altitude. It is $h\,\mathrm{m}$ above the ground $t\,\mathrm{s}$ after beginning its dive, where $h = 8t^2 - 80t + 250$. Find

    a its least altitude during this operation

    b the interval of time during which it was losing height.

4 An open tank is to be constructed with a square base and vertical sides so as to contain $500\,\mathrm{m}^3$ of water.

    a Given that the length of the side of the square base is $x\,\mathrm{m}$, find expressions, in terms of x, for the height of the tank, and for the external surface area of the tank.

b Find the value of x required so that the area of sheet metal used in constructing the tank is a minimum. (Remember to show that the value of x found gives a *minimum* value.)

c Find the minimum area of metal.

5 An open cylinder has radius r cm and volume 27π cm³.

a Find an expression for the external surface area, S cm², in terms of r.

b Find the value of r which makes $\dfrac{dS}{dr} = 0$ and prove that this value of r gives a minimum value of S.

c Hence find the minimum surface area, leaving π in the answer.

6 An open rectangular box is to be made with an external area of 1620 cm². The ratio of the lengths of the sides of the base of the box is 3:5.

Let the length of the shorter side of the base be $3x$ cm, the height of the box h cm and its volume V cm³.

a Draw a sketch of the box and write down an expression for the longer side of the base of the box in terms of x.

b Show that $V = 15hx^2$.

c Show that the external area of the box is given by $(16hx + 15x^2)$ cm².

d Show that $h = \dfrac{1620 - 15x^2}{16x}$ and hence express V in terms of x only.

e Use differentiation to find the value of x which makes the volume of the box a maximum.

f Hence find the maximum volume.

7 This diagram represents the end view of the outer cover of a match box, AB and EF being gummed together, and assumed to be the same length.

If the total length of edge (ABCDEF) is 12 cm, calculate the lengths of AB and BC which will give the maximum possible cross-section area.

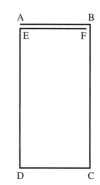

8 This diagram represents a rectangular sheet of metal 8 cm by 5 cm.

Equal squares of side x cm are removed from each corner, and the edges are then turned up to make an open box of volume V cm³.

Show that $V = 40x - 26x^2 + 4x^3$.

Hence find the maximum possible volume, and the corresponding value of x.

9 Repeat Question 8 when the dimensions of the sheet of metal are 8 cm by 3 cm, showing that, in this case, $V = 24x - 22x^2 + 4x^3$.

10 A sealed cylindrical tin is of height h cm and radius r cm.
The area of its total outer surface is A cm^2 and its volume is V cm^3.

   a Find expressions for A and V in terms of r and h.

   b Taking $A = 24\pi$, find an expression for h in terms of r, and hence an expression for V in terms of r.

   c Find the value of r which will make V a maximum.

11 A sweet manufacturer estimates that if it sets the price of a box of speciality chocolates at £p it will sell n boxes per year, where $n = 1000(84 + 12p - p^2)$, for $2.5 \leqslant p \leqslant 15$.

   a Find the price that will maximise the number of boxes sold.

   b Write down the revenue received by selling n boxes at price £p.

   c Hence show that the price that will maximise the manufacturer's revenue is £10.50, to the nearest 50 pence.

15.4 Sketching curves 2

In Chapter 7, curve sketching was approached by transformation of curves. In this section, further techniques are introduced. Ask these questions when sketching a graph.

- Is the graph a standard function or the transformation of one?
- Where does the graph cross the axes?

  Putting $x = 0$ will give the intercept(s) on the y-axis.
  Putting $y = 0$ will give the intercept(s) on the x-axis.

- Does the graph have symmetry?

  Is it symmetrical about the y-axis?
  If all powers of x are even, the graph will be symmetrical about the y-axis.

- What happens for large positive and negative values of x?

  If $f(x)$ is a polynomial in x then the highest power of x will determine its behaviour as $x \to \pm\infty$.

  (Read $x \to \pm\infty$ as 'x tends to plus or minus infinity'.)

  If the highest power term is ax^n then ax^n is the dominant term and

  n even $a > 0$ $f(x) \to +\infty$ as $x \to +\infty$ $a < 0$ $f(x) \to -\infty$ as $x \to +\infty$

  $f(x) \to +\infty$ as $x \to -\infty$ $f(x) \to -\infty$ as $x \to -\infty$

  n odd $a > 0$ $f(x) \to +\infty$ as $x \to +\infty$ $a < 0$ $f(x) \to -\infty$ as $x \to +\infty$

  $f(x) \to -\infty$ as $x \to -\infty$ $f(x) \to +\infty$ as $x \to -\infty$

- Does the function have stationary values, i.e. points where the gradient is zero?

Example 6 $f(x) = 2x^3 - 9x^2 + 12x - 4$

a Show that $2x - 1$ is a factor of $f(x)$.

b Find where the curve $y = f(x)$ crosses the axes.

c State the behaviour of $f(x)$ as $x \to \pm\infty$.

d Find any stationary points of $y = f(x)$.

e Sketch the curve.

f Using the sketch, or otherwise, determine the range of values of k for which $f(x) = k$ has three distinct real roots.

Solution   **a** $f\left(\tfrac{1}{2}\right) = 2 \times \left(\tfrac{1}{2}\right)^3 - 9 \times \left(\tfrac{1}{2}\right)^2 + 12 \times \tfrac{1}{2} - 4$

$\qquad\qquad = \tfrac{1}{4} - \tfrac{9}{4} + 6 - 4 = 0$

$\qquad \therefore \quad 2x - 1$ is a factor of $f(x)$.

$2x - 1 = 0 \Rightarrow x = \tfrac{1}{2}$

$f\left(\tfrac{1}{2}\right) = 0 \Rightarrow 2x - 1$ is a factor of $f(x)$

b $y = 2x^3 - 9x^2 + 12x - 4$

When $x = 0$, $y = -4$, so the curve crosses the y-axis at $(0, -4)$.

The curve crosses the x-axis when $y = 0$.

$\qquad y = 2x^3 - 9x^2 + 12x - 4$

$\qquad\quad = (2x - 1)(x^2 - 4x + 4)$

$\qquad\quad = (2x - 1)(x - 2)^2$

When $(2x - 1)(x - 2)^2 = 0$

$\qquad x = \tfrac{1}{2}$ or $x = 2$ (double root)

$x = 2$ is a *double* root so the curve will meet the x-axis in *two* coincident points, i.e. will touch the x-axis, at $x = 2$.

$\therefore \ y = f(x)$ crosses the x-axis at $\left(\tfrac{1}{2}, 0\right)$ and touches the x-axis at $(2, 0)$.

So the curve crosses the axes at $(0, -4)$ and $\left(\tfrac{1}{2}, 0\right)$ and touches the x-axis at $(2, 0)$.

c   As $x \to +\infty$, $y \to +\infty$

and as $x \to -\infty$, $y \to -\infty$

$2x^3$ is the dominant term and since the power is odd $y \to +\infty$ for large +ve values of x and $y \to -\infty$ for large −ve values of x.

d $f'(x) = 6x^2 - 18x + 12$

At a stationary value, $f'(x) = 0$

$\qquad 6x^2 - 18x + 12 = 0$

$\qquad\quad x^2 - 3x + 2 = 0$

$\qquad (x - 1)(x - 2) = 0$

$\Rightarrow \qquad\qquad x = 1$ or $x = 2$

Differentiate $f(x)$ to find stationary points where gradient $f'(x) = 0$ $\left(\text{use } f'(x) \text{ or } \dfrac{dy}{dx}\right)$.

Stationary points occur when $x = 1$ and $x = 2$.

$\qquad\qquad f'(x) = 6x^2 - 18x + 12$

$\Rightarrow \qquad\quad f''(x) = 12x - 18$

Consider the value of $f''(x)$ to determine the nature of the stationary points.

When $x = 1$ $f(x) = 2 - 9 + 12 - 4$

$\qquad\qquad\qquad = 1$

and $\qquad f''(x) = -6 < 0$

$\therefore \ (1, 1)$ is a maximum point.

$f''(x) < 0 \Rightarrow$ maximum point

When $x = 2$ $f(x) = 2 \times 2^3 - 9 \times 2^2 + 12 \times 2 - 4$

$\qquad\qquad\qquad = 16 - 36 + 24 - 4$

$\qquad\qquad\qquad = 0$

and $\qquad f''(x) = 6 > 0$

$\therefore \ (2, 0)$ is a minimum point.

$f''(x) > 0 \Rightarrow$ minimum point

e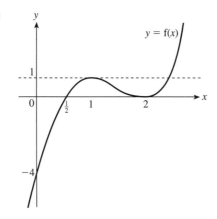

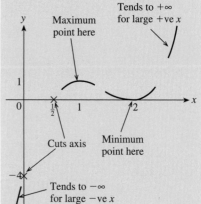

Assemble the information for the sketch.

f f(x) = k has three distinct real roots if y = k cuts y = f(x) in three distinct points.

From the sketch, it can be seen that, for 0 < k < 1, there are three distinct real roots.

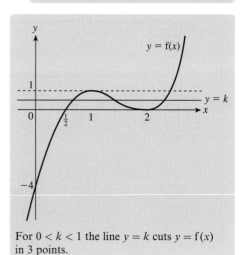

For 0 < k < 1 the line y = k cuts y = f(x) in 3 points.

Note k = 0 or k = 1 will give two distinct real roots
k < 0 or k > 1 will give one real root.

Note All cubics of the form $y = ax^3 + bx^2 + cx + d$ $(a \neq 0)$ have graphs of one of the following forms.

Example 7 $y = x^4 - 2x^2$

    a   State any symmetry of the curve.

    b   State the behaviour of y for large values of x.

    c   Find where the curve meets the axes.

    d   Find any stationary points of the curve.

    e   Sketch the curve.

    f   Find the range of values of k for which $x^4 - 2x^2 - k = 0$ has no real roots.

Solution **a** Let $f(x) = x^4 - 2x^2$

       All powers of $f(x)$ are even so the curve $y = f(x)$

       is symmetrical about the y-axis.

                                 > All powers of x are even.

  b For large x, both +ve and −ve, y is large and +ve.

                                   > x^4 is the dominant term and this is always +ve.

  c When $x = 0$, $y = 0$.

       When $y = 0$, $x^4 - 2x^2 = 0$

       $x^2(x^2 - 2) = 0$

       So $x = 0$ (double root) or $x = \pm\sqrt{2}$.

       So the curve cuts the x-axis at $(-\sqrt{2}, 0)$

       and $(\sqrt{2}, 0)$, touches the x-axis at $(0, 0)$

       and cuts the y-axis at $(0, 0)$.

                       > $x = 0$ is a double root, so the curve will meet the x-axis in two coincident points, i.e. will touch the x-axis, at $x = 0$.

  d $y = x^4 - 2x^2$

       $y' = 4x^3 - 4x$

       > Differentiate to find stationary points where gradient $y' = 0$

       $y' = 0$ when $4x^3 - 4x = 0$

                   $x(x^2 - 1) = 0$

       $\therefore$ $x = 0$ or $x = \pm 1$

                   $y'' = 12x^2 - 4$

       > Consider the value of y'' to determine the nature of the stationary points.

       When $x = 0$, $y = 0$, and $y'' = -4 < 0$.

       $\therefore$ $(0, 0)$ is a maximum point.

       When $x = \pm 1$, $y = -1$, and $y'' = 8 > 0$

       $\therefore$ $(1, -1)$ and $(-1, -1)$ are both
          minimum points.

       > Since y and y'' contain only even powers of x, the same value will be obtained when $x = +1$ or $x = -1$ is substituted.

  e

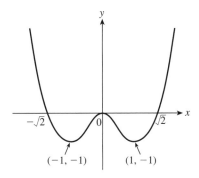

       > Assemble data for the sketch.

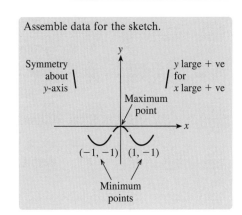

f For $k < -1$, the line $y = k$ will not intersect the curve $y = x^4 - 2x^2$

The intersection of $y = k$ and $y = x^4 - 2x^2$ gives the roots of $x^4 - 2x^2 = k$, i.e. $x^4 - 2x^2 - k = 0$.

∴ for $k < -1$, $x^4 - 2x^2 - k = 0$ will have no real roots.

Note   All quartics of the form $y = ax^4 + bx^3 + cx^2 + dx + e \ (a \neq 0)$ have graphs of one of the following forms.

Exercise 15D

1 The sketch shows the curve $y = 4x^3 - 3x$. The curve cuts the x-axis at the origin and at A and D. There is a maximum point at B and a minimum at C.

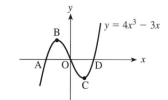

a Find the coordinates of A, B, C and D.

b Copy the sketch and explain how the sketch shows that the equation $4x^3 - 3x = \frac{1}{4}$ has exactly three distinct real roots.

c State the number of distinct real roots of $4x^3 - 3x = c$ when $c = 1$ and $c = 5$.

d Find the range of values of k given that $4x^3 - 3x = k$ has three distinct real roots.

2 For these curves, find where the curve meets the axes, state any symmetry, and the behaviour as $x \to \pm\infty$. Find the coordinates of any stationary points, determine whether they are maximum or minimum points or points of inflexion, and then sketch each curve.

a $y = 3x^2 - x^3$ **b** $y = x^3 - 6x^2$

c $y = x^4 - 10x^2 + 9$ **d** $y = 3x^5 - 5x^3$

3 For these curves, find where the curve meets the axes, and the coordinates of any stationary points and determine their nature. State the behaviour of y as $x \to \pm\infty$ and then sketch each curve.

a $y = (x - 3)(x^2 + 3x + 6)$ **b** $y = 12x - x^3$

c $y = (2 - x)^3$ **d** $y = (x - 1)(x^2 + 4)$

4 a Given that $y = 4x^4 - x^2$ sketch the curve.

b Find the value(s) of k for which $4x^4 - x^2 = k$ has four, three, two, one and no distinct real roots.

5 a Given that $y = 8x^3 - x^4$ sketch the curve.

b Find the value(s) of c for which $8x^3 - x^4 - c = 0$ has three, two, one and no distinct real roots.

**Exercise 15E
(Review)**

1 Find the coordinates of the points on these curves at which the gradient is zero, and determine whether the points are local maxima or local minima, giving reasons.

  a $y = x^3 - 3x$ **b** $y = x^2(x^2 - 8)$ **c** $y = 6\sqrt{x} - 3x$ **d** $y = \dfrac{x^2 + 3}{\sqrt{x}}$

2 Find the range of values of a for which $y = x - \dfrac{a}{x}$ has no stationary points.

3 For each of these functions, decide whether the function is always decreasing, the function is always increasing, or the function is sometimes increasing and sometimes decreasing.

  a $y = x^3 - 3x + 1$

  b $y = x^3 + 3x + 1$

  c $y = 1 - 3x - x^3$

4 Find the maximum or minimum values of these functions, first by differentiation and then by completing the square (see Example 28 page 57).

  a $y = x^2 - 2x + 7$ **b** $y = 3 + 4x - x^2$

  c $y = 2x^2 + x + 1$ **d** $y = 4 + 6x - 2x^2$

5 A circular pipe has outer diameter 4 cm and thickness t cm.

  a Show that the area of cross-section, A cm^2, is given by $A = \pi(4t - t^2)$.

  b Find the rate of increase of A with respect to t when $t = \frac{1}{4}$ and when $t = \frac{1}{2}$, leaving π in the answer.

6 The radius, r cm, of a circle at time t s is given by $r = \dfrac{t^4}{4} + \dfrac{4}{t}$.

  Find the rate at which the radius is changing with respect to time when $t = 1$ and when $t = 2$. State whether the radius is increasing or decreasing in each case.

7 Given that $y = 2x^3 - 15x^2 + 36x - 7$, find the range of values of x for which y is

  a an increasing function   **b** a decreasing function.

8 Prove that there are two points on the curve $y = 2x^2 - x^4$ at which y has a maximum value, and one point at which y has a minimum value.
Give the equations of the tangents to the curve at these three points.

9 For each of these curves, find where the curve meets the axes, state the behaviour as $x \to \pm\infty$, find the coordinates and type of any stationary points, and then sketch the curve.

  a $y = x^3 - 2x^2 + x$ **b** $y = x^2(x - 2)^2$

  c $y = (x + 1)^2(2 - x)$ **d** $y = x^4 + 4x$

10 A stone is thrown vertically upwards and after t seconds its height h metres above its starting point is given by $h = 80t - 5t^2$.

  Find the greatest height reached by the stone and the time when this occurs.

11 A spherical balloon is such that its radius r cm at time t minutes is given by
$r = t^3 - 15t^2 + 48t + 110$ for $0 \leqslant t \leqslant 12$.

   a Find the minimum value of the radius in the first 12 minutes and the time when this occurs.

   b After 12 minutes the balloon bursts. Find its radius just before it bursts.

12 A solid rectangular block has a square base. Find its maximum volume if the sum of the height and any one side of the base is 12 cm.

13 A man wishes to fence in a rectangular enclosure of area 128 m².
One side of the enclosure is formed by part of a brick wall already in position.
What is the least possible length of fencing required for the other three sides?

14 Some water is gradually evaporating from an open vessel, so that the volume V, measured in cm³ after t days, is given by $V = 24(a - t^{\frac{2}{3}} - t^{\frac{1}{3}})$, where a is constant.

   a Given that the water evaporates completely after exactly 27 days, find a and hence the initial volume of water in the vessel.

   b Find the rate of evaporation after 8 days.

15 The stamp counter at a post office is open from 9 am to 3 pm daily.
The average queuing time q, measured in minutes, at time x hours after opening time is modelled by the equation $q = \frac{1}{4}(-x^3 + 6x^2 + 20)$.

   a Show that, using this model, the longest average queuing time occurs around 1 pm and find how long it is.

   b Find when the average queuing time is rising at its maximum rate, and show that this maximum rate of increase is 3 minutes/hour.

16 The mass m of a caterpillar, measured in milligrams, is

$$-\frac{t^3}{6} + \frac{9t^2}{4} + 5t + 10$$

where t is the time since hatching, measured in days. After n days, the caterpillar reaches its maximum mass, stops eating and pupates.

   a Find the value of n and the caterpillar's maximum mass to the nearest mg.

   b Find the caterpillar's maximum growth rate, in mg per day correct to the nearest mg, and when it occurs.

   A 'Test yourself' exercise (Ty15) and an 'Extension exercise' (Ext15) can be found on the CD-ROM.

➤ # Key points

Increasing and decreasing functions

A function, $f(x)$ is **increasing** for $a < x < b$ if $f'(x) > 0$ for $a < x < b$.

A function, $f(x)$ is **decreasing** for $a < x < b$ if $f'(x) < 0$ for $a < x < b$.

Stationary points

At a stationary point $\dfrac{dy}{dx} = 0$, i.e. the gradient is zero.

$\dfrac{dy}{dx} = 0$ and $\dfrac{d^2y}{dx^2} < 0 \Rightarrow$ the point is a local **maximum**.

$\dfrac{dy}{dx} = 0$ and $\dfrac{d^2y}{dx^2} > 0 \Rightarrow$ the point is a local **minimum**.

$\dfrac{dy}{dx} = 0$ and $\dfrac{d^2y}{dx^2} = 0 \Rightarrow$ the point could be a **point of inflexion**

or a local maximum *or* minimum.

The nature of a stationary point can be determined by the above or by considering the gradient of the curve on either side of the stationary point.

To solve maximum and minimum problems, express the quantity (e.g. y) to be maximised or minimised in terms of one variable (e.g. x) and find where $\dfrac{dy}{dx} = 0$.

Curve sketching

Points to consider
- Standard function (or transformation of one)
- Where does the curve cross the axes?
- Symmetry
- Behaviour as $x \to \pm\infty$
- Stationary values, $\dfrac{dy}{dx} = 0$

16 Trigonometric functions

About 2200 years ago, Greek mathematicians began to tabulate trigonometric ratios to enable astronomers to solve right-angled triangles. Over the following 2000 years developments in trigonometry have made it possible to master a huge range of topics connected with astronomy, sea navigation and space exploration, as well as to solve a myriad of everyday problems requiring measurement.

After working through this chapter you should be

- *able to work with angles measured in radians*
- *able to convert radians to degrees and vice versa*
- *familiar with sine, cosine and tangent functions, their graphs, symmetries and periodicity*
- *familiar with and able to use $\tan \theta = \dfrac{\sin \theta}{\cos \theta}$ and $\sin^2 \theta + \cos^2 \theta = 1$*
- *able to solve simple trigonometric equations in a given interval, expressed in either degrees or radians.*

A calculator or computer may be used for checking sketches and to become familiar with the power of the technology. It is important to understand the techniques of curve sketching with and without technological aids.

16.1 Radians

Angles can be measured in various units such as degrees, radians and right angles.

One complete turn $= 360° =$ four right angles $= 2\pi$ radians.

Radian measure is essential in higher level mathematics, particularly for calculus.

Definition of a radian

An arc of length r on a circle of radius r subtends an angle of 1 radian at the centre of the circle. So, if arc length, r, subtends 1 radian, the circumference, $2\pi r$, subtends 2π radians.

So $360° = 2\pi$ rad

$$\boxed{\begin{aligned} \pi \,\textbf{rad} &= \textbf{180}° \\ \textbf{1 rad} &= \frac{\textbf{180}°}{\pi} \approx \textbf{57.3}° \\ \textbf{1}° &= \frac{\pi}{\textbf{180}} \,\textbf{rad} \end{aligned}}$$

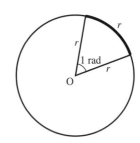

Converting radians to degrees

Using π rad $= 180°$, simple fractions and multiples of π can easily be converted to degrees.

$$\frac{\pi}{2}\,\text{rad} = \frac{1}{2} \times 180° = 90° \qquad \frac{\pi}{3}\,\text{rad} = \frac{1}{3} \times 180° = 60°$$

$$\frac{\pi}{4}\,\text{rad} = 45° \qquad \frac{\pi}{6}\,\text{rad} = 30° \qquad 3\pi\,\text{rad} = 540°$$

Example 1 Convert $\dfrac{7\pi}{15}$ rad to degrees.

$$1\,\text{rad} = \frac{180°}{\pi}$$

$$\therefore \quad \frac{7\pi}{15}\,\text{rad} = \frac{7\pi}{15} \times \frac{\overset{12}{\cancel{180°}}}{\cancel{\pi}}$$

$$= 84°$$

Converting degrees to radians

Some angles can easily be expressed as fractions or multiples of $180°$ ($=\pi$ rad).

$$30° = \tfrac{1}{6} \times 180° = \tfrac{\pi}{6}\,\text{rad} \qquad 150° = \tfrac{5}{6} \times 180° = \tfrac{5}{6}\pi\,\text{rad} \qquad 720° = 4\pi\,\text{rad}$$

Example 2 Convert $48°$ to radians.

$$1° = \frac{\pi}{180}\,\text{rad}$$

$$\therefore \quad 48° = \overset{4}{\cancel{48}} \times \frac{\pi}{\underset{15}{\cancel{180}}}\,\text{rad}$$

$$= \frac{4\pi}{15}\,\text{rad}$$

16.2 Angles of any size

The x and y axes divide the plane into four quadrants.

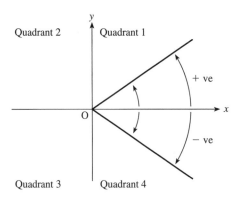

Angles are measured from Ox:

- Positive in an anticlockwise direction
- Negative in a clockwise direction

An angle can be of any size, positive or negative.
Elementary geometry and trigonometry deal with relatively small angles, usually less than 360°. There are many situations, however, which involve large angles.
For example, think of the minute hand of a clock. In a year, it turns through 3 153 600 degrees.

Example 3   In which quadrant would an angle of 560° terminate?

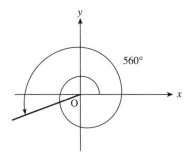

$560° = 360° + 200°$

560° is one complete turn plus 200° anticlockwise.

This would terminate in quadrant 3.

Exercise 16A   *In this exercise, remember that π radians = 180°.*

1   Copy and complete this table.

Degrees	30	45			120	**180**	360
Radians			$\frac{\pi}{3}$	$\frac{\pi}{2}$		**π**	

2   Convert to degrees

a $\frac{\pi}{2}$ rad   **b** $\frac{\pi}{4}$ rad   **c** $\frac{\pi}{3}$ rad

d $\frac{2\pi}{3}$ rad   **e** $\frac{\pi}{6}$ rad   **f** $\frac{3\pi}{2}$ rad

g $\frac{5\pi}{2}$ rad   **h** 4π rad   **i** 5π rad

j $\frac{4\pi}{3}$ rad   **k** $\frac{7\pi}{2}$ rad   **l** $\frac{3\pi}{4}$ rad

3   Convert to radians, leaving π in your answer.

a 360°   **b** 90°   **c** 45°

d 15°   **e** 60°   **f** 120°

g 300°   **h** 270°   **i** 540°

j 30°   **k** 150°   **l** 450°

4   In which quadrants do these angles terminate?

a 100°   **b** $\frac{\pi}{3}$ **c** 70°

d 190°   **e** $\frac{5\pi}{4}$ **f** $-\frac{5\pi}{6}$

g $-\frac{3\pi}{4}$ **h** $\frac{2\pi}{3}$ **i** 330°

j $-80°$ **k** 1000°   **l** $-140°$

16.3 | Trigonometric ratios for any angle

Consider a line segment, OP′, of length r, lying along the positive direction of the x-axis.

Let OP′ rotate through θ to OP and let P have coordinates (x, y).

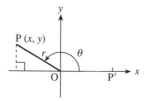

Three trigonometric ratios are defined

$$\sin\theta = \frac{y}{r} \qquad \cos\theta = \frac{x}{r} \qquad \tan\theta = \frac{y}{x}\,(x \neq 0)$$

Note $\quad \tan\theta = \dfrac{\sin\theta}{\cos\theta}$

Since the coordinates, x and y, will sometimes be positive and sometimes negative, the signs of the ratios will also vary. The length, r, is always positive.

To find the trigonometric ratio for any angle it is necessary to find its sign, and its magnitude.

To determine the sign of any trigonometric ratio

By considering the signs of the x and y coordinates in each quadrant, and remembering that the length, r, is always positive, the signs of $\sin\theta$, $\cos\theta$ and $\tan\theta$ can be found.

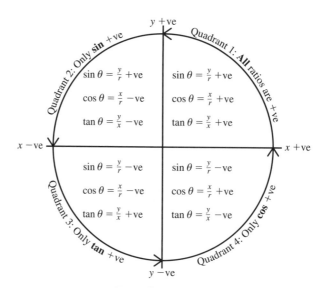

Note To remember in which quadrants ratios are positive, one mnemonic is

All Silly Tom Cats

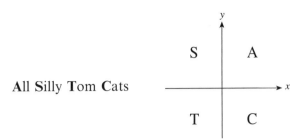

To determine the magnitude of any trigonometric ratio

The magnitude for a ratio of an angle of any size can always be linked to the ratio of an *acute* angle: the acute equivalent angle.

Example 4 Express the given ratios in terms of trigonometric ratios of an acute angle.

a $\cos 122° = \dfrac{x}{r}$

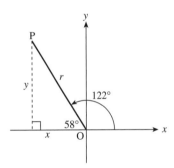

The numerical value of $\frac{x}{r}$ (ignoring the sign) will be the same as for $180° - 122° = 58°$.

$122°$ is in quadrant 2 $\therefore$ $\cos 122°$ is $-$ve.

So, $\cos 122° = -\cos 58°$.

b $\tan 560° = \dfrac{y}{x}$

> $560°$ is one complete turn ($360°$) plus $200°$. $560° - 360° = 200°$.

The numerical value of $\frac{y}{x}$ (ignoring the sign) will be the same as for $200° - 180° = 20°$.

The angle is in quadrant 3 $\therefore$ $\tan 560°$ is $+$ve.

So, $\tan 560° = \tan 20°$.

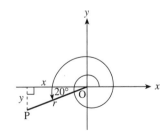

c $\sin(-70°) = \dfrac{y}{r}$

The numerical value of $\dfrac{y}{r}$ (ignoring the sign) will be the same as for 70°.

The angle is in quadrant 4 ∴ $\sin(-70°)$ is −ve.

So, $\sin(-70°) = -\sin 70°$.

Note In each of the above examples, a line is drawn from P perpendicular to the *x*-axis.

To link the ratio for any angle with its acute equivalent angle:

■ **Find the sign using 'All Silly Tom Cats'.**

S	A
T	C

■ **Find the acute equivalent angle by rotating through the given angle to OP, drawing the perpendicular from P to the *x*-axis.**

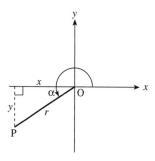

The acute equivalent angle, α, is the angle between OP and the *x*-axis.

A quick method for tackling examples is to think 'Which quadrant?' and 'How far from the *x*-axis?'

$\cos 122°$: quadrant 2 ∴ −ve
122° is 58° from *x*-axis ∴ $\cos 122° = -\cos 58°$

$\tan 200°$: quadrant 3 ∴ +ve
200° is 20° from *x*-axis ∴ $\tan 200° = \tan 20°$

$\sin(-70°)$: quadrant 4 ∴ −ve
−70° is 70° from *x*-axis ∴ $\sin(-70°) = -\sin 70°$

For every acute angle there are four angles between $0°$ and $360°$ with the same numerical ratio.

Two will be positive, and two negative. For example,

$$\sin 37° = \sin(180° - 37°) = \sin 143°$$

$$= -\sin(180° + 37°) = -\sin 217°$$

$$= -\sin(360° - 37°) = -\sin 323°$$

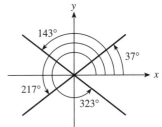

Each angle is $37°$ from the x-axis.

Special cases

For $\theta = 0°$, P has coordinates $(r, 0)$. So

$$\sin\theta = \frac{0}{r} = 0 \qquad \cos\theta = \frac{r}{r} = 1 \qquad \tan\theta = \frac{0}{r} = 0$$

For $\theta = 90°$, P has coordinates $(0, r)$. So

$$\sin\theta = \frac{r}{r} = 1 \qquad \cos\theta = \frac{0}{r} = 0$$

$\tan 90°$ is not defined because division by zero is not defined.

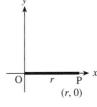

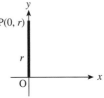

16.4 Graphs of trigonometric functions

Plotting graphs of $\sin\theta$, $\cos\theta$ and $\tan\theta$ shows clearly how the functions vary.

Using the definition $\sin\theta = \dfrac{y}{r}$ and $\cos\theta = \dfrac{x}{r}$ and taking a circle of unit radius so that $r = 1$ gives $\sin\theta = y$ and $\cos\theta = x$.

Then rotating OP through θ and plotting the y coordinate of P for each value of θ gives the sine curve.

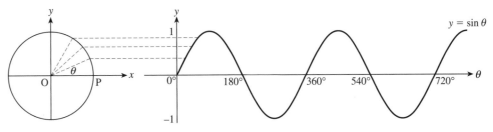

The graph repeats itself after every $360°$, so the function $\sin\theta$ is periodic of period $360°$.

The graph of $\cos\theta$ may be drawn in a similar way. It also has period 360°.

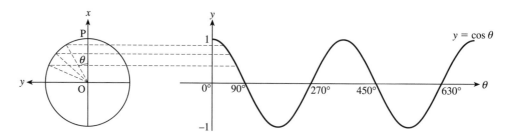

Since $\cos\theta = x$ the axes should be rotated through 90°. Rotating OP through θ and plotting the x coordinate for each value of θ gives the cosine curve.

Note The sine and cosine curves are translations of each other.

For the graph of $\tan\theta$ the tangent to the unit circle at (1, 0) is drawn and OP is extended to meet the tangent at Q. The y coordinate at Q is equal to $\tan\theta$.

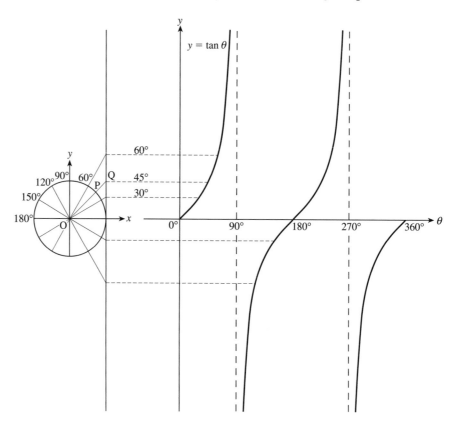

For $\tan\theta$, the period is 180°.

It is important to be familiar with the graphs of $y = \sin\theta$, $y = \cos\theta$ and $y = \tan\theta$.

$y = \sin\theta$

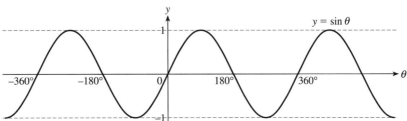

- $\sin 0° = 0$ so the curve passes through the origin.

- Maximum value of $\sin\theta$ is 1.

- Minimum value of $\sin\theta$ is -1.

- Graph repeats every $360°$ (2π radians), i.e. $\sin\theta$ is periodic of period $360°$ (2π radians):

  $$\sin\theta = \sin(\theta + 360°) = \sin(\theta + 2 \times 360°) = \cdots = \sin(\theta + 360n°)$$

  > $n \in \mathbb{Z}$, i.e. n is an integer.

  In radians: $\sin\theta = \sin(\theta + 2\pi) = \sin(\theta + 4\pi) = \cdots = \sin(\theta + 2n\pi)$

- $\sin\theta = -\sin(-\theta)$
  $\therefore$ $\sin\theta$ is an odd function. The graph has $180°$ rotational symmetry about the origin.

$y = \cos\theta$

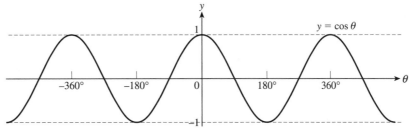

- $\cos 0° = 1$ so the curve cuts the y-axis at $(0, 1)$.

- Maximum value of $\cos\theta$ is 1.

- Minimum value of $\cos\theta$ is -1.

- Graph repeats every $360°$ (2π radians), i.e. $\cos\theta$ is periodic of period $360°$ (2π radians):

  $$\cos\theta = \cos(\theta + 360°) = \cos(\theta + 2 \times 360°) = \cdots = \cos(\theta + 360n°)$$

  > $n \in \mathbb{Z}$.

  In radians: $\cos\theta = \cos(\theta + 2\pi) = \cos(\theta + 4\pi) = \cdots = \cos(\theta + 2n\pi)$

- $\cos\theta = \cos(-\theta)$
  $\therefore$ $\cos\theta$ is an even function. The graph has the y-axis as an axis of symmetry.

$y = \tan \theta$

- $\tan 0° = 0$ so curve passes through the origin $(0, 0)$.

- No maximum or minimum values.

- Asymptotes at $\theta = \pm 90°$, $\pm 270°$, $\pm 450°$, ...

  In radians: $\theta = \pm \frac{1}{2}\pi$, $\pm \frac{3}{2}\pi$, $\pm \frac{5}{2}\pi$, ...

  That is, at all odd multiples of $90°(\frac{1}{2}\pi$ radians).

  This can be summed up by saying:

  asymptotes occur at

  $\theta = (2n + 1)90°$ or $\theta = (2n + 1)\frac{\pi}{2}$.

Note $(2n + 1)$, where n is an integer, will give an odd number.

- Graph repeats every $180°$ (π radians)
  i.e. $\tan \theta$ is periodic of period $180°$ (π radians).

  So $\tan \theta = \tan(\theta + 180°) = \tan(\theta + 2 \times 180°) = \cdots = \tan(\theta + 180n°)$ $n \in \mathbb{Z}$.

  In radians: $\tan \theta = \tan(\theta + \pi) = \tan(\theta + 2\pi) = \cdots = \tan(\theta + n\pi)$

- $\tan \theta = -\tan(-\theta)$
  $\therefore$ $\tan \theta$ is an odd function.
  The graph has $180°$ rotational symmetry about the origin.

Example 5   *This example illustrates how the graphs can be used to find any number of angles with a particular ratio.*

Find three angles with the same trigonometric ratio as

a $\sin 135°$ **b** $\cos 118°$ **c** $\tan(-37°)$

Solution   **a** $135°$ is $45°$ from $180°$.   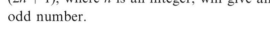 Locate 135° on a sketch of the sine graph.

So, by symmetry, $\sin 45°$ has the same value as $\sin 135°$.

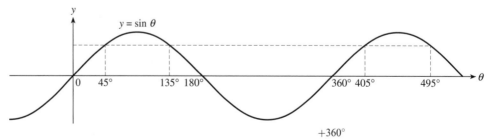

The graph then repeats every $360°$: $\sin 45° = \sin 135° = \sin 405° = \sin 495°$

So the sines of $45°$, $135°$, $405°$ and $495°$ all have the same value.

b 118° is 28° from 90°.

Locate 135° on a sketch of the sine graph.

By symmetry $\cos(270° - 28°) = \cos 242°$ has the same value as $\cos 118°$.

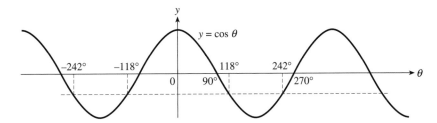

The graph is symmetrical about the y-axis so $\cos(-118°)$ and $\cos(-242°)$ have the same value as $\cos 118°$.

Alternatively, since the graph repeats every 360°, 360° could be subtracted from, or added to, the two values already found.

$$\cos(118° - 360°) = \cos(-242°) \qquad \cos(242° - 360°) = \cos(-118°)$$

So the cosines of $-242°$, $-118°$, $118°$ and $242°$ all have the same value.

c The tan graph repeats every 180°. So given one angle, others can be found by adding or subtracting multiples of 180°. So

$$\overset{-180°}{\overset{\longleftarrow}{}} \quad \overset{+180°}{\overset{\longrightarrow}{}} \quad \overset{+180°}{\overset{\longrightarrow}{}}$$
$$\tan(-217°) = \tan(-37°) = \tan 143° = \tan 323°$$

So the tangents of $-37°$, $143°$, $-217°$ and $323°$ all have the same value.

Note By adding or subtracting multiples of the period, an infinite number of possible solutions can be found to questions like those in Example 5.

Transformations of trigonometric graphs

The graphs of $\sin\theta$, $\cos\theta$ and $\tan\theta$ can be transformed using the methods of Chapter 7 pages 115–118.

A graphical calculator or computer package is a great help for studying these transformations and for checking sketches. Try to resist the temptation to use the calculator before attempting a sketch.

Example 6 For the given functions, sketch the graphs and state the period.

a $y = 4\cos x$ $0° \leqslant x \leqslant 360°$

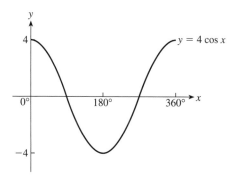

Period: $360°$

The curve $y = \cos x$ undergoes a stretch parallel to the y-axis.

Stretching $y = \cos x$ parallel to the y-axis, scale factor 4, gives $y = 4\cos x$.

The period $y = \cos x$ is $360°$. Stretching parallel to the y-axis will not change the period, so the period of $y = 4\cos x$ is also $360°$.

b $y = \sin 3x$ $-\pi \leqslant x \leqslant \pi$

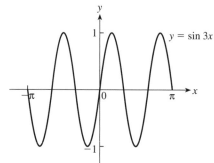

Period: $\dfrac{2\pi}{3}$

The curve $y = \sin x$ is 'stretched' parallel to the x-axis. Stretching parallel to the x-axis, scale factor $\frac{1}{3}$, gives $y = \sin 3x$.

The graph is compressed towards the y-axis. Three periods of $y = \sin 3x$ fit into one of $y = \sin x$. The period of $y = \sin x$ is 2π, so the period of $y = \sin 3x$ is $\frac{2\pi}{3}$.

c $y = \sin(x + 30°)$ $-180° \leqslant x \leqslant 180°$

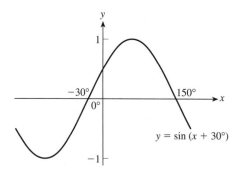

Period: $360°$

The curve $y = \sin x$ is moved $30°$ to the left, i.e. a translation $\begin{pmatrix} -30° \\ 0 \end{pmatrix}$.

This gives $y = \sin(x + 30°)$.

The period is the same as for $y = \sin x$, i.e. $360°$.

Note $\sin kx$ and $\cos kx$ have period $\dfrac{360°}{k}\left(\dfrac{2\pi}{k}\ \text{radians}\right)$.

$\tan kx$ has period $\dfrac{180°}{k}\left(\dfrac{\pi}{k}\ \text{radians}\right)$.

16.5   Special triangles: Trigonometric ratios of 30°, 45° and 60°

For 30° and 60° draw an **equilateral triangle** with sides of 2 units. Split it in half. Work out the height using Pythagoras' theorem. From this half equilateral triangle the sin, cos and tan of 30° and 60° can then be written down.

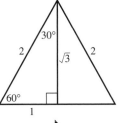

For 45° draw an **isosceles right-angled triangle** with equal sides of 1 unit. Work out the length of the hypotenuse by Pythagoras' theorem.

From this triangle the sin, cos and tan of 45° can then be written down.

θ in degrees	$\sin \theta$	$\cos \theta$	$\tan \theta$	θ in radians
0°	0	1	0	0
30°	$\dfrac{1}{2}$	$\dfrac{\sqrt{3}}{2}$	$\dfrac{1}{\sqrt{3}}$	$\dfrac{\pi}{6}$
45°	$\dfrac{1}{\sqrt{2}}$	$\dfrac{1}{\sqrt{2}}$	1	$\dfrac{\pi}{4}$
60°	$\dfrac{\sqrt{3}}{2}$	$\dfrac{1}{2}$	$\sqrt{3}$	$\dfrac{\pi}{3}$
90°	1	0	Undefined	$\dfrac{\pi}{2}$

The values in the table should be at your fingertips. It is not necessary to learn them although, with use, many will become familiar. Learn ways of working them out.

Example 7   *For this example, the answers are to be given in surds. So, a calculator, unless it is a very sophisticated one, is not helpful.*

Write down, leaving surds in the answer

a $\sin 240°$

b $\tan(-45°)$

c $\cos 330°$

Solution   **a**   240° is in quadrant 3.

$\therefore$ $\sin 240°$ is $-$ve.

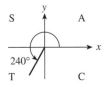

240° is 60° from the x-axis so

$\sin 240° = -\sin 60°$

$\sin 60° = \dfrac{\sqrt{3}}{2}$

So $\sin 240° = -\dfrac{\sqrt{3}}{2}$.

Alternatively, $\sin 240°$ is $-$ve and has the same numerical value as $\sin 60°$.

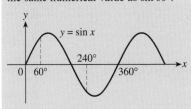

From the special triangle, $\sin 60° = \dfrac{\sqrt{3}}{2}$.

b −45° is in quadrant 4.

∴ tan (−45°) is −ve.

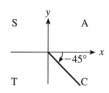

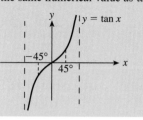

Alternatively, tan (−45°) is −ve and has the same numerical value as tan 45°.

−45° is 45° from the *x*-axis so

tan (−45°) = −tan 45°

tan 45° = 1

So tan (−45°) = −1.

From the special triangle, tan 45° = 1.

c 330° is in quadrant 4.

∴ cos 330° is +ve.

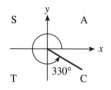

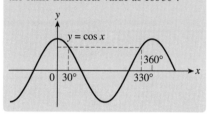

Alternatively, cos 330° is +ve and has the same numerical value as cos 30°.

330° is 30° from the *x*-axis so

cos 330° = cos 30°

$$\cos 30° = \frac{\sqrt{3}}{2}$$

So $\cos 330° = \dfrac{\sqrt{3}}{2}$.

From the special triangle, $\cos 30° = \dfrac{\sqrt{3}}{2}$.

Exercise 16B

1 Express in terms of the trigonometric ratios of acute angles

a sin 170°	**b** tan 300°	**c** cos 200°	**d** sin (−50°)
e cos (−20°)	**f** sin 325°	**g** tan (−140°)	**h** cos 164°
i tan 143°	**j** cos (−130°)	**k** sin 250°	**l** tan (−50°)
m cos 293°	**n** sin (−230°)	**o** sin 1000°	**p** tan 904°

2 Using the special triangles, write down the values of

a sin 30°	**b** cos 30°	**c** cos 45°	**d** tan 30°
e tan 45°	**f** cos 135°	**g** sin 330°	**h** tan 300°
i sin 225°	**j** cos (−60°)	**k** tan (−150°)	**l** sin (−300°)
m sin 405°	**n** cos (−135°)	**o** tan 210°	**p** sin 660°

3 Write down the values of

a sin 270°	**b** cos 0°	**c** tan 180°	**d** sin 0°
e tan 2π	**f** cos $\frac{\pi}{2}$	**g** sin π	**h** cos π

4 Write down the values of these, leaving surds in the answers where appropriate.

a $\cos\dfrac{\pi}{4}$ **b** $\sin\dfrac{\pi}{6}$ **c** $\sin\dfrac{2\pi}{3}$ **d** $\tan\pi$

e $\sin\dfrac{3\pi}{4}$ **f** $\tan\dfrac{\pi}{3}$ **g** $\sin\left(-\dfrac{\pi}{3}\right)$ **h** $\cos\dfrac{\pi}{3}$

i $\tan\dfrac{\pi}{4}$ **j** $\cos\dfrac{11\pi}{6}$ **k** $\tan\dfrac{5\pi}{6}$ **l** $\sin\dfrac{4\pi}{3}$

5 Using a sketch of the graph of the trigonometric function, or otherwise, find all values of θ, where

a $\sin\theta = \sin 30°$ $0° \leqslant \theta \leqslant 720°$ **b** $\cos\theta = \cos 140°$ $0° \leqslant \theta \leqslant 720°$

c $\tan\theta = \tan(-20°)$ $-360° \leqslant \theta \leqslant 360°$ **d** $\sin\theta = \sin\left(-\dfrac{\pi}{4}\right)$ $0 \leqslant \theta \leqslant 4\pi$

e $\cos\theta = \cos\pi$ $-2\pi \leqslant \theta \leqslant 2\pi$ **f** $\tan\theta = \tan\dfrac{\pi}{3}$ $0 \leqslant \theta \leqslant 4\pi$

6 Draw sketches of $\sin\theta$, $\cos\theta$ and $\tan\theta$, on separate axes, for $0° \leqslant \theta \leqslant 360°$.

7 These graphs can be sketched by transformations of the graph of $\sin\theta$.
Draw a sketch of each graph for $0° \leqslant \theta \leqslant 360°$, state the transformation of $\sin\theta$ required, and state the period of the graph.

a $1 + \sin\theta$ **b** $\sin(\theta + 60°)$ **c** $\sin 2\theta$

d $\sin\frac{1}{2}\theta$ **e** $-\sin\theta$ **f** $\sin(-\theta)$

8 These graphs can be sketched by transformations of the graphs of $\sin x$, $\cos x$ and $\tan x$. Draw the sketch of each graph for $-360° \leqslant x \leqslant 360°$, state the transformation of the original graph required and the period of the graph.

a $y = 3\sin x$ **b** $y = 1 + \tan x$ **c** $y = \cos 2x$

d $y = \cos(x - 20°)$ **e** $y = \tan(-x)$ **f** $y = \sin(x + 60°)$

9 These graphs can be sketched by transformations of the graphs of $\sin x$, $\cos x$ and $\tan x$. Draw the sketch of each graph for $-2\pi \leqslant x \leqslant 2\pi$, state the transformation of the original graph required and the period of the graph.

a $y = \sin\dfrac{x}{2}$ **b** $y = \cos\dfrac{2x}{3}$ **c** $y = \sin\left(-\dfrac{x}{2}\right)$

d $y = -1 + \cos x$ **e** $y = \cos\left(x - \dfrac{\pi}{4}\right)$ **f** $y = \tan 2x$

10 By sketching $y = \cos x$ and $y = \sin(x + 90°)$, show that $\cos x = \sin(x + 90°)$ for all values of x.

11 For each function, state the period in radians.

a $\sin \theta$ **b** $\cos 2\theta$ **c** $\tan \theta$ **d** $\tan 3\theta$

e $4\cos \theta$ **f** $\sin \dfrac{\theta}{2}$ **g** $\sin k\theta$ **h** $\tan k\theta$

12 Sketch the graph of $y = 4\sin x$ for $-360° \leqslant x \leqslant 360°$. By drawing the line $y = 3$ on the same axes, explain how you know that the equation $4\sin x = 3$ has four roots in the interval $-360° \leqslant x \leqslant 360°$.

State the values, or range of values, of k for which the equation $4\sin x = k$ has 0, 1, 2, 3, 4 or 5 roots in the interval $-360° \leqslant x \leqslant 360°$.

13 Sketch, on the same axes, the graphs of $y = \sin 2x$ and of $y = \dfrac{x}{4}$ for $-\pi \leqslant x \leqslant \pi$.

State the number of roots of the equation $\sin 2x = \dfrac{x}{4}$ for $-\pi \leqslant x \leqslant \pi$.

14 The triangles below are similar to these two special triangles. In each case, state to which special triangle they are similar, the scale factor to obtain the given triangle from the special triangle and the value of x and, where relevant, y.

a 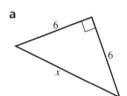   **b** 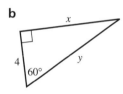   **c**

d 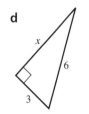   **e** 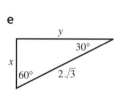   **f** 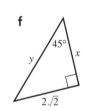   **g**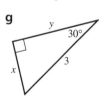

16.6 Trigonometric identities

Two important identities will be proved.

$$\frac{\sin \theta}{\cos \theta} = \tan \theta \qquad \text{and} \qquad \sin^2 \theta + \cos^2 \theta = 1$$

Remember: An identity is true for all values of the variable. The sign $\equiv$ may be used in place of $=$.

Note $(\sin \theta)^2$ is written as $\sin^2 \theta$ and pronounced 'sine squared theta'.
To work out, for example, $\sin^2 42°$ on a calculator, evaluate $\sin 42°$ and square it.

Proof To prove: $\dfrac{\sin\theta}{\cos\theta} = \tan\theta$

State result to be proved, and start with known information.

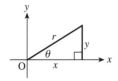

For θ in any quadrant, by definition:

$$\sin\theta = \frac{y}{r} \qquad \cos\theta = \frac{x}{r}$$

$$\therefore \quad \frac{\sin\theta}{\cos\theta} = \frac{y}{r} \div \frac{x}{r} = \frac{y}{r} \times \frac{r}{x} = \frac{y}{x}$$

$$= \tan\theta$$

Start with one side (in this case LHS).
Use known facts to show LHS = RHS.

So $\dfrac{\sin\theta}{\cos\theta} = \tan\theta$.

State conclusion.

Proof To prove: $\sin^2\theta + \cos^2\theta = 1$

State result to be proved.

For any angle θ,

$$\sin^2\theta + \cos^2\theta = \left(\frac{y}{r}\right)^2 + \left(\frac{x}{r}\right)^2$$

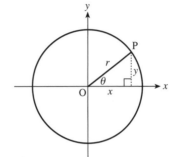

$$= \frac{y^2 + x^2}{r^2}$$

Start with one side (in this case LHS). Use known facts and theorems to show LHS = RHS.

P lies on the circle $\quad y^2 + x^2 = r^2$

$$\therefore \qquad \sin^2\theta + \cos^2\theta = \frac{r^2}{r^2} = 1$$

So $\sin^2\theta + \cos^2\theta = 1$.

State conclusion.

> **For all values of θ:**
> $$\frac{\sin\theta}{\cos\theta} = \tan\theta \qquad \sin^2\theta + \cos^2\theta = 1$$

The identity $\sin^2\theta + \cos^2\theta = 1$ can be rearranged to give

$$\sin^2\theta = 1 - \cos^2\theta \quad \text{or} \quad \cos^2\theta = 1 - \sin^2\theta$$

Two other useful identities:

From the diagram

$$\sin\theta = \frac{b}{c} = \cos(90° - \theta) \qquad\qquad \tan\theta = \frac{b}{a}$$

$$\cos\theta = \frac{a}{c} = \sin(90° - \theta) \qquad \tan(90° - \theta) = \frac{a}{b}$$

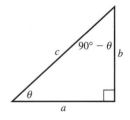

> $$\sin\theta = \cos(90° - \theta) \qquad \cos\theta = \sin(90° - \theta)$$

Note Although the diagram only illustrates the results for $0° < \theta < 90°$, they are true for all values of θ.

Example 8   Prove the identity: $\cos^4 \theta - \sin^4 \theta = \cos^2 \theta - \sin^2 \theta$

To prove: $\cos^4 \theta - \sin^4 \theta = \cos^2 \theta - \sin^2 \theta$ State result to be proved.

$\cos^4 \theta - \sin^4 \theta$ Start with one side (in this case LHS).

$\qquad = (\cos^2 \theta - \sin^2 \theta)(\cos^2 \theta + \sin^2 \theta)$ Factorise using difference of squares.

But $\cos^2 \theta + \sin^2 \theta = 1$

$\therefore \quad \cos^4 \theta - \sin^4 \theta = \cos^2 \theta - \sin^2 \theta$ State conclusion.

Example 9   *In this example, the identity is used in the form $1 - \sin^2 \theta = \cos^2 \theta$.*

If $x = a \sin \theta$, simplify $\dfrac{x}{\sqrt{a^2 - x^2}}$.   Substitute for x.

$\dfrac{x}{\sqrt{a^2 - x^2}} = \dfrac{a \sin \theta}{\sqrt{a^2 - a^2 \sin^2 \theta}}$ Factorise the expression under the root sign.

$\qquad = \dfrac{a \sin \theta}{\sqrt{a^2(1 - \sin^2 \theta)}}$ Use $1 - \sin^2 \theta = \cos^2 \theta$.

$\qquad = \dfrac{a \sin \theta}{\sqrt{a^2 \cos^2 \theta}}$

$\qquad = \dfrac{a \sin \theta}{a \cos \theta}$ Cancel a, use $\dfrac{\sin \theta}{\cos \theta} = \tan \theta$

$\qquad = \tan \theta$

Example 10   *In this example, the identity $\sin^2 \theta + \cos^2 \theta = 1$ will enable θ to be eliminated.*

Eliminate θ from $x = a \sin \theta$, $y = b \cos \theta$.

$x = a \sin \theta \Rightarrow \sin \theta = \dfrac{x}{a}$ Rearrange to give $\sin \theta$ and $\cos \theta$ in terms of x, y, a and b.

$x = b \cos \theta \Rightarrow \cos \theta = \dfrac{y}{b}$

But $\sin^2 \theta + \cos^2 \theta = 1$ Substitute into the identity.

$\therefore \qquad \left(\dfrac{x}{a}\right)^2 + \left(\dfrac{y}{b}\right)^2 = 1$

or $\qquad \dfrac{x^2}{a^2} + \dfrac{y^2}{b^2} = 1$ θ is eliminated.

Example 11   *This example uses an identity and then offers an alternative solution, using Pythagoras' theorem.*

If $\sin\theta = \frac{15}{17}$ and θ is obtuse, find $\cos\theta$.

To find $\cos\theta$, both the sign and the magnitude are needed.

θ is obtuse, so θ is in quadrant 2 $\therefore \cos\theta$ is $-$ve

To find the magnitude, two methods are possible.

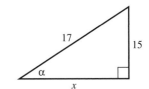

Using the identity $\sin^2\theta + \cos^2\theta = 1$

$$\sin^2\theta + \cos^2\theta = 1$$

Substitute for $\sin\theta$.

$$\left(\tfrac{15}{17}\right)^2 + \cos^2\theta = 1$$

$$\cos^2\theta = 1 - \tfrac{225}{289} = \tfrac{64}{289}$$

$$\therefore \quad \cos\theta = \pm\sqrt{\tfrac{64}{289}} = \pm\tfrac{8}{17}$$

Note: It is not necessary to find θ.

But since $\cos\theta$ is $-$ve, $\cos\theta = -\tfrac{8}{17}$.

Using Pythagoras' theorem

Draw a right-angled triangle with $\sin\alpha = \frac{15}{17}$. α is the acute equivalent angle.

$$x^2 = 17^2 - 15^2$$

$$= (17 - 15)(17 + 15)$$

$$= 2 \times 32$$

$$= 64$$

$$\therefore \quad x = 8$$

But $\cos\theta$ is $-$ve, and $\cos\alpha = \tfrac{8}{17}$ so $\cos\theta = -\tfrac{8}{17}$.

16.7   Solution of trigonometric equations

There are many instances when a solution of an equation, such as $\sin\theta = \frac{1}{4}$, is needed.

One solution can be found by using either the calculator or the special triangles. From this one solution other solutions can be found either by the symmetry of the graph, or by considering the quadrants.

On calculators, the inverse sine function is marked arcsin or $\sin^{-1}$. So, one solution of $\sin\theta = \frac{1}{4}$ is $\theta = \arcsin\frac{1}{4}$ from the calculator. In this book, arcsin, arccos and arctan are used in preference to $\sin^{-1}$, $\cos^{-1}$, $\tan^{-1}$. The inverse trigonometric functions are studied later in the A-level course.

Note   For questions in degrees, give the answer in degrees; for questions in radians give the answer in radians. Answers to trigonometric equations (and to other equations) can be checked on a calculator. For trigonometric equations take care to be in the correct mode.

Answers to Examples 12–19 are given as exact answers, or correct to 1 decimal place if the answer is not exact.

Example 12   Solve $\cos\theta = -\frac{1}{4}$ for $0° < \theta < 360°$.

$$\cos\theta = -\frac{1}{4}$$

$$\arccos\left(-\frac{1}{4}\right) = 104.5°$$

Find one solution on the calculator using $\arccos\left(-\frac{1}{4}\right)$.

So one solution is $\theta = 104.5°$.

$$180° - 104.5° = 75.5°$$

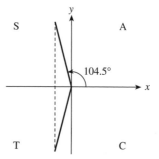

$\cos$ is $-$ve
$\therefore$ solutions are in quadrants 2 and 3, $75.5°$ from the x-axis.

So the second solution is $\theta = 180° + 75.5° = 255.5°$
So $\theta = 104.5°$ or $\theta = 255.5°$.

Example 13   Solve $\sin\left(\theta - \frac{\pi}{4}\right) = \frac{\sqrt{3}}{2}$ for $0 < \theta < 2\pi$.

Find solutions for $\theta - \frac{\pi}{4}$.

$$\sin\left(\theta - \frac{\pi}{4}\right) = \frac{\sqrt{3}}{2}$$

so $\theta - \dfrac{\pi}{4} = \dfrac{\pi}{3}$ or $\theta - \dfrac{\pi}{4} = \pi - \dfrac{\pi}{3}$

$\arcsin\dfrac{\sqrt{3}}{2} = \dfrac{\pi}{3}$ from special triangles.

so $\theta = \dfrac{\pi}{3} + \dfrac{\pi}{4}$ or $\theta = \dfrac{2\pi}{3} + \dfrac{\pi}{4}$

Find other angles which, when $\frac{\pi}{4}$ is added, will still be in range.

$\qquad = \dfrac{7\pi}{12}$ $\qquad = \dfrac{11\pi}{12}$

$\therefore \theta = \dfrac{7\pi}{12}$ or $\dfrac{11\pi}{12}$.

Example 14   Solve $\tan 3\theta = 1$ for $0° \leqslant \theta \leqslant 360°$.

Since $0° \leqslant \theta \leqslant 360°$, it follows that $0° \leqslant 3\theta \leqslant 3 \times 360°$. So all solutions for 3θ up to $3 \times 360° = 1080°$ must be listed.

From the special triangles, $\tan 45° = 1$. So $3\theta = 45°$ is one solution. More values of 3θ can be obtained by adding (or subtracting) multiples of $180°$.

$$3\theta = 45°, \ 45° + 180°, \ 45° + 360°, \ 45° + 540°, \ 45° + 720°, \ 45° + 900°$$

$$= 45°, \ 225°, \ 405°, \ 585°, \ 765°, \ 945°$$

$$\therefore \quad \theta = 15°, \ 75°, \ 135°, \ 195°, \ 255°, \ 315°.$$

Note   For equations such as $\tan 3\theta = 1$ or $\sin\left(\theta - \dfrac{\pi}{4}\right) = \dfrac{\sqrt{3}}{2}$ list solutions for 3θ or $\theta - \dfrac{\pi}{4}$

before dividing by 3 or adding $\dfrac{\pi}{4}$.

Example 15   Find one point of intersection of $y = \sqrt{3}\sin x$ and $y = \cos x$, where x is in radians.

The curves intersect where

$$\sqrt{3}\sin x = \cos x$$

> Dividing both sides by $\cos x$ is permitted because $\cos x = 0$ does not lead to a solution.

$$\therefore \quad \sqrt{3}\tan x = 1$$

$$\tan x = \frac{1}{\sqrt{3}}$$

> Use $\dfrac{\sin x}{\cos x} = \tan x$.

so $x = \arctan\dfrac{1}{\sqrt{3}}$

One value of x is $x = \dfrac{\pi}{6}$.

When $x = \dfrac{\pi}{6}$, $y = \cos\dfrac{\pi}{6} = \dfrac{\sqrt{3}}{2}$

So coordinates of one point of intersection are $\left(\dfrac{\pi}{6}, \dfrac{\sqrt{3}}{2}\right)$.

> Draw special triangle.
> *Note*: $\dfrac{\pi}{6}$ radians $= 30°$.
>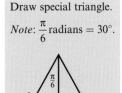

Try illustrating this on a graphic calculator. The point of intersection can be found only approximately.

Note $a\sin x = b\cos x \Rightarrow \tan x = \dfrac{b}{a}$

Solving trigonometric equations which are quadratic

The following examples solve quadratic equations involving trigonometric functions.

Example 16   Solve $2\sin^2\theta - \sin\theta = 0$ for $-180° \leqslant \theta \leqslant 180°$.

> This is a quadratic equation in $\sin\theta$.

$$2\sin^2\theta - \sin\theta = 0$$

$$\sin\theta(2\sin\theta - 1) = 0$$

> $\sin\theta$ could be zero so do not divide through by $\sin\theta$. Put $\sin\theta$, the common factor, outside the bracket.

$$\therefore \qquad \sin\theta = 0 \text{ or } \sin\theta = \tfrac{1}{2}$$

> Use a sketch and the special triangles to find the solution in the given range.

When $\sin\theta = 0$ $\theta = -180°, 0°, 180°$

When $\sin\theta = \tfrac{1}{2}$ $\theta = 30°, 150°$

So $\theta = -180°, 0°, 30°, 150°, 180°$.

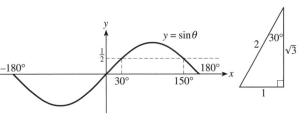

If preferred, Example 16 can be solved by substituting y for $\sin\theta$ to give $2y^2 - y = 0$, and solving for y.

273

Example 17   Solve $3 \sin \theta + \tan \theta = 0$ for $0 \leqslant \theta \leqslant 2\pi$.

$$3 \sin \theta + \tan \theta = 0$$

Sometimes, it can be helpful to express trigonometric equations in terms of $\sin \theta$ and $\cos \theta$ only.

$$3 \sin \theta + \frac{\sin \theta}{\cos \theta} = 0$$

Use $\dfrac{\sin \theta}{\cos \theta} = \tan \theta$ and multiply both sides by $\cos \theta$.

$$3 \sin \theta \cos \theta + \sin \theta = 0$$

$$\sin \theta(3 \cos \theta + 1) = 0$$

$\sin \theta$ is a common factor.

$$\therefore \qquad \sin \theta = 0 \quad \text{or} \quad \cos \theta = -\tfrac{1}{3}$$

If $\sin \theta = 0 \qquad \theta = 0, \pi, 2\pi$

If $\cos \theta = -\tfrac{1}{3} \qquad \theta = 1.9, 2\pi - 1.9$

$$= 1.9, 4.4$$

So $\qquad \theta = 0, 1.9, \pi, 4.4, 2\pi.$

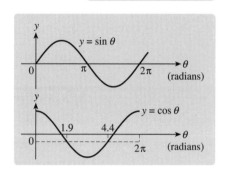

Note   Since the range given for solutions is in radians the answers should be given in radians.

Example 18   Solve $2 \sin x \cos x - 4 \cos x - \sin x + 2 = 0$ for $-\pi < x < \pi$.

$$2 \sin x \cos x - 4 \cos x - \sin x + 2 = 0$$

Try factorising by grouping.

$$2 \cos x(\sin x - 2) - (\sin x - 2) = 0$$

$$(\sin x - 2)(2 \cos x - 1) = 0$$

$(\sin x - 2)$ is a common factor.

so $\qquad\qquad \sin x - 2 = 0 \quad \text{or} \quad \cos x = \tfrac{1}{2}$

When $\sin x - 2 = 0 \qquad \sin x = 2$

$-1 \leqslant \sin x \leqslant 1$ so there is no solution to $\sin x = 2$.

When $\cos x = \tfrac{1}{2} \qquad\qquad x = \pm \dfrac{\pi}{3}$

So solution is $x = \pm \dfrac{\pi}{3}$.

cos +ve $\therefore$ quadrants 1 and 4.

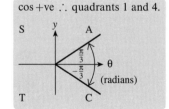

Example 19   Solve $7 - 6\cos^2\theta = 5\sin\theta$, $0° \leqslant \theta \leqslant 360°$.

$$7 - 6\cos^2\theta = 5\sin\theta$$

> By substituting $\cos^2\theta = 1 - \sin^2\theta$ the equation will be a quadratic in $\sin\theta$.

$$7 - 6(1 - \sin^2\theta) = 5\sin\theta$$

$$7 - 6 + 6\sin^2\theta - 5\sin\theta = 0$$

$$6\sin^2\theta - 5\sin\theta + 1 = 0$$

> Put $y = \sin\theta$ if preferred:
> $6y^2 - 5y + 1 = 0$
> $(3y - 1)(2y - 1) = 0$

$$(3\sin\theta - 1)(2\sin\theta - 1) = 0$$

$$\therefore \qquad\qquad \sin\theta = \tfrac{1}{3} \quad \text{or} \ \sin\theta = \tfrac{1}{2}$$

When $\sin\theta = \tfrac{1}{3}$

> Find one value on the calculator: $\theta = \arcsin\tfrac{1}{3} = 19.5°$.
> Find other solutions by symmetry of the graph.

$$\theta = 19.5°$$

or $\theta = 180° - 19.5° = 160.5°$

When $\sin\theta = \tfrac{1}{2}$

$$\theta = 30°$$

or $\theta = 180° - 30° = 150°$

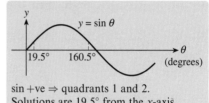

sin +ve $\Rightarrow$ quadrants 1 and 2.
Solutions are 19.5° from the x-axis.

So $\theta = 19.5°, 30°, 150°, 160.5°$.

> *Remember*: From special triangles, $\sin 30° = \tfrac{1}{2}$.

Exercise 16C   *In this exercise, answers should be left in terms of π where appropriate.*

1 If $s = \sin\theta$ and $c = \cos\theta$ simplify

  a $\sqrt{1 - s^2}$ **b** $\sqrt{1 - c^2}$ **c** $\dfrac{s}{\sqrt{1 - s^2}}$ **d** $\dfrac{sc}{\sqrt{1 - s^2}}$

  e $\dfrac{\sqrt{1 - c^2}}{c}$ **f** $\dfrac{c^4 - s^4}{c^2 - s^2}$ **g** $\dfrac{s\sqrt{1 - s^2}}{c\sqrt{1 - c^2}}$

2 Find a solution, where $0 \leqslant x \leqslant \tfrac{\pi}{2}$, to these equations.
   Do *not* use a calculator if the answer can be found from the special triangles.

  a $\sin x = \tfrac{1}{2}$ **b** $\tan x = \sqrt{3}$ **c** $\tan x = 1$ **d** $\tan x = \tfrac{1}{2}$

  e $\sin x = \dfrac{1}{\sqrt{3}}$ **f** $\cos x = \dfrac{1}{\sqrt{3}}$ **g** $\cos x = \dfrac{\sqrt{3}}{2}$ **h** $\tan x = \dfrac{1}{\sqrt{2}}$

3 Find, without using a calculator, the values of

  a $\sin\theta$, $\tan\theta$ if $\cos\theta = \tfrac{4}{5}$ and θ is acute

  b $\cos\theta$, $\tan\theta$ if $\sin\theta = \tfrac{5}{13}$ and θ is obtuse

  c $\sin\theta$, $\cos\theta$ if $\tan\theta = -\tfrac{7}{24}$ and θ is reflex

4 Find the values of θ, from $0°$ to $360°$ inclusive, which satisfy

    a $\cos\theta = \frac{1}{2}$ **b** $\tan\theta = 1$ **c** $\sin\theta = 1$

    d $\cos\theta = -\dfrac{\sqrt{3}}{2}$ **e** $\tan\theta = -\sqrt{3}$ **f** $\sin\theta = 0.6$

    g $\cos(\theta + 60°) = 0.5$ **h** $\tan(\theta - 10°) = -0.1$

5 For $0 \leqslant \theta \leqslant 2\pi$, solve

    a $\tan\theta = \dfrac{1}{\sqrt{3}}$ **b** $\sin\theta = 0.7$ **c** $\cos\left(\theta + \dfrac{\pi}{3}\right) = \dfrac{1}{2}$

    d $\sin\left(\theta - \dfrac{\pi}{6}\right) = 1$ **e** $\sqrt{3}\cos\theta = -1$ **f** $\tan\theta = 0.2$

6 For $0 \leqslant x \leqslant 2\pi$, solve

    a $\sin^2 x = \dfrac{1}{4}$ **b** $\tan^2 x = \dfrac{1}{3}$ **c** $\sin 2x = \dfrac{1}{2}$

    d $\tan 2x = -1$ **e** $\cos 3x = \dfrac{\sqrt{3}}{2}$ **f** $\sin 3x = -1$

7 For $-180° \leqslant \theta \leqslant 180°$, solve

    a $\sin^2 2\theta = 1$ **b** $\tan\theta = 3$ **c** $\tan^2 3\theta = 1$

    d $4\cos 2\theta = 1$ **e** $\sin(2\theta + 30°) = 0.8$ **f** $\tan(3\theta - 45°) = \frac{1}{2}$

8 For $0° \leqslant \theta \leqslant 360°$, solve

    a $\sin\theta(2\sin\theta - 1) = 0$ **b** $(2\cos\theta - 1)(\cos\theta + 1) = 0$

    c $\sin\theta = \cos\theta$ **d** $\tan^2\theta + \tan\theta = 0$

    e $2\cos^2\theta = \cos\theta$ **f** $3\sin^2\theta + \sin\theta = 0$

    g $2\sin^2\theta - \sin\theta - 1 = 0$ **h** $2\cos^2\theta + 3\cos\theta + 1 = 0$

    i $4\cos^3\theta = \cos\theta$ **j** $\tan\theta = \sin\theta$

    k $4\sin^2\theta = 3\cos^2\theta$ **l** $\sin^2\theta = 4\cos^2\theta$

9 For $-\pi \leqslant \theta \leqslant \pi$, solve

    a $3 - 3\cos\theta = 2\sin^2\theta$ **b** $\cos^2\theta + \sin\theta + 1 = 0$

    c $3\sin^2\theta - \sin\theta\cos\theta - 4\cos^2\theta = 0$ **d** $(3\sin\theta - 2)(\tan\theta - 1) = 0$

    e $6\sin\theta\cos\theta - 3\cos\theta + 2\sin\theta - 1 = 0$ **f** $\sin 2\theta = \cos 2\theta$

10 For $-180° \leqslant \theta \leqslant 180°$, solve

   a $(2\sin\theta - 1)(3\sin\theta - 1) = 0$ **b** $(\tan\theta - 2)(\tan\theta + 1) = 0$

   c $(\cos\theta - 1)(2\cos\theta + 1) = 0$ **d** $\tan\theta(\tan\theta - 1) = 0$

   e $(2\sin\theta + 1)(\cos\theta + 1) = 0$ **f** $\tan^2\theta - 4\tan\theta = 0$

   g $4\tan^2\theta + \tan\theta = 0$ **h** $3\sin^3\theta = \sin\theta$

   i $\sin^3\theta = 3\sin\theta$ **j** $3 = 2\sin^2 2\theta + 3\cos 2\theta$

11 For $0 < \theta < 2\pi$, solve

   a $8\cos^3\theta = 1$ **b** $9\tan^4\theta = 1$

   c $2\cos^2\theta - 5\cos\theta + 2 = 0$ **d** $\sin^2\theta + 3\sin\theta + 2 = 0$

   e $3\cos^2\theta = 5(1 - \sin\theta)$ **f** $4\cos^2\theta = 3$

   g $3\cos^2\theta - 13\cos\theta = 3\sin^2\theta - 9$ **h** $\tan 3\theta(\tan 3\theta + 1) = 2$

12 Eliminate θ from these equations.

   a $x = a\cos\theta$, $y = b\sin\theta$ **b** $x = 1 - \sin\theta$, $y = 1 + \cos\theta$

   c $x = 4 - \cos\theta$, $y = 2 + \sin\theta$

13 Prove these identities.

   a $\tan\theta + \dfrac{1}{\tan\theta} = \dfrac{1}{\sin\theta\cos\theta}$ **b** $\dfrac{1}{\sin\theta} + \dfrac{\tan\theta}{\cos\theta} = \dfrac{1}{\sin\theta\cos^2\theta}$

   c $\dfrac{1 - \cos\theta}{\sin\theta} = \dfrac{\sin\theta}{1 + \cos\theta}$ **d** $\left(\dfrac{1}{\cos\theta} + \tan\theta\right)\left(\dfrac{1}{\cos\theta} - \tan\theta\right) = 1$

14 Sketch the graph of $y = 2\cos x$ for $0° \leqslant x \leqslant 360°$.

State the values, or range of values, of k for which the equation $2\cos x = k$ has 0, 1 and 2 roots in the interval $0° \leqslant x \leqslant 360°$.

15 Plot on graph paper, or draw on a graphic calculator, the graphs of $y = 1 - \sin\frac{x}{3}$ and $y = x$ for $0 \leqslant x \leqslant \pi$.
Hence find a solution to $1 - \sin\frac{x}{3} = x$ correct to 1 decimal place.

16 Plot on graph paper, or draw on a graphic calculator, the graphs of $y = \sin(x + 30°)$ and $y = 2 + \tan x$ for $0° \leqslant x \leqslant 180°$. Hence find a solution to $\sin(x + 30°) = 2 + \tan x$, correct to the nearest degree.

1 Sketch $y = \cos 3x$ for $0° \leqslant x \leqslant 180°$, and hence state the number of solutions in the range $0° \leqslant x \leqslant 180°$ for

   a $\cos 3x = \frac{1}{2}$ **b** $\cos 3x = 1$ **c** $\cos 3x = 2$

2 For these functions, sketch the graphs for $0° \leqslant x \leqslant 360°$, describe the required transformation of $y = \sin x$, and state the period.

   a $y = -\sin x$ **b** $y = \sin 2x$ **c** $y = 3\sin x$ **d** $y = \sin(x + 40°)$

   e $y = 2 + \sin x$ **f** $y = \sin(20° - x)$ **g** $y = 1 + \sin\dfrac{x}{2}$

3 Solve these equations for $0° \leqslant x \leqslant 360°$. Do not use a calculator if the answer can be found from the special triangles.

   a $\tan x = \dfrac{1}{\sqrt{3}}$ **b** $\cos x = 0.1$ **c** $\sin 2x = -1$

   d $\tan^2 x = \dfrac{1}{4}$ **e** $\cos(x - 30°) = 0.6$ **f** $\sin x = -\dfrac{1}{\sqrt{2}}$

   g $\tan 3x = 0$ **h** $4\sin\dfrac{x}{2} = 1$

4 Solve these equations for $-\pi < \theta < \pi$.

   a $4\cos^2\theta - 1 = 0$ **b** $2\cos\theta + 3\sin\theta = 0$

   c $\sin\theta\cos\theta - \sin\theta + \cos\theta = 1$ **d** $\tan^2\theta - 3\tan\theta + 2 = 0$

   e $4 - \sin\theta = 6\cos^2\theta$ **f** $\sin^2\theta + 2\cos^2\theta = 2$

   g $3\cos^2\theta + \cos\theta + 2 = 3\sin^2\theta$ **h** $\tan\theta = \cos\theta$

5 Solve these equations.

   a $\sin\left(x - \dfrac{\pi}{2}\right) = \dfrac{3}{4}$ for $0 < x < 2\pi$

   b $\cos(x + 30°) = \dfrac{1}{2}$ for $-180° < x < 180°$

   c $\tan 2x = 1$ for $90° < x < 270°$

   d $4\cos^2 x° + 8\sin x° - 7 = 0$ for $0 < x < 360$

   e $\sin^2\left(x + \dfrac{\pi}{6}\right) = \dfrac{1}{2}$ for $-\pi < x < \pi$

6   Solve $(1 + \cos x)^2 = \frac{1}{4}$ for $-180° \leqslant x \leqslant 180°$.

7   Find one point of intersection of the curve $y = \tan\left(x + \frac{\pi}{5}\right)$ and the line $y = \sqrt{3}$.

8 $f(x) = 2x^3 - 7x^2 + 7x - 2$.

    a   Show that $f(1) = 0$ and hence factorise $f(x)$.

    b   Solve $f(x) = 0$.

    c   Hence, or otherwise, solve for $0 \leqslant x \leqslant 2\pi$
        i $2\tan^3 x - 7\tan^2 x + 7\tan x - 2 = 0$
        ii $2\sin^3 x - 7\sin^2 x + 7\sin x - 2 = 0$
        iii $2\cos^3 x - 7\cos^2 x + 7\cos x - 2 = 0$

9   Simplify, giving exact answers as a single fraction with rational denominator

    a $\sin^2\frac{\pi}{4} + \tan\frac{\pi}{6}$ **b** $\tan\frac{\pi}{6} + \tan\frac{\pi}{3}$ **c** $\sin 45° + \cos 60°$

    d $\dfrac{\sin 120° + \tan 120°}{\sin 150° + \tan 150°}$ **e** $\tan 30° + \dfrac{1}{\cos 30°}$ **f** $2\tan\frac{11\pi}{6} + \dfrac{1}{\cos\frac{11\pi}{6}}$

A 'Test yourself' exercise (Ty16) and an 'Extension exercise' (Ext16) can be found on the CD-ROM.

Key points

Trigonometric ratios

π radians $= 180°$

$\sin\theta = \dfrac{y}{r}$ $\cos\theta = \dfrac{x}{r}$ $\tan\theta = \dfrac{y}{x}$ $x \neq 0$

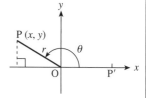

To find the trigonometric ratios of any angle, use either the symmetry of the graph or find

- the **sign** from 'All Silly Tom Cats'

- and the **magnitude** by finding the acute equivalent angles.

Use the special triangles for ratios of $45°$, $30°$, $60°$.

Identities

For all values of θ:

$\dfrac{\sin\theta}{\cos\theta} = \tan\theta$

$\sin^2\theta + \cos^2\theta = 1$ which can be rearranged as $\sin^2\theta = 1 - \cos^2\theta$ or $\cos^2\theta = 1 - \sin^2\theta$

$\sin\theta = \cos(90° - \theta)$ $\cos\theta = \sin(90° - \theta)$

Solving trigonometric equations

- Use identities.
- Recognise quadratic equations.
- $a\sin x = b\cos x \Rightarrow \tan x = \frac{b}{a}$.
- Find one solution with a calculator or using the special triangles and think 'what else?'
- Give answer in degrees if degrees are specified in the question; otherwise use radians.
- Remember that answers to trigonometric equations can be checked on a calculator.
- For equations such as $\cos 3x = 0.2$ or $\sin(x + 40°) = 0.5$, list solutions for $3x$ or $x + 40°$ before dividing by 3 or subtracting $40°$.

Graphs

$y = \sin \theta$

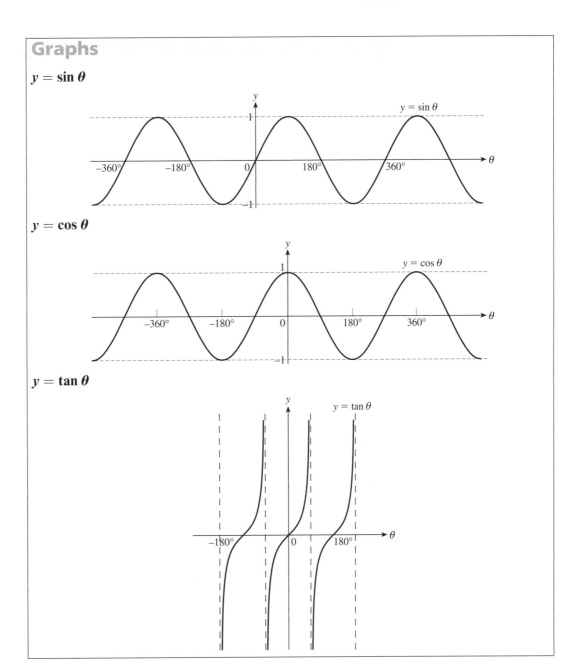

$y = \cos \theta$

$y = \tan \theta$

Proving identities

Use the trigonometric identities on page 280.

For equations or identities it is often helpful, as a first step, to put all the trigonometric functions in terms of sines and/or cosines.

This chapter introduces further applications of trigonometry, some of which you may have met at GCSE.

After working through this chapter you should be able to

■ *find an arc length and the area of a sector using angles measured in radians*
■ *find the area of a triangle in the form $\frac{1}{2}ab\sin C$*
■ *apply the sine and cosine rules.*

17.1   Area of sector and length of arc

Consider a circle, centre O, radius r and an arc, PQ, of the circle subtending an angle of θ radians at the centre of the circle. OPQ is a sector of the circle.

Area of sector OPQ, $A = \dfrac{\theta}{2\pi} \times \pi r^2 = \dfrac{1}{2}r^2\theta$

Length of arc PQ, $s = \dfrac{\theta}{2\pi} \times 2\pi r = r\theta$

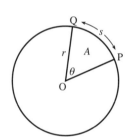

$\dfrac{\theta}{2\pi}$ is the fraction of the circle required.

> **Area of sector,** $A = \frac{1}{2}r^2\theta$
>
> **Length of arc,** $s = r\theta$ **(θ in radians)**

For θ measured in degrees
$$A = \frac{\theta}{360} \times \pi r^2$$
$$s = \frac{\theta}{360} \times 2\pi r$$

Remember π radians $= 180°$

Example 1   The arc, AB, subtends an angle of $\frac{\pi}{3}$ at the centre, O, of a circle radius 6 cm.

Find the length of arc AB and the area of sector OAB.

Solution $s = r\theta$

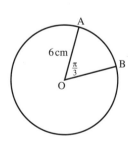

Here, $r = 6$ and $\theta = \dfrac{\pi}{3}$.

So, arc AB $= 6 \times \dfrac{\pi}{3}$ cm $= 2\pi$ cm

$A = \frac{1}{2}r^2\theta$

∴   area of sector OAB $= \dfrac{1}{2} \times 6^2 \times \dfrac{\pi}{3}$ cm^2

$\qquad\qquad\qquad = 6\pi$ cm^2

Note θ must be in radians to use these formulae.

Example 2   The perimeter of the sector of a circle, radius 3 cm, is 8 cm.
Find the area of the sector.

Perimeter $= (3 + 3 + s)$ cm

But perimeter $= 8$ cm

So $s = 2$. Hence the length
of the arc AB, is 2 cm.

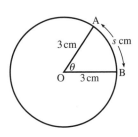

The perimeter of
OAB includes the
radii. $r = 3$ cm.

$$s = r\theta$$

$$2 = 3\theta$$

$$\therefore \qquad \theta = \tfrac{2}{3}$$

θ in radians.

Then $A = \tfrac{1}{2}r^2\theta = \tfrac{1}{2} \times 3 \times 3 \times \tfrac{2}{3} = 3$

$\therefore$ area of sector AOB is 3 cm^2.

Example 3   OAB is a sector of a circle centre O. CD is the tangent to the circle at M.
$OC = OD = a$. Angle $COD = 2\theta$.

Show that the shaded area is given by
$a^2 \cos\theta(\sin\theta - \theta\cos\theta)$

Let the area of $\triangle COD$ be A_t and
the area of the sector OAB be A_s.

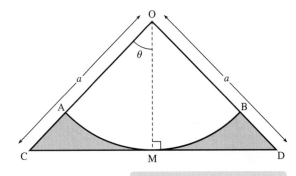

Join O to mid-point M of CD.

Shaded area, $A = A_t - A_s$

$$A_t = CM \times OM$$

$$= a\sin\theta \times a\cos\theta$$

$$= a^2 \sin\theta \cos\theta$$

Angle $OMD = 90°$ because tangent is
perpendicular to radius.

$A_t = \tfrac{1}{2}bh \quad \sin\theta = \dfrac{CM}{a} \quad \cos\theta = \dfrac{OM}{a}$

$$A_s = \tfrac{1}{2}OM^2 \times 2\theta$$

$$= a^2\theta\cos^2\theta$$

Area of the sector $= \tfrac{1}{2}r^2\theta$. Here the angle of the sector is 2θ.

$$A = A_t - A_s$$

$$= a^2 \sin\theta\cos\theta - a^2\theta\cos^2\theta$$

$a^2\cos\theta$ is a common factor.

$$= a^2 \cos\theta(\sin\theta - \theta\cos\theta)$$

Exercise 17A *Unless otherwise stated, all angles in this exercise are measured in radians. Exact answers should be given where appropriate.*

1 A disc makes 100 revolutions in three minutes.
Find the angle through which it turns every second, in radians and in degrees.

2 Find the length of the arcs and the areas of the sectors.

a 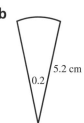 **b** **c**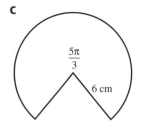

Angles in radians

3 Find the length of an arc which subtends an angle of 0.8 radians at the centre of a circle of radius 10 cm.

4 An arc of a circle subtends an angle of 0.5 radians at the centre.
Given that the length of the arc is 3 cm, find the radius of the circle.

5 Find the angle subtended at the centre of a circle of radius 2.5 cm by an arc 2 cm long.

6 Find the area of a sector containing an angle of 1.5 radians in a circle of radius 2 cm.

7 The radius of a circle is 3 cm. Find the angle contained by a sector of area 18 cm².

8 A round cake with diameter 20 cm is shared equally between six people.
Find the perimeter and the area of the top of each piece.
(Leave π in your answers.)

9 A 1.5 m long pendulum travels from one extreme of its swing to the other extreme in half a second. In this time, the bob at the bottom of the pendulum travels $\frac{\pi}{10}$ m.
Find how long it takes the pendulum to sweep out the equivalent of a complete circle.

10 A coin rolls on its edge along a slope. The coin travels 40 cm and makes n complete revolutions. Show that a face of the coin has an area of $\dfrac{400}{\pi n^2}$ cm².

11 The face of a town hall clock carries a design of concentric circles.
The inner circle has radius 0.5 m and the outer circle has radius 1 m, which is also the length of the minute hand. The region between the circles is shaded.

Find the area of the shaded region swept out by the minute hand in ten minutes.
(Leave π in your answer.)

12 An arc subtends an angle of 1 radian at the centre of a circle, and a sector of area 72 cm^2 is bounded by this arc and the two radii.
Find the radius of the circle.

13 The chord AB of a circle subtends an angle of 60° at the centre.
Find the ratio of the lengths of chord AB to arc AB.

14 A segment is cut off a circle of radius 5 cm by a chord AB, of length 6 cm.
Find the length of the minor arc AB.

15 A circular dartboard is divided into 20 equal sectors, one of which is shown in the diagram.
O is the centre of the circle. The areas for scoring double and treble 20 are marked A and B respectively.
Find the ratio area A:area B in the form n:1, giving n correct to 1 d.p.

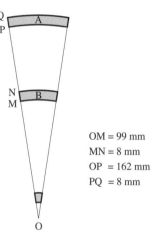

OM = 99 mm
MN = 8 mm
OP = 162 mm
PQ = 8 mm

16 An oar of length l m has the tips of its handle and its blade at A and B respectively. It is pivoted at P, a quarter of the way from A to B.

During a stroke, A travels along an arc of a circle of length $\dfrac{l}{3}$ m.

a Find the area of the sector swept out by PB during a stroke.

The oar is moved so that P is now three-tenths of the way from A to B, and the angle turned by the oar in a stroke is not changed.

b Find the length of the arc along which B travels during a stroke.

17.2 Area of a triangle

The usual notation for labelling triangles is to use capital letters for vertices, the same capitals to represent the size of the angles at the vertices and the corresponding lower case letters to represent the length of sides opposite the vertices. So, for example, the angle at B will be of size B and the side opposite B will be of length b.

This elegantly simple idea was first used by Euler and was one of his many contributions to mathematics.

To prove   Area of triangle $= \frac{1}{2}ab\sin C$

This proof is for an acute-angled triangle, but the formula holds for any triangle.

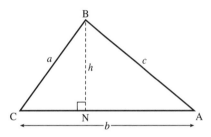

$$\text{Area of } \triangle ABC = \tfrac{1}{2}bh$$

Using area $= \frac{1}{2} \times$ base $\times$ height.

In $\triangle CBN$, $h = a\sin C$

$\sin C = \dfrac{h}{a}$

$\therefore$ Area of $\triangle ABC = \frac{1}{2} \times b \times a\sin C$

$$= \tfrac{1}{2}ab\sin C$$

➤ | **Area of a triangle $= \frac{1}{2}ab\sin C$** |

Similarly: Area $= \frac{1}{2}bc\sin A$
Area $= \frac{1}{2}ac\sin B$

Note   Two sides and the included angle are needed.

Example 4   Find the area of this triangle.

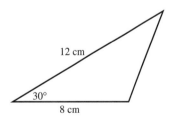

Two sides and the included angle are given.

$a = 8$, $b = 12$, $C = 30°$

Area $= \frac{1}{2}ab\sin C$

$= \frac{1}{2} \times 8 \times 12 \times \sin 30°$

$\sin 30° = \frac{1}{2}$

$= 4 \times 12 \times \frac{1}{2} = 24$

So the area of the triangle is $24\,\text{cm}^2$.

Example 5 Given that chord $AB = 2\sqrt{3}$, show that the area of the minor (shaded) segment ACB of the circle, centre O, radius 2, is $\frac{1}{3}(4\pi - 3\sqrt{3})$.

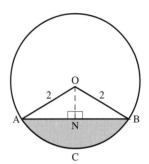

$OA = OB$ (radii) $\therefore$ $\triangle AOB$ is isosceles.

$AB = 2\sqrt{3}$ so $AN = \sqrt{3}$

> ON, the perpendicular from the centre to the chord, bisects AB.

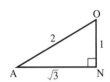

> $\sin A\hat{O}N = \dfrac{\sqrt{3}}{2}$, so from the special triangles $A\hat{O}N = \dfrac{\pi}{3}$ and $ON = 1$.

> ON can also be calculated using Pythagoras' theorem.

$ON = \frac{1}{2}OA$ and $\triangle AON$ is half an equilateral triangle of side 2.

So $A\hat{O}N = \dfrac{\pi}{3}$.

$\therefore$ $A\hat{O}B = \dfrac{2\pi}{3}$

Area of segment ACB = Area of sector OAB − Area of $\triangle AOB$

Area of sector OAB $= \dfrac{1}{2}r^2\theta$

> $\theta = \dfrac{2\pi}{3}$.

$$= \dfrac{1}{2} \times 2^2 \times \dfrac{2\pi}{3}$$

> *Remember*: The angle must be in radians when using this formula.

$$= \dfrac{4\pi}{3}$$

Area of $\triangle OAB = \dfrac{1}{2}ab\sin C$

$$= \dfrac{1}{2} \times 2^2 \sin\dfrac{2\pi}{3}$$

$$= \dfrac{2\sqrt{3}}{2}$$

$$= \sqrt{3}$$

Area of segment ACB $= \dfrac{4\pi}{3} - \sqrt{3}$

$$= \dfrac{1}{3}\left(4\pi - 3\sqrt{3}\right)$$

> Area of segment, S = Area of sector − Area of triangle
> $$= \tfrac{1}{2}r^2\theta - \tfrac{1}{2}r^2\sin\theta$$
> $$= \tfrac{1}{2}r^2(\theta - \sin\theta) \qquad (\theta \text{ in radians})$$

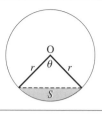

Example 6   Three identical cylinders of radius 3 cm are placed in contact with each other with their axes parallel and a band is placed round the cylinders.

This diagram shows a cross-section.
Calculate the length of the band.

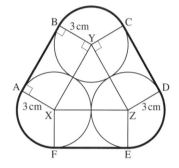

> The band is made up of 3 straight parts and 3 arcs.

Solution   The part of the band AB is a straight line which is a tangent to the circles with centres X and Y at A and B respectively.

$$\text{Y}\hat{\text{B}}\text{A} = \text{X}\hat{\text{A}}\text{B} = 90°$$

> Radius ⊥ tangent.

$$\text{AX} = \text{BY} = 3\,\text{cm}$$

Therefore ABYX is a rectangle and XY = AB.

By symmetry, XY = YZ = ZX. Therefore △XYZ is equilateral.

Also by symmetry, CD = EF = AB.

$$\text{XY} = 2 \times 3\,\text{cm} = 6\,\text{cm}$$

> XY is twice the radius of the cylinder.

So AB = CD = EF = 6 cm.

The part of the band BC is the arc of a circle subtending $\text{B}\hat{\text{Y}}\text{C}$ at the centre Y.

$$\text{B}\hat{\text{Y}}\text{C} = 360° - \text{B}\hat{\text{Y}}\text{X} - \text{C}\hat{\text{Y}}\text{Z} - \text{X}\hat{\text{Y}}\text{Z}$$
$$= 360° - 2 \times 90° - 60° = 120°$$

> $\text{B}\hat{\text{Y}}\text{X} = \text{C}\hat{\text{Y}}\text{Z} = 90°$ (angles of a rectangle).
> $\text{X}\hat{\text{Y}}\text{Z} = 60°$ (△XYZ is equilateral).

So the three arcs BC, DE and FA will together form a circle of radius 3 cm which has circumference $3 \times 2\pi = 6\pi$ cm.

Therefore length of band = 3AB + BC + DE + FA
$$= (3 \times 6 + 6\pi)\,\text{cm}$$
$$= (18 + 6\pi)\,\text{cm}$$

Exercise 17B *Exact answers should be given where appropriate. Otherwise give answers to 3 significant figures.*

1 Find the area of these triangles.

a

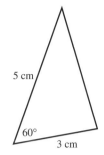

5 cm

60°

3 cm

b

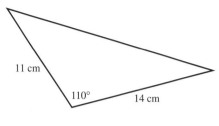

11 cm

110°

14 cm

c

11.4 cm

8.6 cm

d

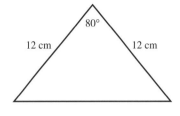

80°

12 cm   12 cm

e

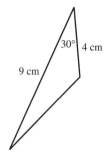

30° 4 cm

9 cm

2 Find the area of the triangle ABC in which

   a $A = 90°$ $B = 27°$ $a = 15$ cm   **b** $C = 90°$ $A = 42°$ $a = 10$

   c $a = b = 12$ cm $c = 4$ cm   **d** $a = b = 12$ cm $A = 80°$

3 The arc, AB, of a sector of a circle, centre O, radius 2 cm, is 4 cm long. Find

   a the area of the sector AOB

   b the area of triangle AOB

   c the area of the minor segment cut off by the line AB.

4 In each diagram, find the area of sector AOB, and the area of triangle AOB, and hence deduce the area of segment ABC.

a

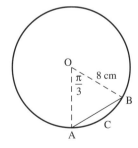

O

$\frac{\pi}{3}$ 8 cm

B

C

A

b

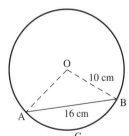

O

10 cm

B

A

16 cm

C

5 A chord AB subtends an angle of $\dfrac{2\pi}{3}$ at O, the centre of a circle with radius 12 cm. Find the area of

   a   sector AOB   **b**   triangle AOB   **c**   the major segment cut off by AB.

6 A chord PQ of a circle with radius r, subtends an angle θ at the centre.

   a   Show that the area of the minor segment cut off by PQ is $\frac{1}{2}r^2(\theta - \sin\theta)$.

   b   Write down the area of the major segment cut off by PQ in terms of r and θ.

   c   Hence show that if the ratio of the areas of the segments is 1:3 then $2\theta - 2\sin\theta = \pi$.

7 The diagram shows the cross-section of a ball, radius r cm, floating in water. The surface of the water touches the ball at A and B. AB subtends an angle of $\dfrac{2\pi}{3}$ at the centre of the ball.

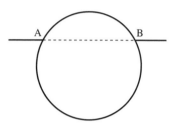

   a   Find the length of AB in terms of r.

   b   Hence find, in terms of r, the circumference of the circle where the ball crosses the surface of the water.

8 Two circles with equal radii r are drawn. Their centres, O and P, are a distance r apart so that the centre of each circle lies on the circumference of the other. The circles intersect at A and B.

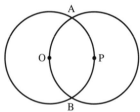

   a   Show that angle AOP $= \dfrac{\pi}{3}$ and find angle AOB.

   b   Hence find the area of the sector PAOB of the circle centre P.

   c   Show that the length of AB is $\sqrt{3}r$ and that the area of overlap of the circles is $2r^2\left(\dfrac{\pi}{3} - \dfrac{\sqrt{3}}{4}\right)$

9 Three circular beer mats of radius 5 cm are placed on a table so that each touches the other two.
Find the area of the space between the mats.

10 The diagram shows a circle, radius r, inscribed and circumscribed by regular hexagons.

a Find exact expressions for the perimeters of both hexagons in terms of r.

b Hence show that π must lie between 3 and $2\sqrt{3}$.

17.3 Solution of triangles

Solution of triangles or solving triangles means finding sides and/or angles given certain data. Right-angled triangles can be solved using trigonometric ratios (sine, cosine and tangent) and Pythagoras' theorem. The sine and cosine rules enable non-right-angled triangles to be solved. It is interesting, before introducing the rules, to discover exactly what is required to specify a triangle. You may prefer to go straight to Section 17.4.

Given data for drawing a triangle, there are cases where

■ no triangle is possible
■ just one triangle is possible
■ more than one triangle is possible.

Exercise 17C *For this exercise, it may be helpful to try to construct the triangles with ruler and compasses or, at least, to plan such a construction.*

1 Consider these sets of data and decide into which category they fall: no triangle ABC possible, just one triangle ABC possible, or more than one triangle ABC possible.

a $A = 60°$ $B = 100°$ $c = 10\,\text{cm}$ **b** $A = 60°$ $B = 100°$ $C = 10°$

2 Repeat Question 1 for these sets of data.

a $a = 8\,\text{cm}$ $A = 60°$ $b = 9\,\text{cm}$ **b** $a = 10\,\text{cm}$ $A = 60°$ $b = 9\,\text{cm}$

c $a = 8\,\text{cm}$ $b = 7\,\text{cm}$ $c = 6\,\text{cm}$ **d** $a = 8\,\text{cm}$ $A = 75°$ $b = 9\,\text{cm}$

e $a = 8\,\text{cm}$ $C = 126°$ $b = 9\,\text{cm}$ **f** $a = 8\,\text{cm}$ $b = 9\,\text{cm}$ $c = 1\,\text{cm}$

g $a = 10\,\text{cm}$ $b = 4\,\text{cm}$ $c = 5\,\text{cm}$ **h** $A = 25°$ $B = 100°$

i $a = 10\,\text{cm}$ $c = 5\,\text{cm}$ $C = 30°$ **j** $a = 10\,\text{cm}$ $c = 8\,\text{cm}$ $A = 90°$

Triangle facts

Some useful facts about triangles should be noted from Exercise 17C.

- The largest angle of a triangle is opposite the longest side and the smallest angle is opposite the shortest side.
- There can only be one obtuse angle in a triangle.
  If there is one, it will be opposite the longest side.
- The sum of the lengths of any two sides of a triangle must exceed the length of the third side.
- There are some sets of data which will result in identical triangles being drawn. All such triangles will be **congruent**. For example,

  Given three sides (SSS): Exercise 17C Question 2c
  Given two sides and the included angle (SAS): Exercise 17C Question 2e
  Given two angles and one corresponding side (AAS): Exercise 17C Question 1a
  Given a right angle, the hypotenuse and one side (RHS): Exercise 17C Question 2j

- When three angles are given all the triangles will be **similar**: Exercise 17C, Question 2h.
- When two sides and an angle (not the included angle) are given there may be 0, 1 or 2 possible triangles. (See Exercise 17C Questions 2a, 2b, 2d, 2i.)

17.4 The sine and cosine rules

The sine and cosine rules provide a method of solving triangles.

- To use the sine rule, an angle and the corresponding side, plus one more side or angle are needed.
- To use the cosine rule, either two sides and the included angle or three sides are needed.

The sine rule

> **The sine rule: For any triangle,**
>
> $$\frac{a}{\sin A} = \frac{b}{\sin B} = \frac{c}{\sin C} \quad \text{or alternatively} \quad \frac{\sin A}{a} = \frac{\sin B}{b} = \frac{\sin C}{c}$$

Proof   To prove the sine rule: $\dfrac{a}{\sin A} = \dfrac{b}{\sin B} = \dfrac{c}{\sin C}$

For an acute-angled triangle
BN is perpendicular to AC.

In $\triangle$ABN   BN $= c \sin A$

In $\triangle$CBN   BN $= a \sin C$

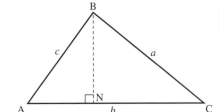

$\sin A = \dfrac{\text{BN}}{c}$

$\sin C = \dfrac{\text{BN}}{a}$

So $\qquad a \sin C = c \sin A$

$\therefore \qquad \dfrac{a}{\sin A} = \dfrac{c}{\sin C}$

Similarly $\qquad \dfrac{a}{\sin A} = \dfrac{b}{\sin B}$

$\therefore \qquad \dfrac{a}{\sin A} = \dfrac{b}{\sin B} = \dfrac{c}{\sin C}$

> The perpendicular from C to AB would give this result.

For an obtuse-angled triangle
BN is perpendicular to AC produced.

> AC produced means AC is extended.

In $\triangle$ABN $\qquad$ BN $= c \sin A$

In $\triangle$CBN $\qquad$ BN $= a \sin (180° - C)$

$\qquad\qquad\qquad = a \sin C$

So $\qquad a \sin C = c \sin A$

$\therefore \qquad \dfrac{a}{\sin A} = \dfrac{c}{\sin C}$

Similarly $\qquad \dfrac{c}{\sin C} = \dfrac{b}{\sin B}$

$\therefore \qquad \dfrac{a}{\sin A} = \dfrac{b}{\sin B} = \dfrac{c}{\sin C}$

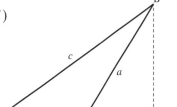

> $\sin A = \dfrac{\text{BN}}{c}$

> $\sin (180° - C) = \sin C$

> The perpendicular from A to BC produced would give this result.

In Examples 7–13, all answers are given exactly or correct to 1 decimal place for lengths and for angles.

Example 7 $\quad$ Given $A = 60°$, $B = 100°$ and $c = 10\,\text{cm}$, find AC.

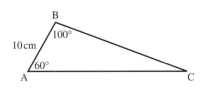

> See Exercise 17c, Question 1a.

$C = 180° - 100° - 60° = 20°$

> Given two angles, the third can be calculated.

Using the sine rule $\qquad \dfrac{c}{\sin C} = \dfrac{b}{\sin B}$

> Substitute $B = 100°$, $c = 10$, and $C = 20°$.

$\qquad\qquad\qquad \dfrac{10}{\sin 20°} = \dfrac{b}{\sin 100°}$

> Multiply both sides by $\sin 100°$

$\qquad\qquad\qquad b = \dfrac{10 \sin 100°}{\sin 20°}$

$\qquad\qquad\qquad\quad = 28.8$

So $AC = 28.8\,\text{cm}$.

Example 8   In this example, given two sides and an angle (not the included angle), more than one triangle may satisfy the data. This is called the 'ambiguous case'.

Given $a = 8$ cm, $A = 60°$ and $b = 9$ cm, find the possible values of B and C and sketch the triangles.

See Exercise 17c, Question 2a.

Using the sine rule

$$\frac{\sin A}{a} = \frac{\sin B}{b}$$

$$\frac{\sin 60°}{8} = \frac{\sin B}{9}$$

$$\sin B = \frac{9 \sin 60°}{8}$$

$$B = \arcsin\left(\frac{9 \sin 60°}{8}\right)$$

$$\therefore \quad B = 77.0° \text{ or } B = 180° - 77.0°$$
$$= 103.0°$$

There are two possible positions for B (marked B_1 and B_2) and $\therefore$ 2 possible values for angle B.

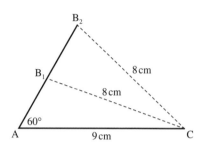

When $B = 77.0°$ $C = 180° - 60° - 77.0° = 43.0°$
When $B = 103.0°$ $C = 180° - 60° - 103.0° = 17.0°$

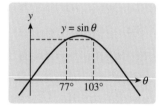

There are two possible triangles.

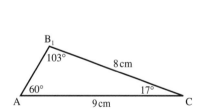

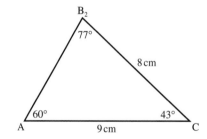

Example 9 How many triangles can be drawn with $a = 10$ cm, $A = 60°$ and $b = 9$ cm?

For each possible triangle find angle B.

See Exercise 17c, Question 2b.

By the sine rule: $\dfrac{\sin B}{b} = \dfrac{\sin A}{a}$

$b = 9$, $A = 60°$, $a = 10$.

$$\dfrac{\sin B}{9} = \dfrac{\sin 60°}{10}$$

$$\sin B = \dfrac{9 \sin 60°}{10}$$

$$B = \arcsin \left(\dfrac{9 \sin 60°}{10} \right)$$

$\therefore \qquad\qquad\quad B = 51.2°$

or $\qquad\qquad\quad B = 180° - 51.2° = 128.8°$

Angle B cannot be $128.8°$, because angle $A = 60°$ and this would make $A + B > 180°$

So only one triangle can be drawn, with angle $B = 51.2°$

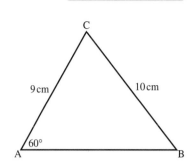

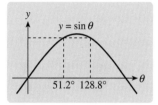

Exercise 17D

1 Use the sine rule to find the values of the unknown sides and angles, to 3 significant figures. In any ambiguous case, give both solutions.

a

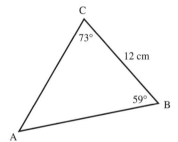

b

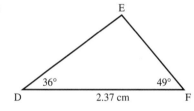

c

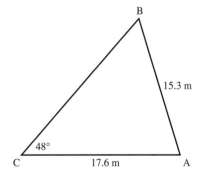

d

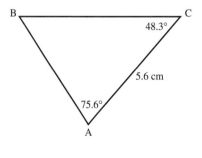

e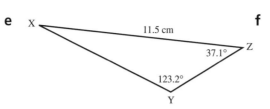

f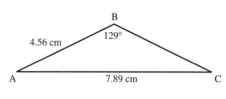

2 Find the values of the unknowns in these triangles, without using a calculator. Give exact answers.

a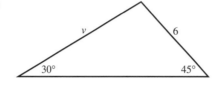

b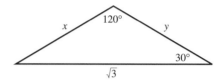

The cosine rule

➤

> **The cosine rule: For any triangle,**
>
> $$a^2 = b^2 + c^2 - 2bc \cos A$$

Proof

To prove the cosine rule: $a^2 = b^2 + c^2 - 2bc \cos A$

For an acute-angled triangle
Draw CN perpendicular to BA.

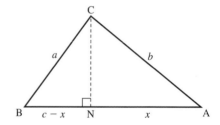

Let $AN = x$, then $\quad BN = c - x$

In $\triangle BCN \qquad CN^2 = a^2 - (c - x)^2$

$$= a^2 - c^2 + 2cx - x^2$$

In $\triangle ACN \qquad CN^2 = b^2 - x^2$ **Using Pythagoras' theorem**

So $\quad a^2 - c^2 + 2cx - x^2 = b^2 - x^2$

$\therefore \qquad\qquad a^2 = b^2 + c^2 - 2cx$

But, in $\triangle ACN \qquad x = b \cos A$ $\cos A = \dfrac{x}{b}$

so $\qquad\qquad a^2 = b^2 + c^2 - 2bc \cos A$

Similarly $\qquad b^2 = a^2 + c^2 - 2ac \cos B$ Putting BN = x gives this result.

and $\qquad\qquad c^2 = a^2 + b^2 - 2ab \cos C$ The perpendicular from B to AC gives this result.

For an obtuse-angled triangle
Draw BN perpendicular to CA produced.

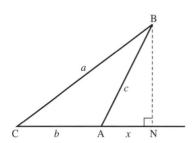

In $\triangle$BCN $\qquad$ $BN^2 = a^2 - (b+x)^2$

$\qquad\qquad\qquad\qquad = a^2 - b^2 - 2bx - x^2$

In $\triangle$BAN $\qquad$ $BN^2 = c^2 - x^2$

So $\quad a^2 - b^2 - 2bx - x^2 = c^2 - x^2$

$\therefore \qquad\qquad\qquad a^2 = b^2 + c^2 + 2bx$

But, in $\triangle$ABN $\qquad$ $x = c\cos(180° - A)$

$\qquad\qquad\qquad\qquad = -c\cos A$

$\boxed{\cos(180° - A) = \dfrac{x}{c}}$

$\boxed{\cos(180° - A) = -\cos A}$

so $\qquad\qquad a^2 = b^2 + c^2 - 2bc\cos A$

Similarly $\qquad b^2 = a^2 + c^2 - 2ac\cos B$

and $\qquad\qquad c^2 = a^2 + b^2 - 2ab\cos C$

> **Rearranging the cosine rule gives** $\cos A = \dfrac{b^2 + c^2 - a^2}{2bc}$

Given three sides of a triangle, the rearrangement can be used to find the angles.

Note For right-angled triangles, the sine rule gives

$$\frac{a}{\sin 90°} = \frac{b}{\sin B}$$

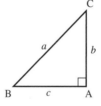

But $\sin 90° = 1$, so $\sin B = \dfrac{b}{a}$.

For right-angled triangles, the cosine rule gives

$$a^2 = b^2 + c^2 - 2bc\cos 90°$$

But $\cos 90° = 0$, so $a^2 = b^2 + c^2$, which is Pythagoras' theorem.

Example 10   Given $a = 8$ cm, $C = 126°$ and $b = 9$ cm, find AB and angle B.

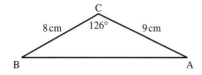

See Exercise 17c, Question 2e.

Two sides and the included angle are given: the cosine rule can be used.

Using the cosine rule

$$c^2 = a^2 + b^2 - 2ab\cos C$$

$$= 8^2 + 9^2 - 2 \times 8 \times 9 \times \cos 126°$$

$$\therefore \quad c = 15.2$$

So   AB $= 15.2$ cm.

Work out c^2 on the calculator and take the square root. State AB correct to 1 d.p. Keep accurate value of c in memory on calculator for use in the sine rule.

Using the sine rule

$$\frac{\sin B}{b} = \frac{\sin C}{c}$$

$$\frac{\sin B}{9} = \frac{\sin 126°}{c}$$

$$\sin B = \frac{9\sin 126°}{c}$$

$$\therefore \quad B = 28.7°$$

$B = \arcsin\left(\dfrac{9\sin 126°}{c}\right)$

Since C is obtuse, B must be acute.

Example 11   In the triangle shown, find Q.

By the cosine rule

$$q^2 = p^2 + r^2 - 2pr\cos Q$$

$$\cos Q = \frac{p^2 + r^2 - q^2}{2pr}$$

so $$\cos Q = \frac{4^2 + 8^2 - 10^2}{2 \times 4 \times 8}$$

$$= \frac{16 + 64 - 100}{2 \times 4 \times 8}$$

$$= \frac{-\overset{5}{\cancel{20}}}{2 \times \cancel{4} \times 8}$$

$$= -\frac{5}{16}$$

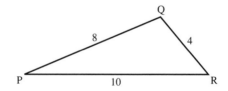

$p = 4$, $q = 10$, $r = 8$.

$\cos Q$ could be evaluated here on a calculator.

$Q = \arccos\left(-\frac{5}{16}\right)$

$\cos Q$ is $-$ve $\therefore$ angle Q is obtuse, so $Q = 108.2°$.

Exercise 17E

1 Use the cosine rule to find the values of the unknown sides and angles, to 3 significant figures.

a

A
60°
8 cm
15 cm
C
B

b

C
8 m
5 m
A
7 m
B

c

E
8 cm
10 cm
D
15 cm
F

d

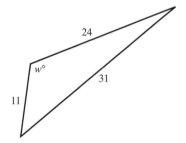

P
63 cm
R
100° 17 cm
Q

e

M
120°
1 cm
2 cm
L
N

2 Find the values of the unknowns in these triangles, without using a calculator. Give exact answers.

a

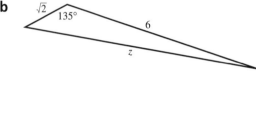

24
$w°$
31
11

b

$\sqrt{2}$
135°
6
z

Example 12 Given $a = 8\,\text{cm}$, $b = 7\,\text{cm}$ and $c = 6\,\text{cm}$, find all angles of the triangle.

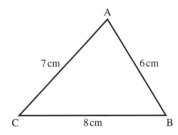

A
7 cm
6 cm
C
8 cm
B

See Exercise 17c, Question 2c.

Given 3 sides, the cosine rule can be used.

$a = 8, b = 7, c = 6$

The cosine rule gives

$$\cos A = \frac{b^2 + c^2 - a^2}{2bc}$$

$$= \frac{7^2 + 6^2 - 8^2}{2 \times 7 \times 6}$$

$$= \frac{49 + 36 - 64}{2 \times 7 \times 6}$$

$$= \frac{21}{2 \times 7 \times 6} = \frac{1}{4}$$

$$\therefore \quad A = \arccos \tfrac{1}{4} = 75.5°$$

Using the sine rule

$$\frac{\sin A}{a} = \frac{\sin B}{b}$$

$$\frac{\sin A}{8} = \frac{\sin B}{7}$$

$$\text{so} \quad \sin B = \frac{7 \sin A}{8}$$

$$= \frac{7 \times \sqrt{15}}{8 \times 4}$$

$$\therefore \quad B = \arcsin \frac{7\sqrt{15}}{32}$$

$$= 57.9°$$

$$C = 180° - A - B$$

$$= 180° - 75.5° - 57.9°$$

$$= 46.6°$$

> No angle is given so the sine rule cannot be used first. Instead, use the cosine rule to find one angle.

> $\cos A$ could be evaluated on the calculator. The method shown can lead to more elegant solutions.

> For $\sin A$ either use the calculator or, since $\cos A = \frac{1}{4}$, draw a triangle with $\cos A = \frac{1}{4}$ and use Pythagoras' theorem to find the third side.
>
> $$\sin A = \frac{\sqrt{15}}{4}$$
>
>

> Accurate values of A and B should be kept on the calculator and used to calculate C.

Example 13   In triangle LMN, with LN $= 6$ cm, MN $= 5$ cm and $N = 40°$, find angles L and M.

Using the cosine rule

$$\text{LM}^2 = 5^2 + 6^2 - 2 \times 5 \times 6 \times \cos 40°$$

$$= 25 + 36 - 60 \cos 40°$$

$$= 61 - 60 \cos 40°$$

$$\therefore \quad \text{LM} = \sqrt{61 - 60 \cos 40°} = 3.877\ldots$$

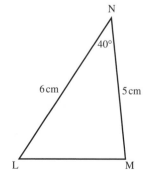

> The sine rule cannot be used as no side and corresponding angle are known. The cosine rule can be used to find LM.

> Store LM in memory for future use.

To find angles L and M, the sine rule can be used. However, when the sine rule is used to find an angle two possibilities are obtained, one acute and one obtuse. The obtuse angle may not be a possible solution.

Angle M (opposite the longest side) could be acute or obtuse.

Angle L (not opposite the longest side) must be acute, so, since it has only one possible value, angle L should be calculated first.

Using the sine rule: $\dfrac{\sin L}{5} = \dfrac{\sin 40°}{LM}$

> Recall LM from calculator memory or use original expression for LM.

$$\sin L = \dfrac{5 \sin 40°}{LM}$$

> L must be acute because it is not opposite the longest side.

$\therefore \qquad\qquad L = 56.0°$

Now, $\qquad\qquad N = 40°$

$\therefore \qquad\qquad M = 180° - 40° - 56.0° = 84.0°$

So angles L and M are 56.0° and 84.0° respectively.

Exercise 17F

Give answers correct to 3 significant figures unless stated otherwise.

1 Find all missing sides and angles in these triangles.

    a $A = 73.2°$ $B = 61.7°$ $c = 171\,\text{cm}$ **b** $a = 10\,\text{cm}$ $b = 12\,\text{cm}$ $c = 9\,\text{cm}$

    c $a = 136$ $B = 102°$ $C = 43°$ **d** $a = 31\,\text{m}$ $b = 42\,\text{m}$ $C = 104°$

    e $A = 28°$ $a = 8.5\,\text{m}$ $b = 14.8\,\text{m}$ **f** $a = 22$ $b = 62$ $c = 48$

2 Two points A and B on a straight coastline are 1 km apart, B being due east of A. If a ship is observed on bearings 167° and 205° from A and B respectively, what is its distance from the coastline?

3 A boat is sailing directly towards a cliff.
The angle of elevation of a point on the top of the cliff and straight ahead of the boat increases from 10° to 15° as the ship sails a distance of 50 m.
Find the height of the cliff.

4 A triangle has sides 10 cm, 11 cm and 15 cm.
By how much does its largest angle differ from a right angle?

5 A ship rounds a headland by sailing first 4 nautical miles on a course of 069° then 5 nautical miles on a course of 295°.
Calculate the distance and bearing of its new position from its original position.

6 A motorist travelling along a straight level road in the direction 053° observes a pylon on a bearing of 037°. 800 m further along the road the bearing of the pylon is 296°. Calculate the distance of the pylon from the road.

7 Two light ropes hang 1.5 m apart from the ceiling of a gym. One is 4 m long and just reaches the ground. The other is 3.8 m long. The ends of the ropes are pulled together so that, with the ropes taut, their free ends touch at P. Find

    a the angle between the free ends of the ropes at P

    b the height of P above the ground.

> *Questions 8–11 are of types that could have been set in the context of astronomy, stage (and other) design, surveying, sailing, engineering etc. A strength of mathematics is that the same problem-solving techniques can apply in so many fields.*

8 Two points, A and B, are marked, on level ground, 30 m apart and in line with a tower. The angles of elevation from A and B are 29° and 37° respectively. Find the height of the tower to the nearest metre.

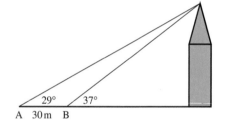

9 From the top of a building the angles of depression of the top and bottom of a tower 100 m tall are 41° and 52°. Assuming that the tower and the building are on level ground, find the height of the building, correct to the nearest metre.

10 From a point A, 25 m due south of a tree, the angle of elevation of the top of the tree is 27°. From a point B the bearing of the tree is 204° and the angle of elevation of its top is 18°. The points A, B and the foot of the tree are on level ground. Find, correct to the nearest 0.1 m

    a the height of the tree

    b the distance from B to the foot of the tree

    c the distance AB.

11 From a point A the bearing of the base of a tower 60 m high is 053° and the angle of elevation of the top of the tower is 16°. From a point B the bearing is 300° and the angle of elevation is 20°. Find the distance AB.

Exercise 17G (Review)

1 Find any missing angles, missing sides and the area of these triangles.

a
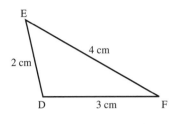

b
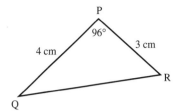

c
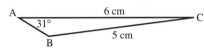

2 Find the area of these triangles, without using a calculator.

a
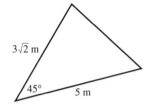

b
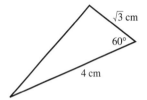

3 An arc AB of a circle centre O, radius 15 cm, subtends an angle of $\frac{2\pi}{3}$ at O. Find the length of arc AB and the area of sector AOB.

4 A chord PQ of a circle centre O, radius 10 cm, subtends an angle of $\frac{\pi}{4}$ at the centre of the circle. Find the length of the major arc PQ and the area of the minor segment cut off by the chord.

5 The lengths of the sides of a triangle are 10 cm, x cm and $(x-2)$ cm. The side of length $(x-2)$ cm is opposite an angle of 60°. Find x.

6 In triangle XYZ, $x = 29$ cm, $y = 21$ cm and $z = 20$ cm. Find

  a the area of the triangle

  b the length of the perpendicular from X to YZ.

7 The area of a sector of a circle, diameter 7 cm, is 18.375 cm². Find, without using a calculator, the length of the arc of the sector.

8 A flower bed is designed as follows: Triangle DEF is an equilateral triangle of side 8 m. Arcs EF, FD and DE have radii 8 m and centres D, E and F respectively. Find the area of the flower bed correct to the nearest 0.1 m².

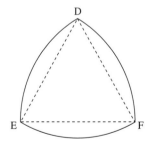

9 O is the centre of a circle radius r. AB and CD subtend angles $\dfrac{3\pi}{4}$ and $\dfrac{\pi}{4}$ at O.

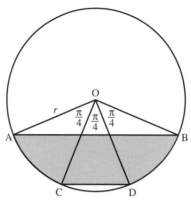

Show that the area shaded is equal to $\frac{1}{4}$ of the area of the circle.

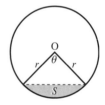 A 'Test yourself' exercise (Ty17) and an 'Extension exercise' (Ext17) can be found on the CD-ROM.

➤ Key points

Applications of trigonometry

When θ is measured in radians

■ area of sector $= \frac{1}{2}r^2\theta$

■ length of arc $= r\theta$

Area of triangle $= \frac{1}{2}ab\sin C$

Area of segment $S =$ Area of sector $-$ Area of triangle

$$= \tfrac{1}{2}r^2(\theta - \sin\theta)$$

Sine rule

$$\frac{a}{\sin A} = \frac{b}{\sin B} = \frac{c}{\sin C} \quad \text{or alternatively} \quad \frac{\sin A}{a} = \frac{\sin B}{b} = \frac{\sin C}{c}$$

(Remember the ambiguous case: if $\sin A = k < 1$, there may be two possible values for A.)

Cosine rule

$$a^2 = b^2 + c^2 - 2bc\cos A \quad \text{or alternatively} \quad \cos A = \frac{b^2 + c^2 - a^2}{2bc}$$

In multistage problems, avoid rounding errors by using full calculator values throughout.

Exam practice 3

1. A manufacturer needs to make a thin metal plate in the shape of a circular sector with perimeter 20 cm. The figure shows such a sector with radius r cm, angle θ radians and area A cm^2.

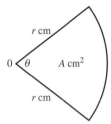

   a Prove that $A = 25 - (r - 5)^2$.

   Given that r can vary,

   b deduce the value of r for which A is a maximum and state the maximum value of A.

   Edexcel specimen

2. A circle C has equation

$$x^2 + y^2 - 10x + 6y - 15 = 0.$$

   a Find the coordinates of the centre of C.

   b Find the radius of C.

   Edexcel June 2001

3. Find the values of θ, to 1 decimal place, in the interval $-180 \leqslant \theta < 180$ for which

$$2\sin^2\theta° - 2\sin\theta° = \cos^2\theta°.$$

   Edexcel Jan 2002

4. $f(x) \equiv 2x^3 + x^2 - 8x - 4$.

   a Show that $(2x + 1)$ is a factor of $f(x)$.

   b Factorise $f(x)$ completely.

   c Hence find the values of x for which $f(x) = 0$.

   London May 1995

5. Find in degrees to 1 decimal place, the values of x which lie in the interval $-180° \leqslant x \leqslant 180°$ and satisfy the equation $\sin 2x = -0.57$.

   London May 1995

6. $f(x) = px^3 + 6x^2 + 12x + q$.

   Given that the remainder when $f(x)$ is divided by $(x - 1)$ is equal to the remainder when $f(x)$ is divided by $(2x + 1)$,

   a find the value of p.

   Given also that $q = 3$, and p has the value found in part **a**,

   b find the value of the remainder.

   Edexcel June 2003

7 There is a straight path of length $70\,\text{m}$ from the point A to the point B. The points are joined also by a railway track in the form of an arc of the circle whose centre is C and whose radius is $44\,\text{m}$, as shown in the figure.

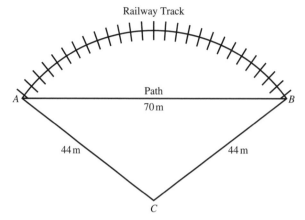

Railway Track

Path
70 m

44 m   44 m

C

a Show that the size, to 2 decimal places, of $\angle ACB$ is 1.84 radians.

Calculate

b the length of the railway track,

c the shortest distance from C to the path,

d the area of the region bounded by the railway track and the path.

London June 1996

8 Find, in degrees, the values of θ in the interval $0 \leqslant \theta \leqslant 360°$ for which

$$4\sin^2\theta - 2\sin\theta = 4\cos^2\theta - 1.$$

State which of your answers are exact and which are given to a degree of accuracy of your choice, which you should give.

Edexcel specimen

9 A circle C has centre $(3, 4)$ and radius $3\sqrt{2}$. A straight line l has equation $y = x + 3$.

a Write down an equation of the circle C.

b Calculate the exact coordinates of the two points where the line l intersects C, giving your answers in surds.

c Find the distance between these two points.

Edexcel Jan 2002

10 a Expand $(1 - 2x)^{10}$ in ascending powers of x up to and including the term in x^3, simplifying each coefficient in the expansion.

b Use your expansion to find an approximation to $(0.98)^{10}$, stating clearly the substitution which you have used for x.

London Jan 1997

11

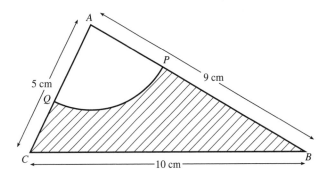

Triangle ABC has $AB = 9\,\text{cm}$, $BC = 10\,\text{cm}$ and $CA = 5\,\text{cm}$.

A circle, centre A and radius $3\,\text{cm}$, intersects AB and AC at P and Q respectively, as shown in the figure.

a Show that, to 3 decimal places, $\angle BAC = 1.504$ radians.

Calculate,

b the area, in cm^2, of the sector APQ,

c the area, in cm^2, of the shaded region $BPQC$,

d the perimeter, in cm, of the shaded region $BPQC$.

Edexcel Jan 2001

12 Given that $f(x) = 15 - 7x - 2x^2$,

a find the coordinates of all the points at which the graph of $y = f(x)$ crosses the coordinates axes.

b Sketch the graph of $y = f(x)$.

c Calculate the coordinates of the stationary point of $f(x)$.

Edexcel May 2002

13 The coefficient of x^2 in the binomial expansion of $\left(1 + \dfrac{x}{2}\right)^n$, where n is a positive integer, is 7.

a Find the value of n.

b Using the value of n found in **a**, find the coefficient of x^4.

London June 1998

14 a Write down the first four terms of the binomial expansion, in ascending powers of x, of $(1 + 3x)^n$, where $n > 2$.

Given that the coefficient of x^3 in this expansion is ten times the coefficient of x^2,

b find the value of n,

c find the coefficient of x^4 in the expansion.

Edexcel June 2002

15 $f(x) \equiv ax^3 + bx^2 - 7x + 14$, where a and b are constants.

Given that when $f(x)$ is divided by $(x - 1)$ the remainder is 9,

a write down an equation connecting a and b.

Given also that $(x + 2)$ is a factor of $f(x)$,

b find the values of a and b.

Edexcel June 2001

16 Given that
$$(2 - x)^{13} \equiv A + Bx + Cx^2 + \ldots,$$
find the values of the integers A, B and C.

London June 1996

17 a Expand $(3 + 2x)^4$ in ascending powers of x, giving each coefficient as an integer.

b Hence, or otherwise, write down the expansion of $(3 - 2x)^4$ in ascending powers of x.

c Hence by choosing a suitable value for x show that $(3 + 2\sqrt{2})^4 + (3 - 2\sqrt{2})^4$ is an integer and state its value.

London Jan 1998

18 Find, in degrees to the nearest tenth of a degree, the values of x for which
$$\sin x \tan x = 4, \quad 0 \leqslant x < 360°.$$

Edexcel Mock paper 1

19 The function f, defined for $x \in \mathbb{R}$, $x > 0$, is such that
$$f'(x) = x^2 - 2 + \frac{1}{x^2}.$$

a Find the value of $f''(x)$ at $x = 4$.

b Given that $f(3) = 0$, find $f(x)$.

c Prove that f is an increasing function.

Edexcel June 2001

20 Expand $\left(x - \dfrac{1}{x} \right)^5$, simplifying the coefficients.

London May 1995

21 On a journey, the average speed of a car is $v \, \text{m s}^{-1}$. For $v \geqslant 5$, the cost per kilometre, C pence, of the journey is modelled by
$$C = \frac{160}{v} + \frac{v^2}{100}$$

Using this model,

a show, by calculus, that there is a value of v for which C has a stationary value, and find this value of v.

b Justify that this value of v gives a minimum value of C.

c Find the minimum value of C and hence find the minimum cost of a 250 km car journey.

Edexcel Jan 2003

22 i Solve, for $0° < x < 180°$, the equation
$$\sin (2x + 50°) = 0.6,$$
giving your answers to 1 decimal place.

ii In the triangle ABC, $AC = 18$ cm, $\angle ABC = 60°$ and $\sin A = \frac{1}{3}$.

a Use the sine rule to show that $BC = 4\sqrt{3}$.

b Find the exact value of $\cos A$.

Edexcel Nov 2002

23 An architect is drawing up plans for a mini-theatre. The diagram shows the plan of the base which consists of a rectangle of length $2y$ metres and width $2x$ metres and a semicircle of radius x metres which is placed with one side of the rectangle as diameter.

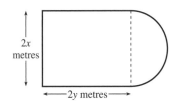

Find in terms of x and y, expressions for

a the perimeter of the base, **b** the area of the base.

The architect decides the base should have a perimeter of 100 metres.

c Show that the area A square metres of the base is given by
$$A = 100x - 2x^2 - \tfrac{1}{2}\pi x^2.$$

d Given that x can vary, find the value of x for which $\dfrac{\mathrm{d}A}{\mathrm{d}x} = 0$ and determine the corresponding value of y, giving your answers to 2 significant figures.

e Find the maximum value of A and explain why this value is a maximum.

Edexcel Specimen paper

24 $f(x) \equiv 2x^3 + 5x^2 - 8x - 15$.

a Show that $(x + 3)$ is a factor of $f(x)$.

b Hence factorise $f(x)$ as the product of a linear factor and a quadratic factor.

c Find, to 2 decimal places, the two other values of x for which $f(x) = 0$.

Edexcel Specimen paper

25 For the curve C with equation $y = x^4 - 8x^2 + 3$,

a find $\dfrac{\mathrm{d}y}{\mathrm{d}x}$,

b find the coordinates of each of the stationary points,

c determine the nature of each stationary point.

The point A, on the curve C, has x-coordinate 1.

d Find an equation for the normal to C at A, giving your answer in the form $ax + by + c = 0$, where a, b and c are integers.

Edexcel June 2003

26 $f(n) = n^3 + pn^2 + 11n + 9$, where p is a constant.

a Given that $f(n)$ has a remainder of 3 when it is divided by $(n + 2)$, prove that $p = 6$.

b Show that $f(n)$ can be written in the form $(n + 2)(n + q)(n + r) + 3$, where q and r are integers to be found.

c Hence show that $f(n)$ is divisible by 3 for all positive integer values of n.

Edexcel Jan 2003

27 Find, in degrees, the value of θ in the interval $0 \leqslant \theta < 360°$ for which
$$2\cos^2\theta - \cos\theta - 1 = \sin^2\theta.$$

Give your answers to 1 decimal place where appropriate.

Edexcel June 2003

28 The first three terms in the expansion, in ascending powers of x, of $(1 + px)^n$, are $1 - 18x + 36p^2x^2$. Given that n is a positive integer, find the value of n and the value of p.

Edexcel Jan 2003

29 $f(x) = x^3 + ax^2 + bx - 10$, where a and b are constants.

When $f(x)$ is divided by $(x - 3)$, the remainder is 14.

When $f(x)$ is divided by $(x + 1)$, the remainder is -18.

  a Find the value of a and the value of b.

  b Show that $(x - 2)$ is a factor of $f(x)$.

Edexcel May 2002

30 **a** Sketch, for $0 \leqslant x \leqslant 360°$, the graph of $y = \sin(x + 30°)$.

  b Write down the coordinates of the points at which the graph meets the axes.

  c Solve, for $0 \leqslant x < 360°$, the equation

$$\sin(x + 30°) = -\tfrac{1}{2}.$$

Edexcel Jan 2003

31 **a** Using the factor theorem, show that $(x + 3)$ is a factor of

$$x^3 - 3x^2 - 10x + 24.$$

  b Factorise $x^3 - 3x^2 - 10x + 24$ completely.

Edexcel Nov 2002

32 Find all values of θ in the interval $0 \leqslant \theta < 360$ for which

  a $\cos(\theta + 75)° = 0.5$

  b $\sin 2\theta° = 0.7$, giving your answers to one decimal place.

Edexcel Jan 2001

33 A circle C has equation
$$x^2 + y^2 - 6x + 8y - 75 = 0.$$

  a Write down the coordinates of the centre of C, and calculate the radius of C.

A second circle has centre at the point (15, 12) and radius 10.

  b Sketch both circles on a single diagram and find the coordinates of the point where they touch.

Edexcel June 2003

34 The circle C has equation $x^2 + y^2 - 8x - 16y - 209 = 0$.

  a Find the coordinates of the centre of C and the radius of C.

The point $P(x, y)$ lies on C.

  b Find, in terms of x and y, the gradient of the tangent to C at P.

  c Hence, or otherwise, find an equation of the tangent to C at the point (21, 8).

Edexcel May 2002

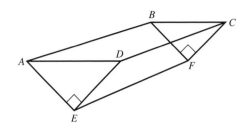

35 The diagram shows an open tank for storing water, *ABCDEF*. The sides *ABFE* and *CDEF* are rectangles. The triangular ends, *ADE* and *BCF* are isosceles and $\angle AED = \angle BFC = 90°$. The ends *ADE* and *BCF* are vertical and *EF* is horizontal. Given that $AD = x$ metres,

a show that the area of $\triangle ADE$ is $\frac{1}{4}x^2$ m².

Given also that the capacity of the container is 4000 m³ and that the total area of the two triangular and two rectangular sides of the container is S m²,

b show that $S = \dfrac{x^2}{2} + \dfrac{16\,000\,\sqrt{2}}{x}$.

Given that x can vary,

c use calculus to find the minimum value of S,

d justify that the value of S you have found is a minimum.

London June 1998

36

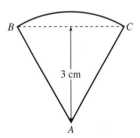

The shape of a badge is a sector *ABC* of a circle with centre *A* and radius *AB*, as shown in the diagram. The triangle *ABC* is equilateral and has perpendicular height 3 cm.

a Find, in surd form, the length of *AB*.

b Find in terms of π, the area of the badge.

c Prove that the perimeter of the badge is $\dfrac{2\sqrt{3}}{3}(\pi + 6)$ cm.

Edexcel May 2002

37 The curve with equation $y = (2x + 1)(x^2 - k)$, where k is a constant, has a stationary point where $x = 1$.

a Determine the value of k.

b Find the coordinates of the stationary points and determine the nature of each.

c Sketch the curve and mark on your sketch the coordinates of the points where the curve crosses the coordinate axes.

Edexcel Mock paper 1

38 The expansion of $(2 - px)^6$ in ascending powers of x, as far as the term in x^2, is

$$64 + Ax + 135x^2.$$

Given that $p > 0$, find the value of p and the value of A.

Edexcel June 2003

Exponentials and logarithms

In previous chapters, we have looked at functions of the form $y = x^n$, such as $y = x$, $y = x^2$, $y = x^{\frac{3}{2}}$. In all these, the variable x was the base and the exponent or index was a rational number. In this chapter, we look at functions where the base is a positive number and the variable is in the exponent. These exponential functions, such as $y = 2^x$, $y = 3^x$, and their inverse functions, logarithmic functions, occur frequently in the physical world. They can be found in such varied fields as: measuring the strength of an earthquake, wave propagation, measuring intensity of sound, and the rate of cooling of an engine.

After working through this chapter you should be

■ able to convert from index to logarithmic form and vice versa

■ able to apply the laws of logarithms

■ familiar with exponential graphs

■ able to solve equations of the form $a^x = b$.

By the beginning of the seventeenth century, there was a pressing need for easier methods of computation, especially in astronomy and navigation. The concept of, for example, multiplying numbers simply by adding (as is the case with logarithms) was important and exciting.

The idea was developed by John Napier, a Scotsman, and independently, but at the same time, by Joost Bürgi, a Swiss. For the following 300 years, logarithms were the main method of computation.

Logarithms have now been superseded by calculators and computers. The concept of logarithms led, however, to the development of new functions which play an important role in higher mathematics.

18.1 Definition of a logarithm

The **logarithm (log)** of a positive number to a given **base** is the **power** to which the base must be raised to equal the given number.

$$a^b = c \Leftrightarrow \log_a c = b$$

Say 'log to the base a of c is b'.

Assume $a > 0$ and $a \neq 1$.

For example, $10^2 = 100 \Leftrightarrow \log_{10} 100 = 2$

Example 1 **a** As $2^5 = 32$, the log to the base 2 of 32 is 5

or $2^5 = 32 \Leftrightarrow \log_2 32 = 5$

b As $10^4 = 10\,000$, the log to the base 10 of 10 000 is 4

or $10^4 = 10\,000 \Leftrightarrow \log_{10} 10\,000 = 4$.

The index form ($10^4 = 10\,000$) and the log form ($\log_{10} 10\,000 = 4$) are equivalent.
The two forms say the same thing in two different ways.
It is important to be able to convert readily from one form to the other.

Exercise 18A *This exercise may be done orally.*

1 What are the bases and logarithms in these statements?

a $10^2 = 100$ **b** $10^{1.6021} \approx 40$ **c** $9 = 3^2$ **d** $4^3 = 64$

e $1 = 2^0$ **f** $8 = (\frac{1}{2})^{-3}$ **g** $a^b = c$

2 Express these statements in logarithmic notation.

a $2^4 = 16$ **b** $27 = 3^3$ **c** $125 = 5^3$

d $10^6 = 1\,000\,000$ **e** $1728 = 12^3$ **f** $64 = 16^{\frac{3}{2}}$

g $10^4 = 10\,000$ **h** $4^0 = 1$ **i** $0.01 = 10^{-2}$

j $\frac{1}{2} = 2^{-1}$ **k** $9^{\frac{3}{2}} = 27$ **l** $8^{-\frac{2}{3}} = \frac{1}{4}$

m $81 = (\frac{1}{3})^{-4}$ **n** $e^0 = 1$ **o** $16^{-\frac{1}{4}} = \frac{1}{2}$

p $(\frac{1}{8})^0 = 1$ **q** $27 = 81^{\frac{3}{4}}$ **r** $4 = (\frac{1}{16})^{-\frac{1}{2}}$

s $c = a^5$ **t** $a^3 = b$ **u** $p^q = r$

3 Express these in index notation.

a $\log_2 32 = 5$ **b** $\log_3 9 = 2$ **c** $2 = \log_5 25$

d $\log_{10} 100\,000 = 5$ **e** $7 = \log_2 128$ **f** $\log_9 1 = 0$

g $-2 = \log_3 \frac{1}{9}$ **h** $\log_4 2 = \frac{1}{2}$ **i** $\log_e 1 = 0$

j $\log_{27} 3 = \frac{1}{3}$ **k** $2 = \log_a x$ **l** $\log_3 a = b$

m $\log_a 8 = c$ **n** $y = \log_x z$ **o** $\log_q r = p$

4 Find x.

a $\log_2 64 = x$ **b** $\log_{10} 100 = x$ **c** $\log_{10} 10^7 = x$

d $\log_a a^2 = x$ **e** $\log_8 2 = x$ **f** $\log_4 1 = x$

5 Evaluate these.

a $\log_{27} 3$ **b** $\log_5 125$ **c** $\log_e e^3$

d $\log_e \frac{1}{e}$ **e** $\log_a \sqrt{a}$ **f** $\log_{2a} 4a^2$

18.2 Combining logs

As a log is a power (or index), the rules for combining powers (or indices) apply.

> **For any base c, the rules are**
> - $\log_c ab = \log_c a + \log_c b$
> - $\log_c \dfrac{a}{b} = \log_c a - \log_c b$
> - $\log_c a^n = n \log_c a$

Note: $\log a^n = \log (a^n)$.

To prove these rules, let c be any base ($c > 0$, $c \neq 1$) and let $x = \log_c a$ and $y = \log_c b$

$\log_c a = x \Leftrightarrow a = c^x \qquad$ ①

$\log_c b = y \Leftrightarrow b = c^y \qquad$ ②

To prove $\quad \log_c ab = \log_c a + \log_c b$

$\qquad$ From ① and ② $\qquad ab = c^x c^y$

Note: To *multiply* numbers, *add* indices or logs.

$\qquad\qquad\qquad\qquad\qquad = c^{x+y}$

$\qquad$ So $\qquad\qquad\qquad \log_c ab = x + y$

$\qquad\qquad\qquad\qquad\qquad\qquad = \log_c a + \log_c b$

To prove $\quad \log_c \dfrac{a}{b} = \log_c a - \log_c b$

$\qquad$ From ① and ② $\qquad \dfrac{a}{b} = \dfrac{c^x}{c^y}$

Note: To *divide* numbers, *subtract* indices or logs.

$\qquad\qquad\qquad\qquad\qquad = c^{x-y}$

$\qquad$ So $\qquad\qquad\qquad \log_c \dfrac{a}{b} = x - y$

$\qquad\qquad\qquad\qquad\qquad\qquad = \log_c a - \log_c b$

To prove $\quad \log_c a^n = n \log_c a$

$\qquad$ From ① $\qquad\qquad a^n = (c^x)^n$

Note: To *raise to a power*, *multiply* by the index or log.

$\qquad\qquad\qquad\qquad\qquad = c^{nx}$

$\qquad$ So $\qquad\qquad\qquad \log_c a^n = nx$

$\qquad\qquad\qquad\qquad\qquad\qquad = n \log_c a$

Considering -1, 0 and 1 as powers of a, gives three important results, which are useful when simplifying logs.

> **For any base a ($a > 0$, $a \neq 1$)**
>
> ▪ $a^1 = a \Rightarrow \log_a a = 1$
> ▪ $a^0 = 1 \Rightarrow \log_a 1 = 0$
> ▪ $a^{-1} = \frac{1}{a} \Rightarrow \log_a \left(\frac{1}{a}\right) = -1$

Note When the same base is used throughout a piece of work, the base can be omitted, as here.

Example 2 Express $\log 45 - 2 \log 3$ in terms of a single logarithm.

$$\begin{aligned} \log 45 - 2 \log 3 &= \log 45 - \log 3^2 \\ &= \log 45 - \log 9 \\ &= \log \tfrac{45}{9} \\ &= \log 5 \end{aligned}$$

> $\log a^n = n \log a \Rightarrow 2 \log 3 = \log 3^2$

> $\log \dfrac{a}{b} = \log a - \log b \Rightarrow \log 45 - \log 9 = \log \tfrac{45}{9}$

Example 3 Express $3 \log_{10} a + \frac{1}{2} \log_{10} b - 2$ as the logarithm of a single term.

$$3 \log_{10} a = \log_{10} a^3$$

> Using $\log a^n = n \log a$

$$\tfrac{1}{2} \log_{10} b = \log_{10} b^{\frac{1}{2}} = \log_{10} \sqrt{b}$$

$$2 = \log_{10} 10^2 = \log_{10} 100$$

So $\quad 3 \log_{10} a + \tfrac{1}{2} \log_{10} b - 2 = \log_{10} a^3 + \log_{10} \sqrt{b} - \log_{10} 100$

$$\begin{aligned} &= \log_{10} (a^3 \sqrt{b}) - \log_{10} 100 \\ &= \log_{10} \left(\tfrac{a^3 \sqrt{b}}{100}\right) \end{aligned}$$

> Using $\log ab = \log a + \log b$

> Using $\log \dfrac{a}{b} = \log a - \log b$

Example 4 Express $\log \sqrt{\dfrac{a^4 b}{3c^3}}$ in terms of $\log a$, $\log b$ and $\log c$.

$$\begin{aligned} \log \sqrt{\frac{a^4 b}{3c^3}} &= \log \left(\frac{a^4 b}{3c^3}\right)^{\frac{1}{2}} \\ &= \tfrac{1}{2} \log \left(\frac{a^4 b}{3c^3}\right) \\ &= \tfrac{1}{2} (\log a^4 b - \log 3c^3) \\ &= \tfrac{1}{2} (\log a^4 + \log b - \log 3 - \log c^3) \\ &= \tfrac{1}{2} (4 \log a + \log b - \log 3 - 3 \log c) \end{aligned}$$

> Use index form.

> Using $\log a^n = n \log a$

> Using $\log \dfrac{a}{b} = \log a - \log b$

> Using $\log ab = \log a + \log b$

> Using $\log a^n = n \log a$

Example 5 Simplify $\dfrac{\log 64}{\log 32}$.

$$\frac{\log 64}{\log 32} = \frac{\log 2^6}{\log 2^5}$$

$$= \frac{6\log 2}{5\log 2}$$

$$= \tfrac{6}{5}$$

> 64 and 32 are powers of the same number, 2.
> By expressing them both as powers of 2 and
> using the log rules, the expression can be simplified.

Exercise 18B

In this exercise, the base of the logarithms is omitted unless it has a special bearing on the question. Note: $\lg x$ *means* $\log_{10} x$.

1 Express in terms of $\log a$, $\log b$ and $\log c$

  a $\log ab$ **b** $\log \dfrac{a}{c}$ **c** $\log \dfrac{1}{b}$ **d** $\log a^2 b^{\frac{3}{2}}$

  e $\log \dfrac{1}{b^4}$ **f** $\log \dfrac{a^{\frac{1}{3}}b^4}{c^3}$ **g** $\log \sqrt{a}$ **h** $\log \sqrt[3]{b}$

  i $\log \sqrt{(ab)}$ **j** $\lg(10a)$ **k** $\lg \dfrac{1}{100b^2}$ **l** $\log \sqrt{\left(\dfrac{a}{b}\right)}$

2 Express these as single logarithms.

  a $\log 2 + \log 3$ **b** $\log 18 - \log 9$

  c $\log 4 + 2\log 3 - \log 6$ **d** $3\log 2 + 2\log 3 - 2\log 6$

  e $\log c + \log a$ **f** $\log x + \log y - \log z$

  g $2\log a - \log b$ **h** $2\log a + 3\log b - \log c$

  i $\frac{1}{2}\log x - \frac{1}{2}\log y$ **j** $\log p - \frac{1}{3}\log q$

  k $2 + 3\lg a$ **l** $1 + \lg a - \frac{1}{2}\lg b$

  m $2\lg a - 3 - \lg 2c$ **n** $3\lg x - \frac{1}{2}\lg y + 1$

  o $a\log b - b\log a$

3 Simplify

  a $\lg 1000$ **b** $\frac{1}{2}\log_3 81$ **c** $\frac{1}{3}\log_2 64$ **d** $-\log_2 \frac{1}{2}$

  e $\frac{1}{3}\log 8$ **f** $\frac{1}{2}\log 49$ **g** $-\frac{1}{2}\log 4$ **h** $3\log 3 - \log 27$

  i $5\log 2 - \log 32$ **j** $\dfrac{\log 8}{\log 2}$ **k** $\dfrac{\log 81}{\log 9}$ **l** $\dfrac{\log 49}{\log 343}$

18.3   Exponentials and their graphs

The word 'exponential', when applied to an increase, means 'more and more rapid'. Exponential graphs are very steep, steeper eventually than any polynomial graph.

Compare $y = x^2$ and $y = 2^x$.

For $y = x^2$, the base, x, is the variable; the power (or exponent), 2, is constant.

For $y = 2^x$, the base, 2, is constant; the power (or exponent), x, is the variable.

$y = x^2$ is a quadratic function.
$y = 2^x$ is an exponential function.

Since $y = 2^x \Leftrightarrow x = \log_2 y$, 'taking a log' and 'raising to a power' are the reverse processes of each other: one 'undoes' the other. (Pairs of functions such as these are called 'inverse' functions of each other.)

Try an example on a calculator:

$$\log_{10} 20 = 1.301\ldots \text{ and } 10^{1.301\ldots} = 20$$

An **exponential function** has the form $y = a^x$. This is not meaningful for all values of a. Negative values of a are excluded. Imagine plotting $y = (-4)^x$. The function could be evaluated for integer values of x but not for non-integers. Also when $a = 0$ or $a = 1$ a^x has a constant value. So $y = a^x$ is restricted to $a > 0$, $a \neq 1$.

When $x = 0$, $y = 1$ for all values of a, so all graphs of $y = a^x$ will pass through the point $(1, 0)$. Also all values of y will be positive.

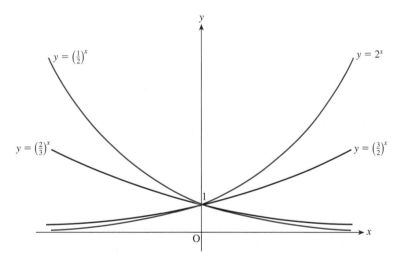

Note   Comparing $y = 2^x$ and $y = (\frac{1}{2})^x = 2^{-x}$, x has been replaced by $(-x)$.

The graphs are reflections of each other in the y-axis.

Note   The transformations of Chapter 7 can be applied to exponential curves. For example $y = 2^{x+3}$ would be a translation of 3 units to the left of $y = 2^x$. When $x = 0$, $y = 2^3 = 8$, so the graph cuts the y-axis at $(0, 8)$.

Notice from the graph of $y = a^x$ that $y > 0$ for all values of x, i.e. 'raising to a power' always gives a *positive* result.

So for the inverse process, 'taking the log', one can only take a log of a *positive* number. The log of a negative number is undefined.

For both exponential functions and log functions, the base, a, is greater than 0 and not equal to 1, i.e. $a > 0$, $a \neq 1$.

Graphs of log functions will be introduced later in the A-level course.

Notice also that since 'taking a log' and 'raising to a power' are inverse processes

$$a^x = N \Leftrightarrow x = \log_a N$$

So $\qquad N = a^{\log_a N}$

or $\qquad x = \log_a a^x$

This gives two more important results:

> **Log functions and exponential functions are inverses of each other. One 'undoes' the other.**
>
> **For any base a $(a > 0, a \neq 1)$**
> - $\log_a a^x = x$
> - $a^{\log_a N} = N$

The most important bases for logarithms are 10 and e, a number which is introduced in detail later in the A-level course. The integer powers of 10 are convenient in the decimal system, while e arises naturally in many areas of mathematics and science. Logs to base e, abbreviated to ln, are called natural or Napierian logarithms (after John Napier).

Example 6   Simplify   **a** $3^{\log_3 4}$ **b** $\log_2 2^7$ **c** $\ln e^{2x+1}$

a $3^{\log_3 4} = 4$

Using $a^{\log_a N} = N$

b $\log_2 2^7 = 7$

Using $\log_a a^x = x$

c $\ln e^{2x+1} = \log_e e^{2x+1}$

Remembering ln is $\log_e$

$\qquad = 2x + 1$

Using $\log_a a^x = x$

To change the base of a log

Calculators only evaluate $\log_{10}$ (lg or log) and $\log_e$ (ln), but it is possible to change the base of a logarithm so that any log can be expressed in one of these two bases (or in any other base).

$$\log_a b = x \Leftrightarrow a^x = b$$

Taking logs to base c of both sides of $a^x = b$

> Doing the same to both sides does not affect the equality.

$$\log_c a^x = \log_c b$$

$$x \log_c a = \log_c b$$

$$x = \frac{\log_c b}{\log_c a}$$

So $\log_a b = \dfrac{\log_c b}{\log_c a}$

> **To change the base of a log use** $\log_a b = \dfrac{\log_c b}{\log_c a}$

Example 7   Find $\log_2 7$ by changing the base of the log.

$$\log_2 7 = \frac{\log_{10} 7}{\log_{10} 2}$$

$$= 2.81 \ (3 \text{ s.f.})$$

> On a calculator, $\log_2$ cannot be evaluated directly, but $\log_{10}$ (lg) or $\log_e$ (ln) can. Here, base 10 has been used and the two logs have been divided on the calculator.
> Check mentally: $2^2 = 4$, $2^3 = 8$, so the answer should lie between 2 and 3.

Exercise 18C   *Note:* ln *means* $\log_e$

1   Evaluate or simplify

    a $\log_a a^2$ **b** $x^{\log_x y}$ **c** $\log_3 3^5$

    d $\log_2 2^n$ **e** $x^{\log_x 7}$ **f** $\log_a a^n$

    g $3^{\log_3 5}$ **h** $\ln e^5$ **i** $\ln e^x$

    j $\log_7 1$ **k** $\log_5 5^{2x+1}$ **l** $\ln e^e$

    m $\log_3 (\log_4 4^3)$ **n** $\log_x (\log_y y^x)$

2   Evaluate, correct to 3 decimal places, by changing the base of the log

    a $\log_6 7$ **b** $\log_3 \frac{1}{2}$ **c** $\log_{0.5} 9$ **d** $\log_2 3$

    e $\log_{0.25} 6$ **f** $\log_8 72.4$ **g** $\log_5 4$ **h** $\log_{0.7} 0.1$

Solving equations involving indices and logs

Example 8   Solve $3^{2x+1} = 80$.

$$3^{2x+1} = 80$$

Taking logs to base 10 of both sides

$$(2x+1)\log_{10} 3 = \log_{10} 80$$

$$2x+1 = \frac{\log_{10} 80}{\log_{10} 3}$$

$$2x = \frac{\log_{10} 80}{\log_{10} 3} - 1$$

$$x = \frac{1}{2}\left(\frac{\log_{10} 80}{\log_{10} 3} - 1\right)$$

$$= 1.49 \ (3 \text{ s.f.})$$

Example 9   Solve $3^{2x+1} - 3^x - 24 = 0$.

> $3^{2x+1} = 3^{2x} \times 3^1 = (3^x)^2 \times 3$

Let $y = 3^x$

Then $3y^2 - y - 24 = 0$

$$(3y+8)(y-3) = 0$$

$$\Rightarrow \qquad y = -\tfrac{8}{3} \text{ or } y = 3$$

> If the substitution $y = 3^x$ is not used, the equation factorises as $(3^{x+1}+8)(3^x-3) = 0$ or $(3 \times 3^x + 8)(3^x - 3) = 0$.

Substituting for y

$$3^x = -\tfrac{8}{3} \text{ or } 3^x = 3$$

There is no solution for $3^x < 0$ but $3^x = 3$ gives $x = 1$.

Example 10   Find the smallest integer value of n such that $1 - 0.7^n > 0.998$.

$$1 - 0.7^n > 0.998$$

$$\Rightarrow \qquad 0.7^n < 1 - 0.998$$

> Rearranging

$$< 0.002$$

Taking logs to base 10 of both sides

$$n\log_{10} 0.7 < \log_{10} 0.002$$

> Divide by $\log 0.7$

$$n > \frac{\log_{10} 0.002}{\log_{10} 0.7}$$

> The inequality sign must be reversed as $\log_{10} 0.7$ is negative.

$$\therefore \qquad n > 17.4\ldots$$

The smallest integer n which satifies this inequality is 18.

Example 11   Solve the simultaneous equations $\log_2 x + \log_2 y = 3$ and $\log_4 x - \log_4 y = -\frac{1}{2}$.

$$\log_2 x + \log_2 y = 3 \quad \text{①}$$

$$\log_4 x - \log_4 y = -\tfrac{1}{2} \quad \text{②}$$

From ① $\log_2 xy = 3$

$$2^3 = xy$$

So $xy = 8 \quad \text{③}$

From ② $\log_4 \dfrac{x}{y} = -\dfrac{1}{2}$

$$4^{-\frac{1}{2}} = \dfrac{x}{y}$$

So $\dfrac{x}{y} = \dfrac{1}{2} \quad \text{④}$

③ $\times$ ④ gives $x^2 = 4$

x must be positive for its log to exist.

$\therefore \quad x = 2$ and $y = 4$.

> Simultaneous equations are usually solved by addition or subtraction. Here the equations are multiplied.

Example 12   Solve $3 + 2\log_2 x = \log_2(10x - 3)$

$$3 + 2\log_2 x = \log_2(10x - 3)$$

> Use $n\log a = \log a^n$.

$$\log_2 x^2 - \log_2(10x - 3) = -3$$

> Combine log terms using $\log a - \log b = \log \frac{a}{b}$

$$\log_2 \dfrac{x^2}{10x - 3} = -3$$

Therefore $\dfrac{x^2}{10x - 3} = 2^{-3}$

> Use $\log_a c = b \Leftrightarrow a^b = c$.

$$\dfrac{x^2}{10x - 3} = \dfrac{1}{8}$$

> Cross-mulitply.

So $8x^2 = 10x - 3$

$$8x^2 - 10x + 3 = 0$$

$$(2x - 1)(4x - 3) = 0$$

Therefore $x = \tfrac{1}{2}$ or $x = \tfrac{3}{4}$

> Since only logs of positive quantities are defined check that the solutions do not lead to logs of negative quantities. These values of x are valid solutions.

In this exercise, use logs only when necessary.

1 Solve for x.

a $2^x = 64$ **b** $2^x = \frac{1}{32}$ **c** $3^x = 64$ **d** $(0.2)^x = 125$

e $3^{4x} = 2$ **f** $3^{x-2} = 27$ **g** $3^{x-2} = 28$ **h** $5^{3x} = 7$

i $7^{-x} = 1$ **j** $\log_2 x = 4$ **k** $\log_x 4 = 2$ **l** $\log_4 2 = x$

2 Solve for x.

a $2^{x+1} = 5^x$ **b** $7^{x-3} = 4^{2x}$ **c** $2^x \times 2^{x+1} = 10$

d $\dfrac{1}{2^x} = 6$ **e** $\left(\dfrac{2}{3}\right)^x = \dfrac{1}{16}$ **f** $7^x = 6^{-x}$

3 Solve for x.

a $2^{2x} - 2^x - 6 = 0$ **b** $2^{2x} - 6 \times 2^x + 8 = 0$

c $3^{2x} - 10 \times 3^x + 9 = 0$ **d** $3^{2x+1} - 26 \times 3^x - 9 = 0$

e $2(2^{2x}) - 5(2^x) + 2 = 0$ **f** $3^{2x+1} - 4(3^x) + 1 = 0$

g $\log_2 (x^2 + 4) = 2 + \log_2 x$ **h** $3 + \log_2 (2x - 5) = \log_2 (5x - 7)$

i $9^x - 2 \times 3^{x+1} + 8 = 0$ **j** $4^x - 6(2^x) + 5 = 0$

4 Solve, giving the answers exactly or correct to 3 significant figures.

a $4^x > 64$ **b** $4^x < 100$ **c** $3^{2x-7} > 1000$

d $0.5^x < 0.02$ **e** $0.4^{2x+1} < 256$ **f** $0.7^{1-x} < 48$

5 Find the smallest integer n such that $3^n > 1\,234\,567$.

6 Find the largest integer n such that $0.2^n > 0.000\,05$.

7 Solve the equation $1 - 2\log_2 x = \log_2 (5x - 12)$

8 Given that $\log_3 x = k$, find, in terms of k,

a $\log_3 (x^2)$ **b** $\log_3 (9x)$ **c** $\log_3 \left(\dfrac{x^{\frac{1}{2}}}{3}\right)$

d Hence, or otherwise, solve $\log_3 (x^2) + \log_3 \left(\dfrac{x^{\frac{1}{2}}}{3}\right) - \log_3 (9x) = 0$

9 Show that the equation $\log_3 (3x - 1) - 2\log_3 x = 1$ has no real roots.

10 Sketch $y = 2^x$ and, on the same axes, sketch

    a $y = 2^{-x}$ **b** $y = 2^{2x}$ **c** $y = 3 \times 2^x$

    In each case describe the transformation of $y = 2^x$ and state the coordinates of the point of intersection of the curve with the y-axis.

11 **a** Find the coordinates of the point of intersection of the graphs of $y = 3^{2x}$ and $y = 3^{x+1}$.

    b Sketch the graphs showing clearly their point of intersection.

Exercise 18E (Review)

In this exercise, for Question 6, remember that lg *stands for* $\log_{10}$.

1 Express in logarithmic notation

    a $2^5 = 32$ **b** $10^2 = 100$ **c** $a^b = c$

    d $p^3 = q$ **e** $3 = 27^{\frac{1}{3}}$ **f** $\frac{1}{3} = 3^{-1}$

2 Express in index notation

    a $\log_2 8 = 3$ **b** $\log_6 36 = 2$ **c** $\log_a b = c$

    d $\log_d c^4 = f$ **e** $\log_p 8 = 4$ **f** $q = \log_c 3$

3 Evaluate

    a $\log_2 128$ **b** $\log_{10} 1000$ **c** $\log_p p^4$

    d $\log_2 16$ **e** $\log_{16} 2$ **f** $\log_e \dfrac{1}{e^2}$

4 Express in terms of $\log a$, $\log b$ and $\log c$

    a $\log \dfrac{ab^2}{c}$ **b** $\log \sqrt{ab}$ **c** $\log \dfrac{a^4}{b^2 c}$ **d** $\log a^2 b^3 c^4$

5 Express each of these as a single logarithm.

    a $\log(a^2 b) - \log(b^2 a)$ **b** $\frac{1}{2}\log x + 3\log y$

    c $\log 5 + 2\log 10 - 3\log 2$ **d** $a\log x + a\log x^2$

6 Simplify

    a $\lg 75 + 2\lg 2 - \lg 3$ **b** $\lg 1\,000\,000$ **c** $\frac{1}{2}\log 64$

    d $5\log 3 - \log 81$ **e** $\dfrac{\log 16}{\log 2}$ **f** $\dfrac{\log 5}{\log 125}$

7 Solve these.

    a $\log_{10}(n^2 - 90n) = 3$ **b** $9^x = 27^{\frac{3}{4}}$

    c $3^x = 4$ **d** $5^{2x+1} = 25$

    e $5^{2x} - 6 \times 5^x + 5 = 0$ **f** $\log_2(y^2 + 7y) = 3$

8 Solve, giving the answers exactly or to 3 significant figures.

    a $2^x < 50$ **b** $2^x > 1024$

9 The roots of the equation $2\log_4 x - \log_4(7x-3) = -\frac{1}{2}$ are α and β, where $\alpha < \beta$.

    a Find α and β and show that $\log_4 \alpha = -\frac{1}{2}$.

    b Calculate $\log_4 \beta$ to 3 significant figures.

10 Sketch, on the same axes, the graphs of $y = 3 \times 4^x$ and $y = 4^{-x}$.
Find, to 3 significant figures, the x coordinates of their point of intersection.

11 Many phenomena – from stock market prices to census data to heat capacities of chemicals – obey Benford's law. This states that for a set of numerical data, the proportion of numbers starting with the digit D is approximately $\log_{10}\left(1 + \dfrac{1}{D}\right)$

    a Show that Benford's law predicts that around 30% of numbers will start with a 1, and around 18% with a 2. What proportion of numbers does the law predict will start with a 9?

    b Show that $\displaystyle\sum_{D=1}^{9} \log_{10}\left(1 + \frac{1}{D}\right) = 1$.

A 'Test yourself' exercise (Ty18) and an 'Extension exercise' (Ext18) can be found on the CD-ROM.

Key points

Exponentials

An exponential function has the form $y = a^x$, $a > 0$, $a \neq 1$.

All graphs of exponential functions pass through the point (0, 1).

Logarithms

- For *any* base c

    $\log_c ab = \log_c a + \log_c b$ $\log_c \dfrac{a}{b} = \log_c a - \log_c b$ $\log_c a^n = n\log_c a$

-   $\log_a a = 1$ $\log_a 1 = 0$ $\log_a\left(\frac{1}{a}\right) = -1$

- To change the base of a log, use $\log_a b = \dfrac{\log_c b}{\log_c a}$

- Log functions and exponential functions are inverses of each other. One 'undoes' the other.

    $N = a^{\log_a N}$ $x = \log_a a^x$

- To solve an equation of the type $a^x = b$, take logs of both sides.

19 Further integration and the trapezium rule

The problems of finding gradients of curves and of finding areas under curves may not, at first sight, seem connected, but calculus solves both these problems. As we have seen, differentiating gives the gradient of a curve; in this chapter we will show how integration gives the area under a curve. This powerful tool enables, for example, the area of an aeroplane's wing to be calculated precisely and is widely used in numerous fields, such as economics and engineering.

After working through this chapter you should be able to

- evaluate a definite integral
- apply integration to finding areas
- apply the trapezium rule for finding approximate areas.

19.1 Indefinite and definite integrals

The integrals in Chapter 10 resulted in *functions* of the variable used.

There was always a constant of integration which could or could not be found, depending on information given.

Such integrals are called **indefinite integrals**.

$$\int 2x\,dx = \underbrace{x^2 + c}_{\text{a function of } x}$$

There are situations where an integral has to be evaluated numerically.

Example 1   Given that the rate of change of displacement, s m, of a particle at time t s is given by

$$\frac{ds}{dt} = 3t^2 + 4t$$

find the total displacement from $t = 1$ to $t = 3$.

Solution
$$s = \int (3t^2 + 4t)\,dt$$

Integrate $\frac{ds}{dt}$ to find s in terms of t.

$$= t^3 + 2t^2 + c$$

When $t = 3$ $s_3 = 3^3 + 2 \times 3^2 + c$

When $t = 1$ $s_1 = 1^3 + 2 \times 1^2 + c$

Difference in displacement is the difference between s evaluated at $t = 3$ and at $t = 1$.

$$\therefore \quad s_3 - s_1 = 3^3 + 2 \times 3^2 + c - (1^3 + 2 \times 1^3 + c)$$
$$= 27 + 18 - 1 - 2$$
$$= 42$$

So displacement is 42 metres.

Note   The value of the integral was calculated at $t = 3$ and $t = 1$ and the difference found. The constant, c, cancelled out.

There is a special notation for evaluating an integral in this way.

Upper limit of integration

Square brackets

$$\int_1^3 (3t^2 + 4t)\,\mathrm{d}t = \left[t^3 + 2t^2 \right]_1^3$$

Lower limit of integration

Substitute 3 first and subtract value when 1 is substituted.
Note: c does not appear.

$$= 3^3 + 2 \times 3^2 - (1^3 + 2 \times 1^2)$$
$$= 27 + 18 - 1 - 2$$
$$= 42$$

An integral evaluated in this way is called a **definite integral**. Such an integral is used when the *difference* in value between the integral at its lower and upper limits is required.

So if $\int f(x)\mathrm{d}x = F(x) + c$

then $\int_a^b f(x)\mathrm{d}x = (F(b) + c) - (F(a) + c)$

$$= F(b) - F(a)$$

19.2   Area 'under' a curve

Consider a continuous curve $y = f(x)$. To find the area 'under' the curve (more precisely the area between the curve and the x-axis), bounded by the lines $x = a$ and $x = b$, several methods could be used.

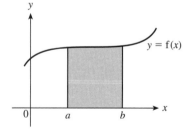

For a rough estimate, the graph could be plotted on graph paper and squares counted. Alternatively the area could be split into trapeziums or rectangles and their areas summed. (See Section 19.5.)

To find an area using integration

Consider the area bounded by the curve $y = f(x)$, the x-axis, $x = a$ and $x = b$.

Let A be the area from the line $x = a$ up to PN where P is the point (x, y) on the curve.

(Just as y depends on x, the area, A, also depends on x. The question is, how?)

Consider a point $Q(x + \delta x, y + \delta y)$ close to P, with ordinate QM.

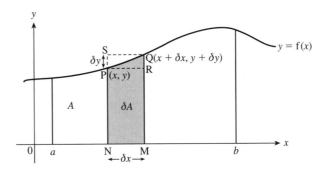

The shaded area represents a small increase in area A. Let this increase be δA.

From the diagram,

Area NMRP $< \delta A <$ Area NMQS

$$y \delta x < \delta A < (y + \delta y) \delta x$$

> Divide through by δx.

$\therefore \qquad y < \dfrac{\delta A}{\delta x} < y + \delta y$

As $\delta x \to 0$, $\delta y \to 0$ and so $\dfrac{\delta A}{\delta x}$ lies between y and something which can be made as close to y as one wants.

$\therefore \qquad \displaystyle\lim_{\delta x \to 0} \frac{\delta A}{\delta x} = y$

But $\quad \displaystyle\lim_{\delta x \to 0} \frac{\delta A}{\delta x} = \frac{\mathrm{d}A}{\mathrm{d}x}$

$\therefore \qquad \dfrac{\mathrm{d}A}{\mathrm{d}x} = y$

Integrating both sides with respect to x

$$A = \int y \, \mathrm{d}x$$

So the area 'under' a curve can be found by integrating the equation of the curve.

Suppose $\displaystyle\int y \, \mathrm{d}x = \mathrm{F}(x) + c$

Then $A = \mathrm{F}(x) + c$

Now A represents the area measured from $x = a$ up to PN where P is the point (x, y) on the curve.

At $x = a$, the area $A = 0$

$\therefore \; 0 = \mathrm{F}(a) + c$

$\quad c = -\mathrm{F}(a)$

So $A = \mathrm{F}(x) - \mathrm{F}(a)$

To find the area up to $x = b$, substitute b for x

So area $= F(b) - F(a)$

$$= \int_a^b y\,dx$$

$$A = \int_a^b y\,dx$$

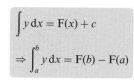

The definite integral, with limits a and b, gives the area bounded by the curve $y = f(x)$, the x-axis, $x = a$ and $x = b$.

*This result, namely that the area under a curve can be found by integration, is a version of the theorem called the **Fundamental Theorem of Calculus.***

Example 2   Find the area bounded by $y = 3x^2$, $x = 2$, $x = 3$ and the x-axis.

$$A = \int_2^3 y\,dx$$

$$= \int_2^3 3x^2\,dx$$

$$= \left[x^3\right]_2^3$$

$$= 3^3 - 2^3$$

$$= 19$$

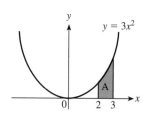

Note   In deriving $A = \int_a^b y\,dx$ the value of y was +ve.
If y is −ve, i.e. the area is below the x-axis, the integral A, will be −ve.
An area, however, is always +ve.

In the diagram, if $\int_a^b y\,dx = A$ then $A < 0$.
In this case, the area shaded $= |A|$.

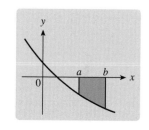

$|A|$, pronounced mod A, means the numerical value of A, regardless of the sign.
So $|6| = 6$ and $|-6| = 6$.

> For area between curve and x-axis use $\displaystyle\int_a^b y\,\mathrm{d}x$
>
> $\displaystyle\int_a^b y\,\mathrm{d}x$ ⟨ +ve for area above x-axis
>   −ve for area below x-axis
>
> Limits a and b are values of x.

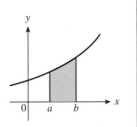

Extension

The area, A, between a curve and the y-axis from $y = c$ to $y = d$ is given by

$$A = \int_c^d x\,\mathrm{d}y, \text{ for } x > 0$$

The derivation is similar to that given on pages 326–328.

⊘ For an example using this look on the CD-ROM (E19.2).

> For area between curve and y-axis use $\displaystyle\int_c^d x\,\mathrm{d}y$
>
> $\displaystyle\int_c^d x\,\mathrm{d}y$ ⟨ +ve for area to right of y-axis
>   −ve for area to left of y-axis
>
> Limits c and d are values of y.

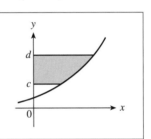

Note   *Differentiating* gives the *gradient*. *Integrating* between limits can give the *area*.

Example 3   The diagram shows a sketch of $y = x^2 - 4x + 3$ with areas A and B, bounded by the curve and the axes, shaded.

a   Find $\displaystyle\int_0^3 (x^2 - 4x + 3)\,\mathrm{d}x$.

b   Find $\displaystyle\int_1^3 (x^2 - 4x + 3)\,\mathrm{d}x$.

c   Deduce the total shaded area.

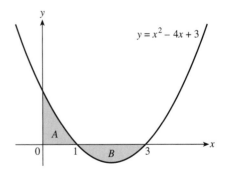

Solution   a $\displaystyle\int_0^3 (x^2 - 4x + 3)\,\mathrm{d}x = \left[\frac{x^3}{3} - 2x^2 + 3x\right]_0^3$

$$= \frac{3^3}{3} - 2 \times 3^2 + 3 \times 3 - 0$$

$$= 9 - 18 + 9$$

$$= 0$$

> Substitute $x = 3$ and $x = 0$.
> Substituting $x = 0$ gives zero.

b $\int_1^3 (x^2 - 4x + 3)\,\mathrm{d}x = \left[\dfrac{x^3}{3} - 2x^2 + 3x \right]_1^3$

Substitute $x = 3$ and subtract value with $x = 1$.

$$= \dfrac{3^3}{3} - 2 \times 3^2 + 3 \times 3 - \left(\tfrac{1}{3} - 2 + 3 \right)$$

$$= 9 - 18 + 9 - \tfrac{4}{3}$$

$$= -\tfrac{4}{3}$$

The integral is $-$ve since the area is below the x-axis.

c From **a**

$$\int_0^3 (x^2 - 4x + 3)\,\mathrm{d}x = 0$$

$\int_0^3 (x^2 - 4x + 3)\,\mathrm{d}x = 0$ so the positive and negative integrals for A and B cancel each other out.

So Area $A = $ Area B.

From **b**   Area $B = \tfrac{4}{3}$

$\therefore$ Total shaded area $= \tfrac{4}{3} + \tfrac{4}{3} = \tfrac{8}{3}$

Note   When calculating areas by integration a sketch is important to make sure that $+$ve and $-$ve integrals are properly interpreted.

Exercise 19A

1   Evaluate

  a $\left[\dfrac{x^4}{4} \right]_1^2$ **b** $\left[x^3 - 3x \right]_{-2}^0$ **c** $\left[3x^3 - 4x \right]_{-1}^1$ **d** $\left[x^3 - \dfrac{1}{x^2} \right]_{-4}^{-3}$

2   Find the value of these definite integrals and draw sketches to show the area found.

  a $\int_0^3 x^2 \,\mathrm{d}x$ **b** $\int_2^5 (2x^2 + 1)\,\mathrm{d}x$ **c** $\int_{-2}^2 x^3 \,\mathrm{d}x$ **d** $\int_1^2 \dfrac{1}{x^2}\,\mathrm{d}x$

3   Evaluate

  a $\int_0^{\frac{3}{2}} (4x - 3)\,\mathrm{d}x$ **b** $\int_{-8}^5 (6 + x)\,\mathrm{d}x$ **c** $\int_2^3 (3x - 1)(x + 2)\,\mathrm{d}x$

  d $\int_2^4 (x^2 - 3x + 4)\,\mathrm{d}x$ **e** $\int_{-1}^1 (3x^2 + 10x - 8)\,\mathrm{d}x$ **f** $\int_{-1}^3 (x^4 + x)\,\mathrm{d}x$

  g $\int_{-1}^1 (2x^4 - 3x^3)\,\mathrm{d}x$ **h** $\int_0^1 (6x^3 - 8x + 1)\,\mathrm{d}x$ **i** $4\int_{-3}^{-2} (3x^2 + 8x)\,\mathrm{d}x$

4 Evaluate

a $\displaystyle\int_1^4 \frac{3}{x^3}\,dx$ **b** $\displaystyle\int_1^4 \sqrt{x}\,dx$ **c** $\displaystyle\int_1^8 \sqrt[3]{x}\,dx$

d $\displaystyle\int_4^9 \left(5 - \frac{2}{\sqrt{x}}\right)dx$ **e** $\displaystyle\int_{-3}^{-2} \frac{3x-2}{x^3}\,dx$ **f** $\displaystyle\int_8^{27} \frac{1}{2\sqrt[3]{x}}\,dx$

g $\displaystyle\int_1^4 \frac{x^2 + x^3}{\sqrt{x}}\,dx$ **h** $\displaystyle\int_1^9 \frac{5x^{\frac{1}{2}} + x}{x^3}\,dx$ **i** $\displaystyle\int_1^{64} \frac{\sqrt{x} + \sqrt[3]{x}}{x}\,dx$

5 Find the shaded areas.

a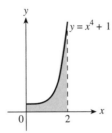

b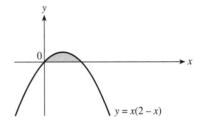

c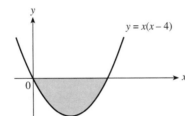

d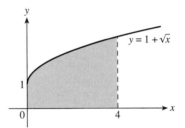

6 Find the areas enclosed by the x-axis and each of these curves and straight lines.

    a $y = 3x^2$, $x = 1$, $x = 3$ **b** $y = x^2 + 2$, $x = -2$, $x = 5$

    c $y = (x-1)(x-2)$, $x = -2$, $x = -1$ **d** $y = \dfrac{3}{x^2}$, $x = 1$, $x = 6$

7 Sketch the curve $y = x^2 - 5x + 6$ and find the area cut off below the x-axis.

8 Find, by integration, the area enclosed by $x + 4y - 20 = 0$ and the axes. Check by another method.

9 Sketch these curves and find the areas enclosed by the curves, the x-axis and the given straight lines.

    a $y = x(4-x)$, $x = 5$ **b** $y = -x^3$, $x = -2$ **c** $y = (x-2)^2$, $x = 1$

10 Find these shaded areas using $A = \int y \, dx$ and a method of subtraction.

a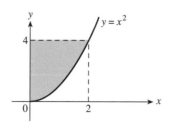

b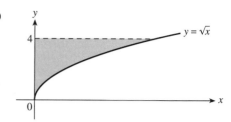

c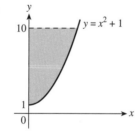

d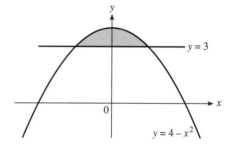

11 Find the shaded areas in parts **a** and **b** of Question 10 using $A = \int x \, dy$ i.e. by considering the area between the curve and the y-axis.

12 Find the area enclosed by the y-axis and these curves and straight lines.

a $x = y^2$, $y = 3$

b $y = x^3$, $y = 1$, $y = 8$

c $x - y^2 - 3 = 0$, $y = -1$, $y = 2$

d $x = \dfrac{1}{\sqrt{y}}$, $y = 2$, $y = 3$

13 Find these shaded areas.

a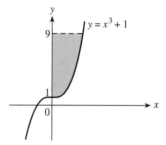

b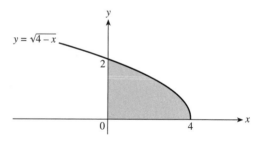

c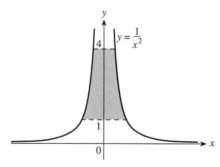

d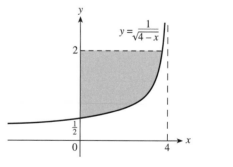

19.3   Area between a curve and a straight line

This example shows how the area between a curve and a straight line can be calculated by subtraction.

Example 5   The diagram shows the graph of $y = 4 - x^2$. The line PQ is the tangent to the curve at P(2, 0) and cuts the y-axis at Q.

Find **a** the coordinates of Q
      b the shaded area.

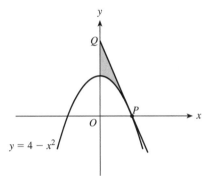

a $y = 4 - x^2$

    So $y' = -2x$

    At P, where $x = 2$, $y' = (-2) \times 2 = -4$.

    So PQ passes through (2, 0) and has gradient -4, hence its equation is

$$y - 0 = -4(x - 2)$$
$$y = 8 - 4x$$

> To find the coordinates of Q the equation of PQ is needed.

> Use $y - y_1 = m(x - x_1)$

PQ cuts the y-axis at Q.
When $x = 0$, $y = 8$ so Q is (0, 8)

b   Shaded area = Area of triangle OPQ – Area under curve from $x = 0$ to $x = 2$

$$= \frac{2 \times 8}{2} - \int_0^2 (4 - x^2)\,dx$$

> Area of triangle $= \frac{1}{2} \times$ base $\times$ height

$$= 8 - \left[4x - \frac{x^3}{3} \right]_0^2$$

$$= 8 - \left(4 \times 2 - \frac{2^3}{3} \right)$$

> Evaluate the square bracket. Put the result in brackets and be careful with the signs when the brackets are removed.

$$= 8 - 8 + \frac{8}{5}$$

$$= \frac{8}{3}$$

So the shaded area is $\frac{8}{3}$.

This example shows how the area between a curve and a straight line can be calculated by splitting the area into regions, each of which can be calculated easily. A simpler and more elegant method, which would work for this example, involves subtracting the equations of the curve and line (see Section 19.4).

Example 5   The sketch shows the curve $y = x^2 - 3x$ and the straight line $y = x$ intersecting at the origin O and at the point L. Find the coordinates of L and the area of the shaded region.

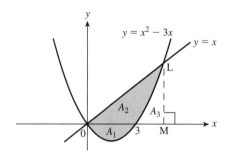

> To sketch $y = x^2 - 3x$, find where the curve cuts the x-axis, i.e. where $y = 0$.
> $x^2 - 3x = 0 \Rightarrow x(x - 3) \Rightarrow x = 0$ or $x = 3$.

> The perpendicular LM from L to the x-axis is shown.

Solution $y = x^2 - 3x$ intersects $y = x$ where

$$x^2 - 3x = x$$

$$x^2 - 4x = 0$$

$$x(x - 4) = 0$$

i.e. where $x = 0$ or $x = 4$

$x = 0$ at the origin, $x = 4$ at L.

So the coordinates of L are (4, 4).

Consider the three regions marked A_1, A_2, and A_3.

Let the shaded area be R.

Then the area required, R, is given by

$$R = A_1 + A_2$$

and $A_2 = \triangle \text{LOM} - A_3$.

> A_3 can be calculated by $\int_3^4 (x^2 - 3x)\, dx$.
>
> The value of A_3 can be subtracted from the area of $\triangle$LOM to find A_2.

$$\int_0^3 (x^2 - 3x)\, dx = \left[\frac{x^3}{3} - \frac{3x^2}{2} \right]_0^3$$

$$= \frac{27}{3} - \frac{27}{2}$$

$$= \frac{54 - 81}{6} = -\frac{9}{2}$$

> A_1 can be found from $\int_0^3 (x^2 - 3x)\, dx$
>
> The integral will be $-$ve as the area is below the x-axis.

So area $A_1 = \dfrac{9}{2}$

The integral is negative but the area is positive.

Area of $\triangle \text{LOM} = \dfrac{1}{2} \times 4 \times 4 = 8$

Area $A_3 = \displaystyle\int_3^4 (x^2 - 3x)\, \mathrm{d}x$

$\qquad = \left[\dfrac{x^3}{3} - \dfrac{3x^2}{2} \right]_3^4$

$\qquad = \dfrac{64}{3} - 24 - \left(9 - \dfrac{27}{2} \right)$

$\qquad = \dfrac{11}{6}$

$R = A_1 + A_2$ and $A_2 = \triangle \text{LOM} - A_3$

So $R = A_1 + \triangle \text{LOM} - A_3$

$\qquad = \dfrac{9}{2} + 8 - \dfrac{11}{6}$

$\qquad = \dfrac{32}{3}$

So the shaded region is $\frac{32}{3}$ square units.

19.4 Area between two curves

To find the area between the curves $y = \mathrm{f}(x)$ and $y = \mathrm{g}(x)$ bounded by $x = a$ and $x = b$ the method of subtraction is used.

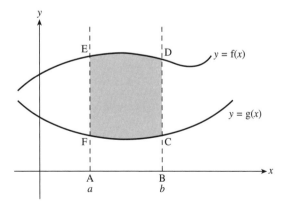

Area required $=$ Area ABDE $-$ Area ABCF

$\qquad = \displaystyle\int_a^b \mathrm{f}(x)\, \mathrm{d}x - \int_a^b \mathrm{g}(x)\, \mathrm{d}x$

$\qquad = \displaystyle\int_a^b \Big(\mathrm{f}(x) - \mathrm{g}(x) \Big)\, \mathrm{d}x$

Even when one, or both, of the curves lies beneath the x-axis the same result, namely $\int (f(x) - g(x))\, dx$, gives the area.

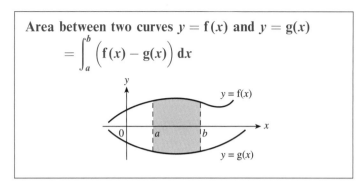

Area between two curves $y = f(x)$ and $y = g(x)$
$$= \int_a^b \left(f(x) - g(x) \right) dx$$

Using this method in Example 5, with $f(x) = x$ and $g(x) = x^2 - 3x$, the area R could be calculated by $\int 4_0(x - (x^2 - 3x))\, dx = \int_0^4 (4x - x^2)\, dx = \dfrac{32}{3}$.

Example 6

Find the area enclosed between the curves $y = 16 - x^2$ and $y = x^2 - 4x$.

The curves intersect where
$$16 - x^2 = x^2 - 4x$$
$$2x^2 - 4x - 16 = 0$$
$$x^2 - 2x - 8 = 0$$
$$(x - 4)(x + 2) = 0$$

i.e. where $x = 4$ or $x = -2$

So the limits of integration are -2 and 4.

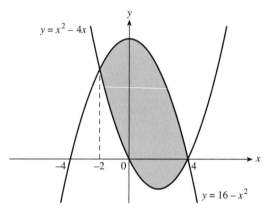

Area required $= \displaystyle\int_{-2}^4 \left(16 - x^2 - (x^2 - 4x) \right) dx$

$= \displaystyle\int 4_{-2}(16 + 4x - 2x^2)\, dx$

$= 2 \displaystyle\int_{-2}^4 (8 + 2x - x^2)\, dx$

Area between curves $= \displaystyle\int_{-2}^4 \left(f(x) - g(x) \right) dx$
The method deals correctly with the area below the x-axis.

Take the common factor 2 outside the integral sign.

$$= 2\left[8x + x^2 - \frac{x^3}{3}\right]_{-2}^{4}$$
$$= 2\left(32 + 16 - \frac{64}{3} - \left(-16 + 4 + \frac{8}{3}\right)\right)$$
$$= 2\left(60 - \frac{72}{3}\right)$$
$$= 72$$

So area is 72 square units.

Exercise 19B

1 Find these shaded areas.

a

b

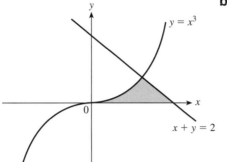
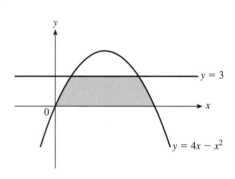

2 The curve C has equation $y = x^2 + 4$. The normal to C at P(1, 5) meets the x-axis at Q.

a Sketch the curve C.

b Find the equation of the normal.

c Find the coordinates of Q.

d Find the area bounded by the x-axis, the y-axis, the curve and the line PQ.

3 The diagram shows the graph of the curve C whose equation is
$y = x(x - 1)(x - 2)$

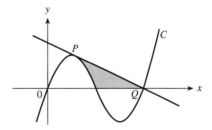

a Find the equation of the tangent at the point P whose x coordinate is $\frac{1}{2}$.

b Show that the point Q(2, 0) lies on the tangent and on C.

c Find the shaded area.

4 a Sketch the curve $y = 2 + x - x^2$. On the sketch draw the tangent at P(1, 2) to the curve. The tangent at P meets the x and y axes at L and M respectively.

  b Find the coordinates of L and M.

  c Find the area bounded by the curve, the tangent and the x-axis.

  d Find the area bounded by the curve, the tangent and the y-axis.

5 For these curves and straight lines, find their points of intersection, draw a sketch on the same axes and find the area of the segment cut off from each curve by the corresponding straight line.

  a $y = x^2 - 4x + 6$, $y = 3$ **b** $y = 7 - x - x^2$, $y = 5$

  c $y = \frac{1}{2}x^2$, $y = 2x$ **d** $y = 3x^2$, $3x + y - 6 = 0$

  e $y = (x + 1)(x - 2)$, $x - y + 1 = 0$ **f** $y = \sqrt{x} + 3$, $2y = x + 6$

6 The diagram shows a sketch of $f(x) = x^2 - 3x + 2$.

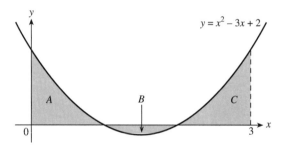

  a Find $\displaystyle\int_0^3 f(x)\,dx$.

  b Find the sum of the areas A, B and C.

  c Explain the difference between the answers to parts **a** and **b**.

7 Show that the line $y = 3x + 5$ is a tangent to the curve $y = x^3 + 3$ at the point $(-1, 2)$. Find the coordinates of the other point of intersection of the line and the curve. Draw a sketch of the line and the curve and find the area enclosed by them.

8 Sketch the graph of $y = x(x - 1)(x + 1)$.

  Find $\displaystyle\int_{-1}^1 y\,dx$ and explain the answer.

9 The normal to the curve $y = x^3$ at the point P (1, 1) meets the x-axis at Q. Find the area bounded by the x-axis, the curve $y = x^3$ and PQ.

10 Find the area bounded by the curve $y = 9 - x^2$ and the line $x + y + 11 = 0$.

11 The curves $y = x^2 + 2$ and $y = 4 - x^2$ meet at the points A and B.

   a Find the coordinates of A and B and hence sketch the curves.

   b Find the area enclosed by the two curves.

12 Find these shaded areas.

   a

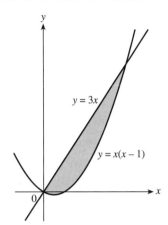

   b

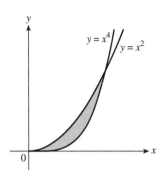

   c

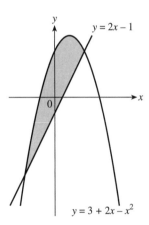

   d

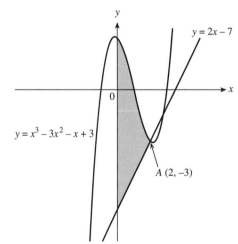

19.5   The trapezium rule

So far in this chapter, definite integrals (those with limits) have been evaluated exactly, in particular to find the area 'under' a curve. There are many functions which cannot be integrated (or cannot be integrated by simple methods). In these situations, methods of approximating the integral are needed. One method is to estimate the integral by finding an approximation to the area.

A simple approach is to approximate the area to a trapezium.

The area of a trapezium $= \dfrac{1}{2} \times$ (sum of the parallel sides) $\times$ (distance between them)

$$= \dfrac{h}{2}(a+b)$$

Example 7   The area A is bounded by $y = \dfrac{1}{1+x^3}$, the axes and $x = 1$.

By approximating the area to a trapezium, find an approximate value of A and

hence, an estimate of the definite integral $\displaystyle\int_0^1 \dfrac{1}{1+x^3}\,\mathrm{d}x$.

Solution   When $x = 0$, $y = \dfrac{1}{1+0} = 1$

When $x = 1$, $y = \dfrac{1}{1+1^3} = \dfrac{1}{2}$

$A \approx \dfrac{1}{2} \times (1 + \tfrac{1}{2}) \times 1$

$\approx \dfrac{3}{4}$

$A = \displaystyle\int_0^1 \dfrac{1}{1+x^3}\,\mathrm{d}x$

Hence $\displaystyle\int_0^1 \dfrac{1}{1+x^3}\,\mathrm{d}x \approx \dfrac{3}{4}$

Note   Because of the curvature of $y = \dfrac{1}{1+x^3}$ the trapezium will give an underestimate of the area A and, hence, an underestimate of the integral.

For a more accurate result, the area to be approximated can be divided into a number of trapeziums and their areas summed.

Formula for the trapezium rule

To find an approximation to the area, A, between $y = \mathrm{f}(x)$, the x-axis, $x = a$ and $x = b$, the area to be estimated is split into n strips of equal width, h.

For n intervals, there are $n + 1$ ordinates, y_0 to y_n.

> *Remember*: The ordinate is another name for the y-coordinate.

The n intervals from $x = a$ to $x = b$ are equal width so $h = \dfrac{b-a}{n}$.

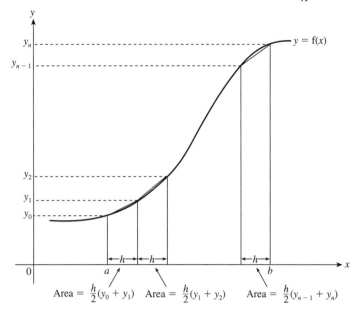

$$\text{Area} = \frac{h}{2}(y_0 + y_1) \quad \text{Area} = \frac{h}{2}(y_1 + y_2) \quad \text{Area} = \frac{h}{2}(y_{n-1} + y_n)$$

Notice that the curve at the top of each strip is replaced by the chord joining the end points so that each strip is a trapezium.

Some trapeziums (e.g. the first two) give an overestimate of the area under the curve.

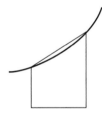

Some trapeziums (e.g. the last one) give an underestimate of the area under the curve.

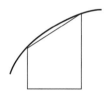

Summing the areas of the trapeziums

> Area of trapezium $= \dfrac{h}{2}(a + b)$

$$A \approx \frac{h}{2}(y_0 + y_1) + \frac{h}{2}(y_1 + y_2) + \frac{h}{2}(y_2 + y_3) + \cdots + \frac{h}{2}(y_{n-2} + y_{n-1}) + \frac{h}{2}(y_{n-1} + y_n)$$

$$\approx \frac{h}{2}(y_0 + y_1 + y_1 + y_2 + y_2 + \ldots + y_{n-1} + y_{n-1} + y_n)$$

$$\approx \frac{h}{2}[y_0 + y_n + 2(y_1 + y_2 + \ldots + y_{n-1})]$$

> Note that y_0 and y_n only appear once. $y_1, y_2, \ldots, y_{n-1}$ appear twice.

> **Trapezium rule**
>
> $$\int_a^b f(x)\,dx \approx \frac{h}{2}[y_0 + y_n + 2(y_1 + y_2 + \cdots + y_{n-1})]$$

This rule can be stated as: Integral $\approx \dfrac{h}{2} \times$ (first $+$ last $+$ $2 \times$ sum of others).

Note A better approximation to the area can be obtained by increasing the number of trapeziums, i.e. by increasing n and reducing h.

Example 8 Find an approximation to the area bounded by the axes, $y = \dfrac{1}{1 + x^2}$ and $x = 2$.

Use the trapezium rule with 5 intervals (i.e. 6 ordinates).

Using the graph, comment on whether the approximation is likely to be an overestimate or an underestimate.

Solution $h = \dfrac{2}{5} = 0.4$

The interval 0 to 2 is divided into five equal intervals. The integral to find this area can be calculated on some graphic calculators.

$y_0 = \dfrac{1}{1 + 0} = 1$

$y_1 = \dfrac{1}{1 + 0.4^2} = 0.8620\ldots$

$y_2 = \dfrac{1}{1 + 0.8^2} = 0.6097\ldots$

$y_3 = \dfrac{1}{1 + 1.2^2} = 0.4098\ldots$

$y_4 = \dfrac{1}{1 + 1.6^2} = 0.2808\ldots$

$y_5 = \dfrac{1}{1 + 2^2} = 0.2$

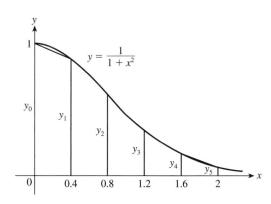

Note The value of each y-coordinate has been worked out on a calculator and included here to make the method clear. Providing the method is clearly shown, the calculation can be completed on a calculator and only the answer written down.

$A \approx \dfrac{h}{2}[y_0 + y_5 + 2(y_1 + y_2 + y_3 + y_4)]$

The trapezium rule.

$\approx \dfrac{0.4}{2}\left[\dfrac{1}{1+0} + \dfrac{1}{1+2^2} + 2\left(\dfrac{1}{1+0.4^2} + \dfrac{1}{1+0.8^2} + \dfrac{1}{1+1.2^2} + \dfrac{1}{1+1.6^2}\right)\right]$

$\approx 0.2[1 + 0.2 + 2(0.8620\ldots + 0.6097\ldots + 0.4098\ldots + 0.2808\ldots)]$

≈ 1.105

The first trapezium is an underestimate of the area and the fifth trapezium is an overestimate, so from the graph it is not possible to comment on whether the total area has been overestimated or underestimated.

The area calculated is given exactly by $\int_0^2 \dfrac{1}{1+x^2}\,dx$. Although this cannot be integrated by the methods covered in this book, it will be integrated later in the A-level course. Integration gives a value for the area of 1.107, correct to 3 decimal places, so the trapezium rule gave, in fact, an underestimate.

The following example shows how an area can be approximated even when the function cannot be expressed algebraically.

It also illustrates that the area may give useful practical information. If there are units on the axes then the units of the quantity represented by the area under the curve will be those obtained by 'multiplying' the units on the axes. For example, the area under a graph of speed (in km/h) against time (in h) will represent the distance travelled (in km).

$$\frac{\text{km}}{\text{hour}} \quad \times \quad \text{hour} \quad = \quad \text{km}$$

$$\uparrow \qquad\qquad \uparrow \qquad\qquad \uparrow$$

Speed in kmh^{-1} Time in h   Distance in km
(y-axis)   (x-axis)   (area)

Example 9 The areas, in square centimetres, of the cross-sections of a bone 18 cm long, measured at 2 cm intervals are

$$4,\ 3.9,\ 3.7,\ 3.7,\ 3.6,\ 3.5,\ 3.4,\ 3.5,\ 3.6,\ 3.6$$

Estimate the volume of the bone.

Solution The shaded area is an estimate of the volume, V.

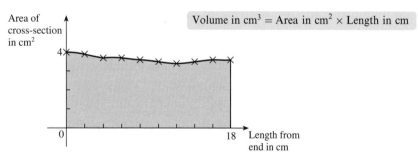

The shaded area can be estimated by applying the trapezium rule.

$$V \approx \frac{2}{2}[4 + 3.6 + 2(3.9 + 3.7 + \cdots + 3.6)]$$

$$= 65.4$$

Intervals of 2 cm
$h = 2$
$y_0 = 4,\ y_9 = 3.6$

So an estimate for the volume of the bone is $65.4\,\text{cm}^3$.

Note The area under a curve represents the quantity obtained when the units on the axes are multiplied.

All the integrals in this exercise can be evaluated, and therefore checked, on a graphic calculator. The calculator will give a more accurate value than these results.

1 The diagram shows a sketch of $y = x^2$. The trapezium rule is to be used, with 3 intervals, to find an approximation to the area between the curve, the x-axis, $x = 2$ and $x = 5$.

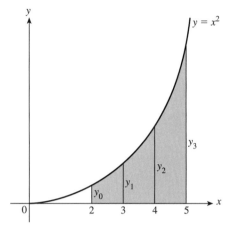

a Complete the table below and hence find an approximation to the area.

n	0	1	2	3
x	2	3	4	5
y_n	4			

b Calculate the exact area by integration. Explain, with reference to the graph, why the trapezium rule gives an overestimate of the area.

2 The diagram shows a sketch of $y = \sqrt{64 - x^3}$. The trapezium rule is to be used, with 6 intervals, to find an approximation to the area between the curve, the y-axis and $x = 3$.

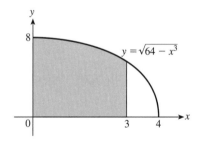

c Verify that the answer obtained by the trapezium rule is within 1.3% of the exact value.

Complete the table below, entering the results to three decimal places. Hence find an approximation to the area, giving the answer to one decimal place.

n	0	1	2	3	4	5	6
x	0	0.5	1	1.5	2	2.5	3
y_n	8				7.483		

3 Use the trapezium rule with 4 intervals to find an approximation to the area bounded by the curve $y = 2^x$, the x-axis, $x = -2$ and $x = 2$.

Explain, with the aid of a sketch, whether the approximation is an overestimate or underestimate.

4 Water is flowing into a tank. The rate of flow, in litres per minute, is recorded at one-minute intervals over a ten-minute period. Here are the results.

Time (min)	0	1	2	3	4	5	6	7	8	9	10
Rate of flow (l/m)	0	16.1	20.3	21.7	30	30	30	28.2	27.1	26.1	24.1

Use the trapezium rule to estimate the increase in volume of water in the tank over the ten-minute period.

5 A motorist travels along a straight road. At three-second intervals, her speed in m s^{-1} is recorded.

Time (s)	0	3	6	9	12	15	18
Speed (m s^{-1})	0	2.8	4.1	5.2	6.3	6.5	7.7

Use the trapezium rule to estimate the distance travelled for the 18 seconds during which the measurements were made.

6 a Tabulate, correct to 3 decimal places, the value of the function $f(x) = \sqrt{1 + x^2}$ for values of x from 0 to 0.8 at intervals of 0.1.

b Use the values found in part **a** to estimate $\int_0^{0.8} f(x)\,dx$ by the trapezium rule, using all the ordinates.

7 Find approximations to these integrals using the trapezium rule.

a $\int_1^7 \dfrac{1}{x}\,dx$ using 6 intervals

b $\int_0^1 \dfrac{1}{x+4}\,dx$ using 4 intervals

c $\int_{-1}^1 \sqrt{1+x}\,dx$ using 4 intervals

d $\int_0^2 \dfrac{1}{1+x^3}\,dx$ using 4 intervals

8 a Use the binomial theorem to expand $(1 + x^3)^{10}$ in ascending powers of x, up to and including the term in x^9.

b Estimate $\int_0^{0.2} (1 + x^3)^{10}\,dx$, correct to 3 decimal places

    i by integration of the expansion in part **a**

    ii by the trapezium rule with 5 ordinates (4 intervals)

Exercise 19D (Review)

1 Find the value of these integrals and draw sketches to show the area found.

a $\displaystyle\int_{-2}^{2} (x^2 + 4)\,dx$ **b** $\displaystyle\int_{0}^{4} (x^3 + 1)\,dx$ **c** $\displaystyle\int_{2}^{3} (x - 2)(x - 3)\,dx$

2 Evaluate

a $\displaystyle\int_{-2}^{3} \left(\frac{3x^2}{2} + \frac{1}{x^2}\right) dx$ **b** $\displaystyle\int_{0}^{1} \frac{4x^2 + x}{x^{\frac{1}{2}}}\,dx$

c $\displaystyle\int_{0}^{32} \frac{2}{3} x^{-\frac{1}{5}}\,dx$ **d** $\displaystyle\int_{0}^{64} (\sqrt{x} - \sqrt[3]{x})\,dx$

3 Sketch the curve $y = 8x - x^2$ and find the area bounded by the curve and the x-axis.

4 Sketch the curve $y = x^3 - 1$ and find the area bounded by the curve and the axes.

5 Find these shaded areas.

a
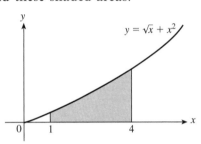

b
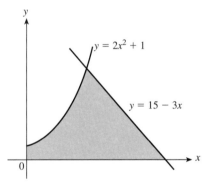

c
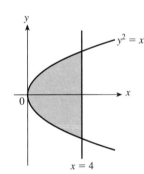

d
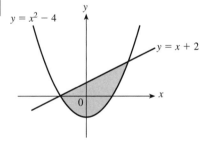

6 a Calculate $\displaystyle\int_{-1}^{1} x(x^2 - 1)\,dx$.

b Find the area bounded by the curve $y = x(x^2 - 1)$ and the x-axis
  i between $x = -1$ and $x = 0$ **ii** between $x = 0$ and $x = 1$

7 Find $\displaystyle\int_{0}^{4} x^2\,dx$

a exactly by integration

b approximately by the trapezium rule with
  i 4 intervals **ii** 8 intervals

c Verify that the answer obtained by the trapezium rule in **b ii** is within 0.8% of the exact value.

8 Use the trapezium rule, with 5 intervals, to estimate

$$\int_1^2 \frac{1}{\sqrt{x}+x}\,dx$$

giving your answer correct to 2 decimal places.

9 The values of a function f(x) are given in the table below.

x	1	2	3	4	5	6	7
f(x)	0.9	1.1	1.4	1.5	1.0	0.8	0.4

Find approximate values, using the trapezium rule, for

a $\displaystyle\int_1^7 f(x)\,dx$ **b** $\displaystyle\int_1^7 \frac{1}{f(x)}\,dx$

10 The line $y = \frac{1}{2}x + 1$ meets the curve $y = \frac{1}{4}(7x - x^2)$ at the points A and B.

   a Find the coordinates of A and B.

   b Find the length of AB.

   c Prove that the upper segment of the curve cut off by the line has area $1\frac{1}{8}$.

11 Find the ratio of area
A to area B.

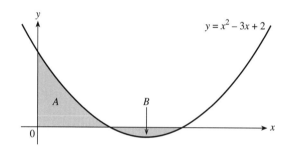

12 Find the area of the region lying between $y = 2x^2$ and $y = x^2$ from the origin up to the line $x = 4$.

13 Find these shaded areas.

a
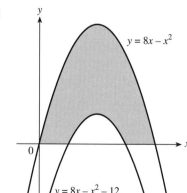

b
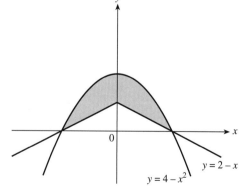

A 'Test yourself' exercise (Ty19) and an 'Extension exercise' (Ext19) can be found on the CD-ROM.

Key points

Indefinite and definite integrals

An **indefinite integral** results in a function being obtained. Remember to add a constant.

A **definite integral** has limits, and results in a value being obtained.

$$\text{If } \int f(x)\,dx = F(x) + c \text{ then } \int_a^b f(x)\,dx = F(b) - F(a)$$

Area

- For area between curve and x-axis use $\int_a^b y\,dx$

$$\int_a^b y\,dx \begin{cases} \text{+ve for area above } x\text{-axis} \\ \text{−ve for area below } x\text{-axis} \end{cases}$$

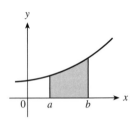

Limits a and b are values of x.

- Area between two curves $y = f(x)$ and $y = g(x) = \int_a^b \Big(f(x) - g(x) \Big)\,dx$

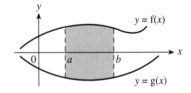

Trapezium rule

For estimating integrals, use the **trapezium rule** for n equal intervals ($n+1$ ordinates) of width h.

$$\int_a^b f(x)\,dx \approx \frac{h}{2}\left[y_0 + y_n + 2(y_1 + y_2 + \cdots + y_{n-1}) \right]$$

20 Geometric series

In Chapter 8, we met the idea of a series, the name given to the sum of a sequence of terms. In an arithmetic series, consecutive terms differ by a constant amount; in a geometric series, consecutive terms have a common ratio. The values at the end of each year of a sum of money invested at a constant rate of compound interest form a geometric sequence, as do the number of cells in an organism at the end of each hour, when, for example, the number of cells doubles every hour. There are numerous situations in many fields, including finance, biology and statistics, where geometric series arise.

After working through this chapter you should

- *know what a geometric series is*
- *be able to find the general term and the sum to n terms of a geometric series*
- *be able to prove the formula for the sum of n terms of a geometric series:*

$$S_n = \frac{a(1 - r^n)}{1 - r}$$

- *be able to find the sum to infinity of a convergent geometric series, including the use of $|r| < 1$.*

20.1 Geometric sequences and series

A geometric sequence is produced by multiplying each term in a sequence by a constant value (the common ratio) to obtain the next term.

Some geometric sequences converge; those that do not converge are said to be divergent.

Starting with 3, for example, and multiplying each time by 2 would generate the sequence 3, 6, 12, 24, ... The terms of this sequence get larger and larger without limit. This geometric sequence is said to be divergent.

Starting with 3 again, and now multiplying each time by $\frac{1}{3}$ would generate the sequence 3, 1, $\frac{1}{3}$, $\frac{1}{9}$, ... The terms of this sequence get smaller and smaller and closer and closer to zero. This geometric sequence is said to be convergent to zero. By going far enough along the sequence, one can get as close to zero as one chooses.

Starting once more with 3, and now multiplying each time by -1 would generate the geometric sequence 3, -3, 3, -3, ... This is an oscillating sequence, it does not converge.

Of the three geometric sequences above, only the sequence with common ratio $\frac{1}{3}$ converged. For a geometric sequence to converge, its common ratio must be numerically less than one (i.e. between -1 and 1). A negative common ratio, numerically less than one, $-\frac{1}{3}$ for example, would also generate a convergent sequence: 3, -1, $\frac{1}{3}$, $-\frac{1}{9}$, ...

To find out more about the behaviour of sequences look on the CD-ROM (E8.1).

Geometric series

Summing any geometric sequence gives a geometric series. So $3 + 1 + \frac{1}{3} + \frac{1}{9} + \cdots$ is an example of a geometric series. Dividing any term by the previous term will give the common ratio, which in this case is $\frac{1}{3}$.

Example 1 below illustrates that a geometric series, with common ratio numerically less than 1, will converge.

The terms of a geometric sequence or geometric series are said to be in **geometric progression**.

➤ | A geometric series is a series whose consecutive terms have a common ratio.

If the common ratio is r, each term is obtained from the previous term by multiplying by r.

Expressed formally:

➤ | $u_n = r u_{n-1}$ **or** $\dfrac{u_n}{u_{n-1}} = r$

Before introducing formulae for geometric series, Examples 1 and 2 solve some problems without using formulae.

Example 1 For the geometric series given

i find the next two terms
ii find the sum of n terms, S_n, for $n = 1, 2, 3, 4$
iii comment on whether or not the series converges.

Solution **a** $2 + 6 + 18 + 54 + \cdots$

i The common ratio, $r = \frac{6}{2} = 3$.

$\therefore$ 5th term $= 54 \times 3 = 162$

and 6th term $= 162 \times 3 = 486$

> Any term divided by the previous term will give the common ratio.

ii $S_1 = 2$
$S_2 = 2 + 6 = 8$
$S_3 = 2 + 6 + 18 = 26$
$S_4 = 26 + 54 = 80$

> Summing the first term, then the first two terms, then three, and so on, will help determine whether or not the series converges.

iii The terms of the sequence $S_1, S_2, S_3, S_4, \ldots$ get larger and larger without limit. Therefore the series $2 + 6 + 18 + 54 + \cdots$ does not converge.

b $1 + \frac{1}{2} + \frac{1}{4} + \frac{1}{8} + \cdots$

   i The common ratio, $r = \frac{1}{2}$.

   $\therefore$ 5th term $= \frac{1}{8} \times \frac{1}{2} = \frac{1}{16}$

   and 6th term $= \frac{1}{16} \times \frac{1}{2} = \frac{1}{32}$

   ii $S_1 = 1$

   $S_2 = 1 + \frac{1}{2} = 1\frac{1}{2}$

   $S_3 = 1 + \frac{1}{2} + \frac{1}{4} = 1\frac{3}{4}$

   $S_4 = 1\frac{3}{4} + \frac{1}{8} = 1\frac{7}{8}$

   iii The terms of the sequence
   $S_1, S_2, S_3, S_4, \ldots$ get closer to 2.
   By calculating further values of
   S_n it can be seen that, as n gets larger,
   S_n tends to 2.
   Therefore the series
   $1 + \frac{1}{2} + \frac{1}{4} + \frac{1}{8} + \cdots$ converges.

   > Expressed formally:
   > as $n \to \infty$, $S_n \to 2$.

   > The series is said to have a sum to infinity, S_∞. In this case $S_\infty = 2$.
   >
   > By taking enough terms of the series one can get as close to the limit, 2, as one wants.

c $2 - 4 + 8 - 16 + \cdots$

   i The common ratio, $r = -\frac{4}{2} = -2$.

   $\therefore$ 5th term $= -16 \times -2 = 32$

   and 6th term $= 32 \times -2 = -64$

   ii $S_1 = 2$

   $S_2 = 2 - 4 = -2$

   $S_3 = 2 - 4 + 8 = 6$

   $S_4 = 6 - 16 = -10$

   iii The terms of the sequence $S_1, S_2, S_3, S_4, \ldots$ get numerically larger without limit.
   Therefore the series $2 - 4 + 8 - 16 + \cdots$ does not converge.

Example 2   For the geometric series $4 + 12 + 36 + 108 + \cdots$, find

   a   the 8th term

   b   the sum of the first eight terms.

Solution   **a** $r = \frac{12}{4} = 3$

   > To obtain the eighth term, the first term has to be multiplied by r seven times ($8 - 1 = 7$).

1st term	2nd term	3rd term	4th term	$\ldots$	8th term
4	4×3	4×3^2	4×3^3		4×3^7

   $\therefore$ the 8th term is 4×3^7.

   > In general, to obtain the nth term, the first term is multiplied by r, $(n - 1)$ times.

b To find the sum of the first eight terms: Multiply both sides by $r (= 3)$ and line up the terms.

$$S_8 = 4 + 4 \times 3 + 4 \times 3^2 + 4 \times 3^3 + \cdots + 4 \times 3^7 \qquad \text{①}$$

$$3S_8 = \qquad 4 \times 3 + 4 \times 3^2 + 4 \times 3^3 + \cdots + 4 \times 3^7 + 4 \times 3^8 \qquad \text{②}$$

$$\text{②} - \text{①} \Rightarrow 2S_8 = 4 \times 3^8 - 4$$

> Subtracting ① from ② eliminates all but the 1st term of S_8 and the last term of $3S_8$.

$$S_8 = \frac{4 \times 3^8 - 4}{2} = 2 \times 3^8 - 2 = 13\,120$$

Formulae for geometric series

Let the first term be a, the common ratio r, the sum of the first n terms S_n, and the sum to infinity, S_∞.

Formula for the nth term

First term $= a$, second term $= ar$, third term $= ar^2$ and so on. So nth term $= ar^{n-1}$.

> n**th term** $= ar^{n-1}$

Note For a geometric series, $u_n = ru_{n-1}$ or $\dfrac{u_n}{u_{n-1}} = r$.
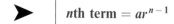

Formula for the sum of a geometric series

$$S_n = a + ar + ar^2 + \cdots + ar^{n-1} \qquad \text{①}$$

$$rS_n = \qquad ar + ar^2 + \cdots + ar^{n-1} + ar^n \qquad \text{②}$$

> Multiply both sides by r and line up the terms.

$$\text{②} - \text{①} \qquad rS_n - S_n = ar^n - a$$

> Subtracting ① from ② eliminates all but two terms.

$$S_n(r - 1) = a(r^n - 1)$$

$$S_n = \frac{a(r^n - 1)}{r - 1} \qquad \text{③}$$

Multiplying numerator and denominator of ③ by -1 gives an alternative formula, which is useful when r is numerically less than 1, i.e. $-1 < r < 1$. This can be written $|r| < 1$. (See note below.)

$$S_n = \frac{a(1 - r^n)}{1 - r}$$

> Two versions of the formula are given to avoid dealing with negative quantities.

> **When** $|r| > 1$ **use** $S_n = \dfrac{a(r^n - 1)}{r - 1}$ **When** $|r| < 1$ **use** $S_n = \dfrac{a(1 - r^n)}{1 - r}$
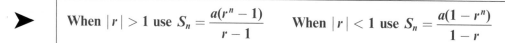

Note $|n|$, pronounced mod n, means the numerical value of n, regardless of the sign. So $|0.2| = 0.2$ and $|-0.2| = 0.2$.
r numerically less than 1 $\Leftrightarrow -1 < r < 1 \Leftrightarrow |r| < 1$
r numerically greater than 1 $\Leftrightarrow r < -1$ or $r > 1 \Leftrightarrow |r| > 1$

Example 3 For the geometric series $3 + 6 + 12 + 24 + \ldots$, find

    a the 12th term

    b the sum of the first 20 terms

    c the least number of terms for which the sum exceeds 10^6.

Solution   **a** $a = 3$, $r = 2$, $n = 12$.

       12th term $= ar^{11} = 3 \times 2^{11} = 6144$

  b $S_{20} = \dfrac{a(r^{20} - 1)}{r - 1}$

        $= \dfrac{3(2^{20} - 1)}{2 - 1} = 3(2^{20} - 1) = 3\,145\,725$

  c $S_n = \dfrac{3(2^n - 1)}{2 - 1}$

> Use the formula to find an expression for the sum of n terms.

       $= 3(2^n - 1)$

For the sum to exceed 10^6

$$3(2^n - 1) > 10^6$$

$$3 \times 2^n - 3 > 10^6$$

$$3 \times 2^n > 10^6 + 3$$

$$2^n > \frac{10^6 + 3}{3}$$

> See second note below.

Taking logs to base 10 (lg) of both sides

> Alternatively, take both logs to base e (ln).

$$n \lg 2 > \lg \frac{10^6 + 3}{3}$$

$$n > \frac{\lg \dfrac{10^6 + 3}{3}}{\lg 2}$$

> Divide both sides by lg 2. $\lg 2 > 0$ so there is no need to reverse the inequality sign.

$$> 18.3\ldots$$

So the least number of terms for which the sum exceeds 10^6 is 19.

> n, the number of terms, must be an integer.

Note If dividing by $\log x$, where $x < 1$, the inequality sign has to be reversed, as, for $x < 1$, $\log x$ is negative.

Note $2^n > \dfrac{10^6 + 3}{3} = 333334.33\ldots$

An alternative method of finding n is to try powers of 2 on the calculator to find which is the first power greater than $333334.33\ldots$

Example 4   Find the value of

$$\sum_{i=10}^{20} 3 \times \left(-\tfrac{2}{3}\right)^i$$

> At first sight, this may not look like a geometric series. Writing out the first few terms will show that it is one.

to 3 significant figures.

$$\sum_{i=10}^{20} 3 \times \left(-\tfrac{2}{3}\right)^i = 3 \times \left(-\tfrac{2}{3}\right)^{10} + 3 \times \left(-\tfrac{2}{3}\right)^{11} + 3 \times \left(-\tfrac{2}{3}\right)^{12} + \cdots + 3 \times \left(-\tfrac{2}{3}\right)^{20}$$

This is a geometric series with $a = 3 \times \left(-\tfrac{2}{3}\right)^{10}$, $r = -\tfrac{2}{3}$ and $n = 11$.

> Take care with signs.
> $\left(-\tfrac{2}{3}\right)^{10}$ is +ve.
> $\left(-\tfrac{2}{3}\right)^{11}$ is −ve.

Using $S_n = \dfrac{a(1 - r^n)}{1 - r}$

$$\sum_{i=10}^{20} 3 \times \left(-\tfrac{2}{3}\right)^i = 3 \times \left(-\tfrac{2}{3}\right)^{10} \dfrac{\left(1 - \left(-\tfrac{2}{3}\right)^{11}\right)}{1 - \left(-\tfrac{2}{3}\right)}$$

> $1 - \left(-\tfrac{2}{3}\right) = 1 + \tfrac{2}{3} = \tfrac{5}{3}$

$$= 3 \times \left(\tfrac{2}{3}\right)^{10} \left(1 + \left(\tfrac{2}{3}\right)^{11}\right) \times \tfrac{3}{5}$$

> Dividing by $\tfrac{5}{3}$ is the same as multiplying by $\tfrac{3}{5}$.

$$= 0.0316$$

Example 5 $(x - 3)$, $(x + 1)$ and $(3x - 5)$ are consecutive terms of a geometric series. Find possible values of x.

Since the terms are consecutive terms of a geometric series

$$r = \frac{x + 1}{x - 3}$$

> The ratio, r, of consecutive terms is constant.
> So for consecutive terms a, b and c
> $\dfrac{b}{a} = \dfrac{c}{b}$

Also
$$r = \frac{3x - 5}{x + 1}$$

So
$$\frac{x + 1}{x - 3} = \frac{3x - 5}{x + 1}$$

> Two equal fractions, so cross multiply.

$$(x + 1)(x + 1) = (x - 3)(3x - 5)$$

$$x^2 + 2x + 1 = 3x^2 - 14x + 15$$

$$2x^2 - 16x + 14 = 0$$

> Divide through by 2.

$$x^2 - 8x + 7 = 0$$

$$(x - 7)(x - 1) = 0$$

$$\Rightarrow \qquad x = 7 \text{ or } x = 1$$

> *Note*: When $x = 7$, terms are 4, 8, 16.
> When $x = 1$, terms are −2, 2, −2.

Example 6 A machine depreciates in value by 5% every year. After how many years would a machine have lost more than half its value?

Let the original value of the machine be £x.

> No value is given. Use algebra.

After one year: value = £$x \times 0.95$

> To reduce by 5%, multiply by $1 - 0.05 = 0.95$.

After n years: value = £$x \times 0.95^n$

If the machine has lost more than half its value,

$$x \times 0.95^n < \frac{x}{2}$$

> Divide through by x.

$$0.95^n < 0.5$$

> Take logs of both sides on calculator to base 10 (log or lg) or base e (ln)

$$n \log 0.95 < \log 0.5$$

$$n > \frac{\log 0.5}{\log 0.95}$$

> Division is by a negative number ($\log 0.95$), so inequality sign must be reversed.

$$\therefore \quad n > 13.5$$

So $n = 14$

> n, the number of years, must be an integer.

So after 14 years the machine loses more than half its value.

Example 7 A building society pays 6% compound interest per annum.

Mr X deposits £1500.

Ms Y decides to invest £2500 at the start of each year.

They do not make any withdrawals.

a How much will be in Mr X's account at the end of the first year, and at the end of five years?

b Find expressions for the amount in Ms Y's account at the end of the first year, the second year, and the third year.

Form a series to find the total value of Ms Y's account after 10 years.

Solution Let £A_n be the amount in an account after n years.

a For Mr X:

$$A_1 = 1500 \times 1.06 = 1590$$

> To increase by 6% multiply by 1.06.

So, after one year, £1590 is in Mr X's account.

$$A_5 = 1500 \times 1.06^5 = 2007.34$$

So after five years, £2007.34 is in Mr X's account.

b For Ms Y:

$$A_1 = 2500 \times 1.06$$

$$A_2 = (2500 \times 1.06 + 2500) \times 1.06$$

> Each year, £2500 is added to the total.

$$= 2500 \times 1.06^2 + 2500 \times 1.06$$

$$A_3 = 2500 \times 1.06^3 + 2500 \times 1.06^2 + 2500 \times 1.06$$

$$A_{10} = 2500 \times 1.06^{10} + 2500 \times 1.06^9 + \cdots + 2500 \times 1.06$$

> A geometric series with $a = 2500 \times 1.06$, $r = 1.06$ and $n = 10$.

$$= \frac{2500 \times 1.06(1.06^{10} - 1)}{1.06 - 1}$$

$$= 34\,929.11$$

So after ten years, £34 929.11 is in Ms Y's account.

Exercise 20A

1 Which of these series are geometric series?
Write down the common ratios of those that are.

a $3 + 9 + 27 + 81$ **b** $1 + \frac{1}{4} + \frac{1}{16} + \frac{1}{64}$

c $-1 + 2 - 4 + 8$ **d** $1 - 1 + 1 - 1$

e $1 + 1\frac{1}{2} + 1\frac{1}{4} + 1\frac{1}{8}$ **f** $a + a^2 + a^3 + a^4$

g $1 + 1.1 + 1.21 + 1.331$ **h** $\frac{1}{2} + \frac{1}{6} + \frac{1}{12} + \frac{1}{36}$

i $2 + 4 - 8 - 16$ **j** $\frac{3}{4} + \frac{9}{2} + 27 + 162$

2 Write down the terms indicated for each of these geometric series.
Do not simplify the answers.

a $5 + 10 + \cdots$; 11th, 20th   **b** $10 + 25 + \cdots$; 7th, 19th

c $\frac{2}{3} + \frac{3}{4} + \cdots$; 12th, nth   **d** $3 - 2 + \cdots$; 8th, nth

e $\frac{2}{7} - \frac{3}{7} + \cdots$; 9th, nth   **f** $3 + 1\frac{1}{2} + \cdots$; 19th, $2n$th

3 Find, giving the answer as a fraction, the sum to n terms of these geometric series with first term a and common ratio r.

a $a = 2$, $r = \frac{1}{3}$ and $n = 5$ **b** $a = 4$, $r = -\frac{1}{2}$ and $n = 7$

c $a = -2$, $r = -\frac{2}{3}$ and $n = 6$ **d** $a = 1$, $r = -2$ and $n = 11$

4 Find the number of terms in these geometric series:

a $2 + 4 + 8 + \cdots + 512$ **b** $81 + 27 + 9 + \cdots + \frac{1}{27}$

c $0.03 + 0.06 + 0.12 + \cdots + 1.92$ **d** $\frac{8}{81} - \frac{4}{27} + \frac{2}{9} - \cdots - 1\frac{11}{16}$

e $5 + 10 + 20 + \cdots + 5 \times 2^n$ **f** $a + ar + ar^2 + \cdots + ar^{n-1}$

5 Find the sums of the geometric series in Question 4. Simplify, but do not evaluate the answers.

6 Evaluate

a $\displaystyle\sum_{1}^{6} 3^r$ **b** $\displaystyle\sum_{5}^{15} 4(1.5^r)$ **c** $\displaystyle\sum_{1}^{5} \frac{1}{2^r}$

7 The third term of a geometric series is 10, and the sixth is 80.
Find the common ratio, the first term and the sum of the first six terms.

8 The third term of a geometric series is 2, and the fifth is 18.
Find two possible values of the common ratio, and the second term in each case.

9 The three numbers, $n - 2$, n, $n + 3$, are consecutive terms of a geometric series.
Find n, and the term after $n + 3$.

10 A man starts saving on 1 April. He saves 1p the first day, 2p the second, 4p the third, and so on, doubling the amount every day.
If he kept on saving under this system until the end of the month (30 days), how much would he have saved?
Give your answer in pounds, correct to 3 significant figures.

11 The first term of a geometric series is 16 and the fifth term is 9.
What is the value of the seventh term?

12 For the geometric series $4 + 2 + 1 + \frac{1}{2} + \cdots$ find the least number of terms for which the sum exceeds 7.998.

13 The second term of a geometric series of positive terms is 100 and the fourth term is 64.
Find the number of terms that must be added if the total is to exceed four times the first term.

14 Write down the terms indicated for both of these geometric series.
Simplify your answers.

a $\dfrac{1}{k^4} + \dfrac{1}{k^3} + \dfrac{1}{k^2} + \cdots$, 13th, nth   **b** $2p^3 + 2p^2 + 2p + \cdots$, 4th, $(n-1)$th

15 The numbers $x - 4$, $x + 2$, $3x + 1$ are in geometric series.
Find the two possible values of the common ratio.

16 Find the sum of the first n terms of the geometric series $5 + 15 + 45 + \cdots$.
What is the smallest number of terms whose total is more than 10^8?

17 Estimates are produced for the number of babies born worldwide each year. The estimates for 2004 and for 2008, given in thousands of births to the nearest thousand, were 130 350 and 137 804 respectively.
Assume that successive yearly estimates are in geometric progression.

    a Find the annual percentage increase in the number of births.

    b Find the estimates for 2006 and 2011 (to the nearest thousand).

    c Find the estimated total number of births between 2004 and 2012 inclusive (to the nearest thousand).

18 The profit made by a supermarket chain in 2004 is £700 million. The managing director wishes to increase this by 2% per year for the next ten years. In fact, the profit increases annually in geometric progression to £765 million in 2009.

    a Find the annual percentage increase in profit from 2004 to 2009, giving the answer correct to 2 significant figures.

    b Find what the annual percentage increase will need to be from 2009 to 2014 if the profit in 2014 is to match the managing director's original target.

The chain again fails to meet the target between 2010 and 2014; instead profits continue to grow at the same annual rate as in 2004 to 2009.

    c Using the answer to **a**, find the total profit (to the nearest million pounds) from 2004 to 2014 inclusive.

19 A finance company offers a high-interest account with these conditions:

    ■ The client must deposit a fixed amount on 1 January each year.
    ■ The client must not make any withdrawals for at least ten years.
    ■ On 31 December each year, the finance company adds 8% to the value of the investment.

A client decides to invest £1500 each year.

    a Write down a series whose sum is the total value of the investment on 31 December of the tenth year, after the addition of interest.

    b Evaluate this series.

20.2 Infinite geometric series

The study of the infinitely large, the infinitely small and infinite series has always fascinated thinkers and led to famous paradoxes.

> To quote from Tom Stoppard in his play Jumpers*:
>
> ... But it was precisely this notion of infinite series which in the sixth century BC led the Greek philospher Zeno to conclude that since an arrow shot towards a target first had to cover half the distance, and then half the remainder after that, and so on ad infinitum, the result was, as I will now demonstrate, that though an arrow is always approaching its target, it never quite gets there, and Saint Sebastian died of fright. ...

Consider the geometric series $1 + \frac{1}{2} + \frac{1}{4} + \frac{1}{8} + \cdots$ of Example 1b on page 353. By considering the sum of the first 1, 2, 3, 4, ... terms it could be seen that the series was tending to 2 and that the more terms that were included the closer their sum was to 2.

The formula for S_n can be used to prove that the limiting value, the sum to infinity, is 2.

In other words, as $n \to \infty$, $S_n \to 2$.

Using the formula $S_n = \dfrac{a(1 - r^n)}{1 - r}$ with $a = 1$ and $r = \frac{1}{2}$:

$$S_n = \frac{1 - \left(\frac{1}{2}\right)^n}{\frac{1}{2}}$$

Multiply numerator and denominator by 2.

$$= 2 - \left(\tfrac{1}{2}\right)^{n-1}$$

$$= 2 - \frac{1}{2^{n-1}}$$

As n increases, 2^{n-1} increases and so $\dfrac{1}{2^{n-1}}$ tends to zero.

By taking large enough values of n, S_n can be made as close to 2 as one wishes.

So the sum to infinity, $S_\infty = 2$.

$S_\infty = \lim\limits_{n \to \infty} S_n = 2$

To obtain a general formula for S_∞, consider

$$S_n = \frac{a(1 - r^n)}{1 - r}$$

$$= \frac{a}{1 - r} - \frac{ar^n}{1 - r}.$$

If r is numerically less than 1 then r^n gets smaller as n increases.

So providing that $|r| < 1$ then, as n tends to infinity, r^n tends to 0 and so the term $\dfrac{ar^n}{1 - r}$ tends to 0.

*Published by Faber and Faber.

$$S_\infty = \frac{a}{1-r} \quad \text{providing} \quad |r| < 1$$

Note The sum to infinity of a geometric series exists if, and only if, $|r| < 1$, i.e. $-1 < r < 1$.

Example 8 *This example offers a proof without words!*

Find the sum to infinity of the geometric series: $1 + \frac{1}{2} + \frac{1}{4} + \frac{1}{8} + \frac{1}{16} + \cdots$.

This diagram should be sufficient to demonstrate that $S_\infty = 2$.

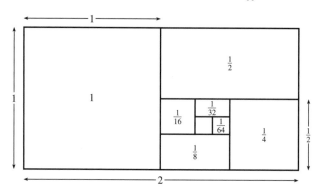

Example 9 Express $0.\dot{7}\dot{2}$ as a fraction by considering the recurring decimal as a geometric series.

$0.\dot{7}\dot{2} = 0.727\,272\,72\ldots$

$= \frac{72}{100} + \frac{72}{10\,000} + \frac{72}{1\,000\,000} + \cdots$

$= \frac{72}{100}\left(1 + \frac{1}{100} + \frac{1}{10\,000} + \cdots\right)$

$= \frac{72}{100} \times \dfrac{1}{1 - \frac{1}{100}}$

$= \frac{72}{100} \times \frac{100}{99}$

$= \frac{72}{99}$

$= \frac{8}{11}$

The series in the brackets is an infinite geometric series with $a = 1, r = \dfrac{1}{100}$.

Use $S_\infty = \dfrac{a}{1-r}$

Example 10 The second term of a geometric series is 16 and its sum to infinity is 100.
Find two possible values of the common ratio and the corresponding first terms.

2nd term $= 16 \quad \Rightarrow \quad ar = 16 \qquad ①$

$S_\infty = 100 \quad \Rightarrow \quad \dfrac{a}{1-r} = 100 \qquad ②$

The two equations in a and r have to be solved simultaneously.

Rearranging ② $\qquad a = 100(1-r)$

Substituting in ①

$$100(1 - r)r = 16$$

Divide both sides by 4.

$$25r - 25r^2 = 4$$

$$25r^2 - 25r + 4 = 0$$

See second note below.

$$(5r - 1)(5r - 4) = 0$$

$$\Rightarrow \qquad r = \tfrac{1}{5} \quad \text{or} \quad r = \tfrac{4}{5}$$

Both values of r are such that $|r| < 1$ so S_∞ does exist.

When $r = \tfrac{1}{5}$, $\quad a = \dfrac{16}{r} = 80$.

When $r = \tfrac{4}{5}$, $\quad a = \dfrac{16}{r} = 20$.

So, either $a = 20$ and $r = \tfrac{4}{5}$, or $a = 80$ and $r = \tfrac{1}{5}$.

Exercise 20B

1 Find the sums of these series **i** to n terms and **ii** to infinity:

a $1 + \tfrac{1}{3} + \tfrac{1}{9} + \tfrac{1}{27} + \cdots$

b $12 + 6 + 3 + 1\tfrac{1}{2} + \cdots$

c $\tfrac{3}{10} + \tfrac{3}{100} + \tfrac{3}{1000} + \tfrac{3}{10\,000} + \cdots$

d $\tfrac{13}{100} + \tfrac{13}{10\,000} + \tfrac{13}{1\,000\,000} + \cdots$

e $0.5 + 0.05 + 0.005 + \cdots$

f $0.54 + 0.0054 + 0.000\,054 + \cdots$

g $1 - \tfrac{1}{2} + \tfrac{1}{4} - \tfrac{1}{8} + \cdots$

h $54 - 18 + 6 - 2 + \cdots$

2 Evaluate these sums to infinity.

a $\displaystyle\sum_{r=1}^{\infty} 0.8^r$

b $\displaystyle\sum_{r=1}^{\infty} \dfrac{5}{3^{r-1}}$

c $\displaystyle\sum_{r=4}^{\infty} \left(\tfrac{3}{4}\right)^r$

d $\displaystyle\sum_{r=6}^{\infty} \dfrac{3p}{2^{2r}}$

3 If the sum to infinity of a geometric series is three times the first term, what is the common ratio?

4 The sum to infinity of a geometric series is 4 and the second term is 1. Find the first, third, and fourth terms.

5 The second term of a geometric series is 24 and its sum to infinity is 100. Find the two possible values of the common ratio and the corresponding first terms.

6 The second term of a geometric series is 11 and the sum to infinity is twice as large as the first term. Find the sum of the first ten terms.

7 The sum to three terms of a geometric series is 61 and the common ratio is 0.8. Find the sum to infinity.

8 The second term of a geometric series is 4 and the sum to infinity is 18.

a Show that one possible value of the common ratio is $\frac{1}{3}$ and find the sum of the first ten terms given this common ratio.

b Find the other possible value of the common ratio. Given this second value, find how many terms must be summed for the total to exceed 17.

**Exercise 20C
(Review)**

1 For these geometric series, find the next two terms, the sum of n terms, state whether or not the series converge and, if so, state the sum to infinity.

a $4 + 2 + 1 + \frac{1}{2} + \cdots$ **b** $3 + 3.3 + 3.63 + 3.993 + \cdots$

c $1 - 4 + 16 - 64 + \cdots$ **d** $60 - 12 + 2.4 - 0.48 + \cdots$

2 Find the sums of these geometric series to ten terms.
Give the answers correct to 3 significant figures.

a $0.1 + 0.2 + \cdots$ **b** $50 + 49 + \cdots$ **c** $16 - 8 + \cdots$

d The geometric series whose second term is 110 and whose fourth term is 133.1

e The geometric series whose fourth term is a thousand and whose ninth term is 0.01

f The geometric series whose first term is 96 and whose fourth term is 40.5

3 The fourth term of a geometric series is -6 and the seventh term is 48.
Find the first three terms of the series.

4 The sum to infinity of a geometric series is four times its first term. Find the common ratio.

5 Evaluate these sums.

a $\displaystyle\sum_{r=1}^{\infty} \frac{3k}{6^r}$ **b** $\displaystyle\sum_{3}^{7}\{(-2)^r + r - 2\}$

6 Find the sum of these geometric series, giving the answers to 3 significant figures.

a $\frac{9}{4} + \frac{3}{2} + 1 + \cdots + \frac{1024}{59\,049}$ **b** $\frac{1}{5} - \frac{3}{20} + \frac{9}{80} - \cdots + \frac{6561}{327\,680}$

7 For the geometric series $\frac{4}{3} + \frac{8}{9} + \frac{16}{27} + \cdots$, find the least number of terms for which the sum exceeds 3.92.

8 For the geometric series $6 + 12.6 + 26.46 + \cdots$, find the greatest number of terms for which the sum does not exceed 1000.

9 Prove that $1 - \dfrac{1}{\sqrt{2}} + \dfrac{1}{2} - \dfrac{1}{2\sqrt{2}} + \cdots = 2 - \sqrt{2}$.

10 Prove that, for $0 < A < \frac{\pi}{2}$,

$$\sin^2 A + \sin^2 A \cos^2 A + \sin^2 A \cos^4 A + \sin^2 A \cos^6 A + \cdots = 1$$

11 Find the difference between the sums to ten terms of the arithmetic series and the geometric series whose first two terms are -2 and 4.

12 Find x and y given that $x + 3y$, 4 and $2x + 2y$ are consecutive terms of a geometric series, and $x + 4y$, 4 and $x - 2y$ are consecutive terms of an arithmetic series.

13 Express these recurring decimals as rational numbers.

a $0.\dot{8}$

b $0.\dot{1}\dot{2}$

c $3.\dot{2}$

d $2.6\dot{9}$

e $1.00\dot{4}$

f $2.9\dot{6}\dot{0}$

14 Show that the sums to infinity of the geometric series $3 + \frac{9}{4} + \frac{27}{16} + \cdots$ and $4 + \frac{8}{3} + \frac{16}{9} + \cdots$ are equal.

15 Show that there are two possible geometric series in each of which the first term is 8, and the sum of the first three terms is 14.

Find the second term and the sum of the first seven terms in each series.

16 A building society offers 6% interest per year on investments.
Someone deposits £4000 in an account and leaves the interest to accumulate.

a Find the total value of the investment (to the nearest pound) after six years.

b Find how many years have elapsed before the investment has at least doubled in value.

c Find how much interest is paid (to the nearest pound) at the end of the second and at the end of the seventh year.

17 The story is told of a poor man who invented the game of chess. He presented it to the king who was so impressed that he offered the inventor any reward he chose. The inventor asked for a grain of rice on the first square of the board, two grains on the second square, four on the third and so on. The king thought this a very modest request; however, when the total was calculated, it came to more than all the grains of rice in the world. Find the total number of grains, giving the answer in standard form, to two significant figures.

A 'Test yourself' exercise (Ty20) and an 'Extension exercise' (Ext20) can be found on the CD-ROM.

Key points

Geometric series

A **geometric series** is a series whose consecutive terms have a common ratio.

If the first term of a geometric series is a and common ratio is r then

nth term $= ar^{n-1}$

$$S_n = \frac{a(1 - r^n)}{1 - r} = \frac{a(r^n - 1)}{r - 1}$$

Use $\quad S_n = \dfrac{a(1 - r^n)}{1 - r}$ for $r < 1$

Use $\quad S_n = \dfrac{a(r^n - 1)}{r - 1}$ for $r > 1$

If $-1 < r < 1$, i.e. $|r| < 1$, the series converges.

$$S_\infty = \frac{a}{1 - r} \qquad \text{for } |r| < 1$$

For a geometric series with terms $u_1, u_2, u_3, \ldots, u_n, \ldots$ the common ratio, r, is the ratio of consecutive terms.

$$u_n = r\, u_{n-1} \text{ or } r = \frac{u_n}{u_{n-1}}$$

1 **a** Using the substitution $u = 2^x$, show that the equation $4^x - 2^{(x+1)} - 15 = 0$ can be written in the form $u^2 - 2u - 15 = 0$.

   b Hence solve the equation $4^x - 2^{(x+1)} - 15 = 0$, giving your answers to 2 decimal places.

   Edexcel Nov 2002

2 Evaluate

   $$\int_1^4 \left(\frac{x}{2} + \frac{1}{x^2} \right) dx.$$

   Edexcel Nov 2002

3

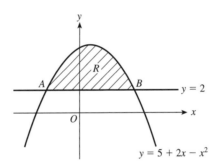

The diagram shows the curve with equation $y = 5 + 2x - x^2$ and the line with equation $y = 2$. The curve and the line intersect at the points A and B.

   a Find the x-coordinates of A and B.

   The shaded region R is bounded by the curve and the line.

   b Find the area of R.

   Edexcel Jan 2001

4 Given that $p = \log_q 16$, express in terms of p,

   a $\log_q 2$,

   b $\log_q (8q)$.

   Edexcel Jan 2002

5 A geometric series has first term 1200. Its sum to infinity is 960.

   a Show that the common ratio of the series is $-\frac{1}{4}$.

   b Find, to 3 decimal places, the difference between the ninth and tenth terms of the series.

   c Write down an expression for the sum of the first n terms of the series.

   Given that n is odd,

   d prove that the sum of the first n terms of the series is

   $$960(1 + 0.25^n).$$

   Edexcel Jan 2003

6 The diagram shows a sketch of part of the curve with equation $y = f(x)$ where $f(x) = -x^3 + 27x - 34$.

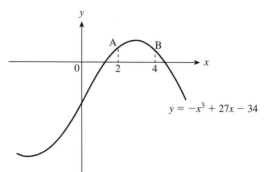

$$y = -x^3 + 27x - 34$$

a Find $\int f(x) \, dx$.

The lines $x = 2$ and $x = 4$ meet the curve at points A and B as shown.

b Find the area of the finite region bounded by the curve and the lines $x = 2$, $x = 4$ and $y = 0$.

c Find the area of the finite region bounded by the curve and the straight line AB.

London May 1995

7 A competitor is running in a 25 km race. For the first 15 km, she runs at a steady rate of 12 km h^{-1}. After completing 15 km, she slows down and it is now observed that she takes 20% longer to complete each kilometre than she took to complete the previous kilometre.

a Find the time, in hours and minutes, the competitor takes to complete the first 16 km of the race.

The time taken to complete the rth kilometre is u_r hours.

b Show that, for $16 \leqslant r \leqslant 25$, $u_r = \dfrac{1}{12}(1.2)^{r-15}$.

c Using the answer to **b**, or otherwise, find the time, to the nearest minute, that she takes to complete the race.

London June 1998

8 $y = 3x^{\frac{1}{2}} - 4x^{-\frac{1}{2}}$, $x > 0$.

a Find $\dfrac{dy}{dx}$.

b Find $\int y \, dx$.

c Hence show that
$$\int_1^3 y \, dx = A + B\sqrt{3}$$
where A and B are integers to be found.

London Jan 1998

9 The nth term of a sequence is u_n, where $u_n = 95\left(\frac{4}{5}\right)^n$, $n = 1, 2, 3, \ldots$

a Find the values of u_1 and u_2.

Giving your answers to 3 significant figures, calculate

b the value of u_{21}

c $\displaystyle\sum_{n=1}^{15} u_n$.

d Find the sum to infinity of the series whose first term is u_1 and whose nth term is u_n.

London May 1995

10 The second term of a geometric series is 80 and the fifth term of the series is 5.12.

a Find the common ratio and the first term of the series.

b Find the sum to infinity of the series, giving your answer as an exact fraction.

c Find the difference between the sum to infinity of the series and the sum of the first 14 terms of the series, giving your answer in the form $a \times 10^n$, where $1 \leqslant a < 10$ and n is an integer.

Edexcel Specimen paper

11

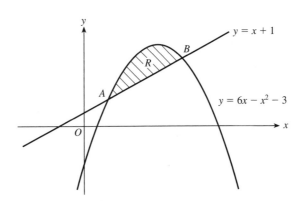

The diagram shows the line with equation $y = x + 1$ and the curve with equation $y = 6x - x^2 - 3$.

The line and the curve intersect at the points A and B, and O is the origin.

a Calculate the coordinates of A and the coordinates of B.

The shaded region R is bounded by the line and the curve.

b Calculate the area of R.

Edexcel Jan 2002

12 a Simplify $\dfrac{x^2 + 4x + 3}{x^2 + x}$

b Find the value of x for which $\log_2(x^2 + 4x + 3) - \log_2(x^2 + x) = 4$.

Edexcel June 2003

13 A measure of the effective voltage, M volts, in an electrical circuit is given by

$$M^2 = \int_0^1 V^2 \mathrm{d}t,$$

where V volts is the voltage at time t seconds. Pairs of values of V and t are given in the following table.

t	0	0.25	0.5	0.75	1
V	-48	207	37	-161	-29
V^2					

Use the trapezium rule with five values of V^2 to estimate the value of M.

Edexcel June 2001

14 A geometric series is $a + ar + ar^2 + \ldots$

a Prove that the sum of the first n terms of this series is given by

$$S_n = \frac{a(1 - r^n)}{1 - r}.$$

The second and fourth terms of the series are 3 and 1.08 respectively.

Given that all terms in the series are positive, find

b the value of r and the value of a,

c the sum to infinity of the series.

Edexcel Jan 2002

15 Given that $f(x) = (2x^{\frac{3}{2}} - 3x^{-\frac{3}{2}})^2 + 5$, $x > 0$,

a find, to 3 significant figures, the value of x for which $f(x) = 5$.

b Show that $f(x)$ may be written in the form $Ax^3 + \dfrac{B}{x^3} + C$, where A, B and C are constants to be found.

c Hence evaluate $\displaystyle\int_1^2 f(x)\mathrm{d}x$.

Edexcel May 2002

16 The following is a table of values for $y = \sqrt{(1 + \sin x)}$, where x is in radians.

x	0	0.5	1	1.5	2
y	1	1.216	p	1.413	q

a Find the value of p and the value of q.

b Use the trapezium rule and all the values of y in the completed table to obtain an estimate of I, where

$$I = \int_0^2 \sqrt{(1 + \sin x)}\mathrm{d}x.$$

Edexcel Jan 2002

17 The third and fourth terms of a geometric series are 6.4 and 5.12 respectively. Find

   a the common ratio of the series,

   b the first term of the series,

   c the sum to infinity of the series.

   d Calculate the difference between the sum to infinity of the series and the sum of the first 25 terms of the series

Edexcel Jan 2001

18 A sequence of numbers $u_1, u_2, \ldots, u_n, \ldots$ is given by the formula

$$u_n = 3\left(\frac{2}{3}\right)^n - 1,$$

where n is a positive integer.

   a Find the values of u_1, u_2 and u_3.

   b Find $\displaystyle\sum_{n=1}^{15} 3\left(\frac{2}{3}\right)^n$, and hence show that $\displaystyle\sum_{n=1}^{15} u_n = -9.014$ to 4 significant figures.

   c Prove that $3u_{n+1} = 2u_n - 1$.

London Jan 1998

19 A geometric series has third term 27 and sixth term 8.

   a Show that the common ratio of the series is $\frac{2}{3}$.

   b Find the first term of the series.

   c Find the sum to infinity of the series.

   d Find, to 3 significant figures, the difference between the sum of the first 10 terms of the series and the sum to infinity of the series.

London June 1997

20 Evaluate $\displaystyle\int_1^4 \frac{2}{x^2}\,\mathrm{d}x$.

Edexcel Mock paper 1

21 The sequence $u_1, u_2, u_3, \ldots, u_n$ is defined by the recurrence relation

$u_{n+1} = pu_n + 5$, $u_1 = 2$, where p is a constant.

Given that $u_3 = 8$,

   a show that one possible value of p is $\frac{1}{2}$ and find the other value of p.

Using $p = \frac{1}{2}$,

   b write down the value of $\log_2 p$.

Given also that $\log_2 q = t$,

   c express $\log_2\left(\dfrac{p^3}{\sqrt{q}}\right)$ in terms of t.

Edexcel Nov 2002

22 a An arithmetic series has first term a and common difference d. Prove that the sum of the first n terms of the series is

$$\tfrac{1}{2}n[2a + (n-1)d].$$

A company made a profit of £54 000 in the year 2001. A model for future performance assumes that yearly profits will increase in an arithmetic sequence with common difference £d. This model predicts total profits of £619 200 for the 9 years 2001 to 2009 inclusive.

b Find the value of d.

Using your value of d,

c find the predicted profit for the year 2011.

An alternative model assumes that the company's yearly profits will increase in a geometric sequence with common ration 1.06. Using this alternative model and again taking the profit in 2001 to be £54 000,

d find the predicted profit for the year 2011.

Edexcel Nov 2002

23

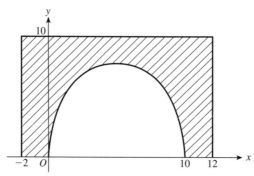

The diagram shows the cross-section of a road tunnel and its concrete surround. The curved section of the tunnel is modelled by the curve with equation

$$y = 8\sqrt{\left(\sin\frac{\pi x}{10}\right)},\text{ in the interval } 0 \leqslant x \leqslant 10.$$

The concrete surround is represented by the shaded area bounded by the curve, the x-axis and the lines $x = -2$, $x = 12$ and $y = 10$. The units on both axes are metres.

a Using this model, copy and complete the table below, giving the values of y to 2 decimal places.

x	0	2	4	6	8	10
y	0	6.13				0

The area of the cross-section of the tunnel is given by $\displaystyle\int_0^{10} y\,\mathrm{d}x$.

b Estimate this area, using the trapezium rule with all the values from your table.

c Deduce an estimate of the cross-sectional area of the concrete surround.

d State, with a reason, whether your answer in part **c** over-estimates or under-estimates the true value.

Edexcel Jan 2003

- The paper is 1 hour 30 minutes long.

- Calculators may be used.

1. The curve C has equation

$$y = 2x + \frac{10}{x}, \quad x > 0.$$

(a) Find $\frac{dy}{dx}$.

(2)

The coordinates of the stationary point on C are $(\sqrt{a}, b\sqrt{a})$, where a and b are integers.

(b) Find the value of a and the value of b.

(4)

(c) Determine the nature of the stationary point.

(4)

___

2. The speed, $v\,\mathrm{m\,s}^{-1}$, of a car at time t seconds is modelled by

$$v = 0.006(400t - 4.9\,t^2), \qquad 0 \leqslant t \leqslant 20.$$

(a) Copy and complete the following table showing the speed of the car at 5 second intervals. Give the values of v to 2 decimal places.

t	0	5	10	15	20
v	0		21.06		36.24

(2)

The distance, s metres, travelled by a car in the first 20 seconds is given by the area under the graph of v against t that lies between the curve, the line $t = 20$ and the t-axis.

(b) Use the trapezium rule, with all the values from your table, to estimate s.

(4)

___

3.

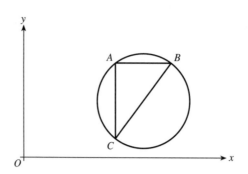

In the figure the coordinates of point A, B and C are $(5, 5)$, $(8, 5)$ and $(5, 1)$ respectively.

(a) Show that $AB^2 + AC^2 = BC^2$.

(2)

A circle is drawn which passes through the points A, B and C.

(b) Find the coordinates of the centre of the circle and the radius of the circle.

(2)

(c) Write down an equation of the circle.

(2)

---

4. The first 3 terms, in ascending powers of x, of the binomial expansion $(1 + bx)^n$ are $1 + 28x + 336x^2$.
Given that n is a positive integer, find the value of n and the value of b.

(8)

---

5. A geometric series has third term 18 and sixth term 486.

(a) Show that the common ratio is 3.

(4)

(b) Find the first term of the series.

(2)

(c) Hence calculate the sum of the first 15 terms.

(2)

6.

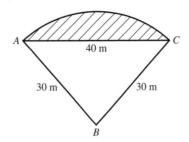

The figure shows a plan of a garden. The triangle ABC is a lawn and the shaded area is a flower-bed.

There is a straight line of length 40 m from the point A to the point C. The points are joined by an arc of the circle whose centre is B and whose radius is 30 m as shown in the figure.

(a) Show that the size of the angle ABC is 1.46 radians to 2 decimal places.

(3)

Calculate, giving your answers to 3 significant figures,

(b) the area of the entire garden,

(2)

(c) the area of the flower-bed.

(4)

---

7. Given that $2\log_3 y = c - \log_3 x$, $x > 0$, $y > 0$,

where c is the constant, and $y = 2$ when $x = 4$,

(a) show that

$$y = 4x^{-\frac{1}{2}}$$

(4)

Given that $\log_3 y = 1 - 2\log_3 x$, $x > 0$, $y > 0$,

(b) show that $yx^a = b$, stating the value of a and the value of b.

(3)

(c) Hence solve the simultaneous equations

$$2\log_3 y = c - \log_3 x, \quad \log_3 y = 1 - 2\log_3 x, \quad x > 0, \ y > 0$$

giving your answers to 3 significant figures.

(3)

8. Find all the values of θ, correct to 1 decimal place, in the interval $0° \leqslant \theta \leqslant 360°$ for which

(a) $\tan(\theta + 30°) = 0.4$

(3)

(b) $\sin\theta + \cos^2\theta = 0.2$

(7)

9.

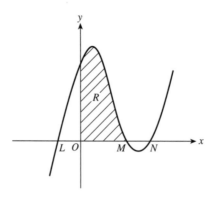

The figure shows the curve C with equation $y = f(x)$.

$$f(x) = x^3 - 4x^2 + x + 6.$$

The curve C crosses the x-axis at L, M and N.

(a) Show that $(x - 3)$ is a factor of $f(x)$

(2)

(b) Hence, or otherwise, factorise $f(x)$ fully.

(4)

The shaded region R is bounded by C, the positive x-axis and the positive y-axis.

(c) Find the area of R.

(5)

END

Answers

1 Surds and Indices

Exercise 1A (p. 6)

1 a 5 b $\frac{1}{2}$ c 48 d $\frac{1}{2}$
  e $\frac{a}{b}$ f 15 g 21 h $\frac{p}{q}$
  i $\frac{1}{4p}$ j $\frac{9a}{2b}$

2 a $2\sqrt{2}$ b $2\sqrt{3}$ c $3\sqrt{3}$
  d $5\sqrt{2}$ e $3\sqrt{5}$ f $11\sqrt{10}$
  g $5\sqrt{3}$ h $4\sqrt{2}$ i $6\sqrt{2}$
  j $7\sqrt{2}$

3 a $\sqrt{18}$ b $\sqrt{12}$ c $\sqrt{80}$ d $\sqrt{24}$
  e $\sqrt{72}$ f $\sqrt{216}$ g $\sqrt{128}$ h $\sqrt{1000}$
  i $\sqrt{175}$ j $\sqrt{392}$

Exercise 1B (p. 8)

1 a $3\sqrt{2}$ b $6\sqrt{3}$ c $4\sqrt{7}$
  d $5\sqrt{10}$ e $28\sqrt{2}$ f 0

2 a $\frac{6}{7} + \frac{2}{7}\sqrt{2}$ b $9 + 4\sqrt{5}$
  c $-1 + \sqrt{2}$ d $4 - 2\sqrt{3}$
  e $-1 - \sqrt{2}$ f $\frac{1}{2} + \frac{3}{4}\sqrt{2}$

3 a $\frac{1}{5}\sqrt{5}$ b $\frac{1}{7}\sqrt{7}$ c $-\frac{1}{2}\sqrt{2}$
  d $\frac{2}{3}\sqrt{3}$ e $\frac{1}{2}\sqrt{6}$ f $\frac{1}{4}\sqrt{2}$
  g $-\frac{1}{2}\sqrt{3}$ h $\frac{3}{8}\sqrt{6}$ i $\sqrt{2} - 1$
  j $2 + \sqrt{3}$
  k $5 + 2\sqrt{6}$
  l $\frac{1}{2}(5 + \sqrt{3} + \sqrt{5} + \sqrt{15})$
  m $-7 + 3\sqrt{6}$
  n $4 + \sqrt{10}$
  o $3 + 2\sqrt{2}$
  p $\sqrt{2}$

4 a 12 b $3\sqrt{3}$ c 6 d 2
  e $3\sqrt{3}$ f $\sqrt{3}$ g $4\sqrt{3}$ h $\frac{1}{6}\sqrt{3}$
  i 30 j 4 k $3\sqrt{3}$ l $11\sqrt{3}$

5 a $6\sqrt{10}$ b 24 c $6\sqrt{6}$
  d $\frac{19}{25}\sqrt{5}$ e $4\sqrt{2}$ f 1

6 $10\sqrt{3}\,\text{cm}$; $18\,\text{cm}^2$; $\sqrt{39}\,\text{cm}$

7 $(6\sqrt{7} + 2\sqrt{21})\,\text{cm}$; $21\sqrt{3}\,\text{cm}^2$; $2\sqrt{21}\,\text{cm}$

8 $(6\sqrt{2} + 5)\,\text{cm}$

9 $2\sqrt{17}\,\text{m}$

10 $5\sqrt{3}$ units

11 $6\sqrt{2}\,\text{cm}$

12 $2\sqrt{6}\,\text{cm}$

13 a $\sqrt{3}$ b 10 c $2 - \sqrt{2}$
  d $4\sqrt{2} + 4$ e $\frac{1}{2}\sqrt{3} + \frac{3}{2}$ f $f - \sqrt{3}$
  g $-\sqrt{2} - 1$

14 a 2.2% b 1%

Exercise 1C (p. 13)

1 a 5 b 3 c 2 d 7
  e $\frac{1}{2}$ f 1 g -2 h -1
  i 16 j 9 k 125 l 343
  m $\frac{1}{8}$ n $\frac{2}{3}$ o $\frac{3}{2}$ p $\frac{2}{3}$

2 a 1 b $\frac{1}{3}$ c 1 d $\frac{1}{4}$
  e $\frac{1}{8}$ f 2 g 9 h 1
  i $\frac{1}{27}$ j $-\frac{1}{6}$ k 1 l $2\frac{1}{4}$
  m 4 n 3 o $4\frac{1}{2}$ p $\frac{5}{9}$

3 a $\frac{1}{2}$ b $\frac{1}{4}$ c $\frac{1}{2}$ d $\frac{1}{8}$
  e $\frac{1}{9}$ f 2 g 2 h 9
  i $1\frac{1}{2}$ j $1\frac{1}{2}$ k $1\frac{1}{2}$ l $\frac{16}{81}$

4 a 0.4 b $\frac{8}{27}$ c $3\frac{3}{8}$ d $\frac{2}{3}$

Exercise 1D (p. 14)

1 a x^{-1} b x^{-2} c $x^{\frac{1}{2}}$ d $x^{\frac{1}{3}}$
  e x^{-4} f x^4 g $x^{-\frac{1}{2}}$ h x^7
  i $x^{-\frac{1}{3}}$ j $x^{\frac{2}{3}}$ k $x^{-\frac{1}{4}}$ l $x^{-\frac{5}{2}}$

2 a $3x^{-1}$ b $\frac{4}{3}x^{-1}$ c $6x^{\frac{1}{2}}$ d $5x^{-3}$
  e $\frac{1}{4}x^{-4}$ f $\frac{1}{3}x^{\frac{1}{2}}$ g $\frac{1}{5}x^{-\frac{1}{2}}$ h $6x^7$
  i $\frac{4}{5}x^{-\frac{1}{3}}$ j $7x^{\frac{2}{3}}$ k $2x^{-\frac{3}{4}}$ l $\frac{8}{3}x^{-\frac{1}{2}}$

Exercise 1E (p. 16)

1 a a^3 b $4b^2$ c x^8
  d y^{11} e $6b^5$ f $24a^6$
  g $15x^5y^3$ h a^2 i x^2
  j y^4 k $64a^6$ l $4a^6$
  m a^{m^2+m} n a^{2m+2} o $2x^{-1}y$
  p u^6 q v^{2x} r v^{3x}
  s v^{6x} t v^{2x^2} u 1
  v $x^{\frac{1}{6}}$ w a^{x-y+z} x $3y^8$
  y p^{-5} z 1

2 a $s^{-2}; \dfrac{1}{s^2}$ b $y^{-5}; \dfrac{1}{y^5}$
  c $t^{-b}; \dfrac{1}{t^b}$ d $x^{p-q}; \dfrac{1}{x^{q-p}}$

3 a 25 b 4 c 1000 d 16
  e 4 f 512 g 4 h 9
  i $\frac{1}{2}$ j 1 k -3 l $-\frac{1}{2}, 1$
  m $-2, -1$ n 19.0 o 0.435
  p 11.2

Exercise 1F Review (p. 17)

1 a 5 b 20 c $3\sqrt{5}$ d $-4\sqrt{5}$
  e 2 f $5\sqrt{5}$ g $\sqrt{10}$

2 a $4\sqrt{5}$ b $4\sqrt{2}$ c $6\sqrt{2}$ d $11\sqrt{5}$

3 a $\frac{1}{3}\sqrt{3}$ b $\frac{1}{2}\sqrt{2}$ c $\frac{4}{7}\sqrt{7}$
  d $\dfrac{4 + \sqrt{10}}{6}$ e $\sqrt{6} - 2$ f $\frac{1}{4}\sqrt{6}$
  g $3(\sqrt{6} + \sqrt{5})$

4 a $-3, 2$ b 49 c 0.511 d 27
  e $\frac{1}{2}$ f -8 g -1 h 0.597

5 $4\sqrt{6}\,\text{cm}$

6 $(14 + 2\sqrt{7})\,\text{cm}$; $(5\sqrt{7} - 2)\,\text{cm}^2$

7 $(15 - 8\sqrt{3})\,\text{cm}$

8 $20\sqrt{13}\,\text{cm}$

9 $12\sqrt[3]{10}\,\text{cm}$; $(6 \times 10^{\frac{2}{3}})\,\text{cm}^2$

10 $x = 3, y = 2$

11 a 1 b 64 c $\frac{1}{4}$ d 2 e 8
  f $\frac{1}{32}$ g 5 h $\frac{8}{27}$ i $\frac{9}{4}$ j $\frac{8}{27}$
  k $\frac{3}{2}$ l 1 m 6 n $\frac{1}{6}$ o 256

12 a 3 b 64 c 7
  d 1 e 1000 f 1

13 a 7^9 b 3^{11} c 80 d 5^7
  e 992 f 2^5 g 8^2 h 222
  i 7^3 j 7^4 k 5^9

14 a a^8 c $5c^3$ d d^9 f $6f^5$
  g $20g^5$ h h^{12} i i^{12} j $9j^2$
  k $4k^6$ l $49l^{12}$ m m^6n^3 n p^6q^{12}
  o $9r^2s^2$ p v^3w^3 q x r y^4
  s z^5

4 a $16\pi^2r^2$ b $4a^6$ c b^4c^6
  d $\dfrac{4a^2}{b^2}$ e $9y^6$ f $\dfrac{10^4a^6}{b^4}$
  g $4a$ h $\frac{9}{16}a^{\frac{1}{2}}$

5 a x^4 b $2y$ c $5ab^2$
  d $6x^3$ e $a^{-2}b^3$ f $-2a$
  g $3xy^3$ h $\dfrac{10}{9x^5y^4}$

6 a $\dfrac{4}{x^2}$ b $\dfrac{1}{3a}$ c $\dfrac{4t^2}{3s}$
  d 1 e $\frac{1}{2}a$ f $9b$
  g $12x$ h $60x^4y^3$ i $4a^4$
  j $\frac{1}{2}b^4$ k $4x^6y^{10}$ l 1
  m $-2b^2c^5$ n $-4l^2m$ o $-2xy^2$
  p $d^{10}e^{15}$ q $60a^3b^3$ r $\dfrac{n}{m}$
  s $-m^5$ t n^6 u $2h^5$
  v $64p^{-6}$ w $\frac{27}{2}a^{-5}$

2 Algebraic expressions

Exercise 2A (p. 21)

1 a 63 **b** 24 **c** $-\frac{3}{4}$
 d $n^2 - 2n$ **e** $9k^2 - 6k$ **f** $a^2 - 1$

2 a 1 **b** 126 **c** $\frac{91}{64}$
 d -7 **e** 2 **f** 0
 g $a^3 + 1$ **h** $27k^3 + 1$

3 a $25 + 10h + h^2$ **b** $10 + h$
 c $2a + h$

Exercise 2B (p. 24)

1 a $x^4 + 2x^3 - x - 5$; 4
 b $-x^5 + 3x^3 + 1$; 5
 c $6x^4 - x^3 - 2x^2$; 4
 d -10; 0

2 a 2; -5 **b** 3; 1
 c -1; 0 **d** 0; -10

3 a $-3y^3 + 2x$
 b $-6x^2 - 2xy - 3y^2$
 c $3x^3 - 7x^2 - 4x$
 d $2ab + 8a^2b - 7ab^2$

4 a $x^2 + 2x + 1$ **b** $x^2 + 4x + 4$
 c $4x^2 + 4x + 1$ **d** $x^2 - 10x + 25$
 e $x^2 + 3x + 2$ **f** $x^2 + 7x + 12$
 g $x^2 - x - 12$ **h** $x^2 - x - 2$
 i $2x^2 + 3x + 1$ **j** $3x^2 - 14x + 8$
 k $16x^2 - 9$ **l** $x^2 - 4$

5 a $2a^2 - 5a - 3$
 b $4d^2 + 11de + 6e^2$
 c $6 - 11x + 4x^2$
 d $-56w^2 + 113w - 56$
 e $9l^2 - 25$
 f $9x^2 + 24x + 16$
 g $4h^2 - 4h - 35$
 h $10x^2 - x - 21$
 i $6x^2 + 11x + 4$
 j $4 - 49x^2$
 k $4y^2 - 12y + 9$
 l $16 - b^4$
 m $x^3 - 3x^2 + 3x - 1$
 n $a - b$
 o $-x^3 - x^2 + 10x - 8$

6 a $4k^2 + 16k + 20$
 b $2m^2 + 6m + 10$
 c $7s^2 + 6s - 7$
 d $2v^2 + 18$
 e $-2y^2 - 24y - 15$
 f $3x^3 + 4x^2 - x - 6$

7 a $x^3 - x^2 + 3x + 4$
 b $-x^3 + 3x^2 + 3x - 6$
 c $x^5 + x^4 - 7x^3 + 7x^2 + 15x - 5$
 d $x^3 + 4x^2 + 2x - 1$
 e $x^4 + 6x^3 + 7x^2 - 6x + 1$
 f $x^3 + 13x^2 - 3x - 10$

8 a 8 **b** 14 **c** 15

9 a 9 **b** 9 **c** 2 **d** -7

Exercise 2C (p. 26)

1 a $\frac{1}{3}$ **b** $\frac{1}{3}$ **c** $\frac{10}{27}$
 d $\frac{2}{13}$ **e** $\frac{5}{7}$ **f** $\frac{3}{8}$
 g $\frac{23}{27}$ **h** $\frac{7}{48}$

2 a $2x$ **b** $\frac{1}{2}b$ **c** a
 d $\frac{2x}{y}$ **e** $\frac{c}{a}$ **f** $\frac{2}{3}p^4$
 g $\frac{3}{2}abc$ **h** $\frac{z^2}{xy}$ **i** $\frac{3x+1}{x+1}$
 j -1 **k** $\frac{2x^2}{x-y}$ **l** $\frac{2}{3x}$

3 a $b^2(1-b)$
 b $a(a+b)$
 c $5a(a-2)$
 d $y^4(y-1)$
 e $9c(2c^2 - d^2)$
 f $4a^2(1-4b)$
 g $27(3x-2)$
 h $3x(x^2 - 2x + 3)$
 i $x(4x^2 + x - 1)$
 j $2a(a^2 - 2a - 1)$
 k $7a(a - ab + 2b^3)$
 l $3x^2y(x^2 - 2xy + 3y^2)$

4 a $\frac{1}{2}gh(g+j)$ **b** $\frac{1}{2}h(a+b)$
 c $\frac{1}{3}lm(l-2m)$ **d** $\frac{1}{6}\pi r^2(8r + 3h)$

5 a 3400 **b** 122 **c** 2900
 d 161.4 **e** 10.5 **f** 1580
 g 30 **h** 27 000

Exercise 2D (p. 31)

1 a $x(x - 36)$ **b** $6x(x - 6)$
 c $2x(x - 2)$ **d** $x(5x - 2)$
 e $x(x + 3)$ **f** $4x(x + 1)$

2 a $(x + 1)(x + 2)$
 b $(x + 7)(x + 1)$
 c $(x + 2)(x + 4)$
 d $(x + 8)(x + 1)$
 e $(x - 1)(x - 3)$
 f $(x - 3)(x - 5)$
 g $(x - 10)(x - 2)$
 h $(x - 4)(x + 1)$
 j $(x - 1)(x + 8)$
 k $(x + 4)(x - 1)$
 l $(x + 9)(x - 10)$
 m $(x + 12)(x - 15)$
 o $(x - 10)(x + 1)$

3 a $(3x + 11)(x + 1)$
 b $(x + 7)(3x + 1)$
 d $(x + 2)(2x + 1)$
 e $2(x + 1)(2x + 1)$
 f $3(x + 1)^2$
 i $(x + 2y)(5x + y)$
 j $(2p - q)(4p - 5q)$
 k $(x - 4)(6x - 1)$
 l $2(q - 6)(3q - 4)$
 m $(2y - 3)(5y - 3)$
 n $(2a - 3)(3a - 2)$
 o $5(c - 1)(c - 2)$
 q $(3z - 2)^2$

s $3(x - 4)(x + 3)$
 t $(5x - 1)(5x - 2)$

4 a $(a - 7)(a + 2)$
 b $(b - 3)(b + 7)$
 c $(c - 2)(c + 4)$
 d $(d - 1)(d + 6)$
 e $(e - 6)(e + 1)$
 f $(2f + 1)(f - 1)$
 g $(2g - 1)(g + 1)$
 i $(x - 3y)(x + 7y)$
 j $(3j - 1)(j - 1)$
 k $(k - 3)(2k + 1)$
 l $(3 - l)(5 + l)$
 m $(1 + 2m)(1 - 4m)$
 n $(n - 2p)(n + 4p)$

5 a $(a + 2)(a - 2)$
 b $(b + 12)(b - 12)$
 c $(c + 3d)(c - 3d)$
 d $(5e + 4f)(5e - 4f)$
 e $(6g + 1)(6g - 1)$
 f $(7h + 8j)(7h - 8j)$
 g $4(2k - 5)(2k + 5)$
 h $25(l + 3)(l - 3)$
 i $3(l + 2)(l - 2)$
 j $2(m + 5n)(m - 5n)$
 k $(3z^2 + a)(3z^2 - a)$
 l $(ef^2g^3 + 11)(ef^2g^3 - 11)$
 m $(9h^2 + 4)(3h + 2)(3h - 2)$
 n $(l^4 + 16)(l^2 + 4)(l + 2)(l - 2)$
 o $(13 + m)(13 - m)$

6 a $(2 - z)(1 + 5z)$
 b $(1 - y)(2 - 5y)$
 c $(2 + w)(1 - 3w)$
 d $(3 - y)(1 + 2y)$
 e $(1 - t)(3 + 5t)$
 f $(1 - 2j)(1 + j)$
 g $(x - 6)(x - 1)$
 h $(1 - x)(2 + x)$
 i $(6 - x)(3 + x)$

7 a $(f - 2)^2$
 b $(x + 1)(x + 2)$
 c $2(k + 4)(k - 4)$
 d $(1 - 4v)(1 + 5v)$
 e $7(x - 2)(x + 1)$
 f $7(1 + 2c)(1 - 2c)$
 i $(1 - 5u)(1 - 4u)$
 j $a(x - 4)(x + 1)$
 k $(s - 1)(19s - 8)$
 l $2l(l - 6)$
 m $(2g + 1)(5g + 4)$
 n $(a - 1)(6a + 25)$

8 a $(m + n)(y + z)$
 b $(x - y)(c + d)$
 c $(5 - n)(x - y)$
 d $(a + b)(a - c)$
 e $(a + b)(5 + b)$
 f $(1 - 2c)(y - 3a)$
 g $(x + 1)(x^3 + 2)$
 h $(y - 1)(y^2 + 1)$
 i $(a + 1)(b - 1)$

9 a 140 **b** 800 **c** -1500
 d 0.4 **e** -400 **f** 14 049

10 a $(t-3)(t+2)$
  b $(2t+5)(3t-2)$
  c $(7-2t)(1+6t)$
  d $(2t-r)(3t-2r)$
  e $(3t-4)(4t-3)$
  g $(2t-3u)(3t+5u)$
  h $12(t+1)(t-2)$
  i $(4t+3)(6t-5)$
  j $(3t+4)(8t-3)$
  k $(1-x)(1+6x)$
  l $(2t-3)(3t-4)$
  m $(4-3t)(5t+2)$
  n $7(1+2c)(1-2c)$
  o $4(3t^2-13t-2)$
  p $(3t+1)(4t-9)$
  q $(t-6)(4t+1)$
  r $(2t-3)^2$
  s $(t-3)(4t-5)$
  t $(t-15)(4t+1)$
  u $4(5t+1)(5t-1)$

11 a $n^2(n+3)(n-3)$
  b $\pi(R+r)(R-r)$
  c $(a+b)(x^2+2)$
  d $(e^2+1)(e^2+3)$
  e $(2-d)(3+d)$
  f $(x^3+2)(2x-1)$
  g $(c+2d)(c-2d)$
  i $(a+b)(2x+y)$
  j $3(4t^2-6t-3)$
  k $4(t-1)^2$
  m $l(1-3l)(1+l)$
  n $(x+7)(x-8)$
  o $6(2+x)(1-x)$

12 a $-3(x+2)$
  b $-(x+3)$
  c $4x$
  d $(x+1)(x+2)(2x+9)$
  e $(N+1)^2(N+2)^2$
  f $(a+b)(a^2-ab+b^2)$
  g $x(x+7)(3x^2+19x-7)$
  h $12(x+2)$
  i $(x+2)(5x+12)$
  j $x(5x+14)$

Exercise 2E (p. 38)

1 a $3abc$ b $8n^2t$ c $12ac$
  d $3e$ e $2a(a-b)$
  f $9y$ g $91x$

2 a $\dfrac{5}{a}$ b $\dfrac{7a}{12}$

  c $\dfrac{1}{x}$ d $\dfrac{5y}{12}$

  e $\dfrac{2b-3a}{ab}$ f $\dfrac{8}{15x}$

  g $\dfrac{1-36a}{12a}$ h $\dfrac{5b+4}{8}$

  i $\dfrac{x}{3}$ j $\dfrac{x^2+z^2}{xyz}$

  k $\dfrac{m^2+n^2}{mn}$ l $\dfrac{2a^3-3b^2}{3a^2}$

  m $\dfrac{15y+31}{3y}$ n $\dfrac{2b^2+a^2-ab}{ab}$

  o $\dfrac{ax-ay}{(a-x)(a-y)}$

3 a $\dfrac{56b-33c}{24}$ b $\dfrac{68-9x}{20}$

  c $\dfrac{13x+9}{15}$ d $\dfrac{27x-89y+17}{18}$

4 a $\dfrac{b}{a}$ b $\dfrac{9x^3}{5z}$ c $3a^2b^2$

  d $\dfrac{2x^2}{3}$ e $\dfrac{6x^2}{y}$ f $\dfrac{3x^2}{2y}$

  g $\dfrac{b}{2c}$ h $\dfrac{a}{d}$ i 1

  j $\dfrac{c}{a}$ k -1 l $3(a-b)$

5 a $\dfrac{a^2}{b^2}$ b $c(1-d)$ c $\dfrac{18y^3}{5}$

  d $\dfrac{y}{x}$ e $\dfrac{2b}{a}$

  f $\dfrac{c}{ac+b}$ g $\dfrac{y}{x-yz}$

  h $\dfrac{a-1}{a}$ i $\dfrac{x^4+x^3}{x^3-1}$

  j $\dfrac{1}{a-b}$

Exercise 2F Review (p. 39)

1 a $9x^2+6x+1$
  b $4x^2-12x+9$
  c $6x^2-7x-3$
  d $16x+9$
  e $11x^2-6x-10$
  f $4+x-5x^2$
  g $10x^2-11$
  h $5x^2-4x-5$

2 a $(a+4)(a+2)$
  b $(a+7)(a-2)$
  c $(3a-7)(2a+1)$
  d $(a+3)(a-3)$
  e $4a(3a-1)$
  f $(2a-5)(2a+5)$
  g $(a+1)(b+1)$
  h $a(a-4)(a-3)$

3 a 4 b $\frac{13}{20}$ c 32 d $\frac{6}{7}$
  e $\frac{105}{16}$ f $\frac{7}{8}$ g -1 h $\frac{30}{7}$
  i $\frac{1}{8}$ j $\frac{15}{4}$ k 4 l $\frac{5}{8}$
  m 147 n -1 o $\frac{7}{24}$ p $\frac{3}{2}$

4 a $3x$ b $x(4x-1)$

  c $(4x-1)^2$ d $4x+\dfrac{1}{x}$

  e $4+\dfrac{2}{x}$ f $\dfrac{3}{4x-1}$

  g 6 h 9

  i $x+1$ j $\dfrac{x(4x-1)}{9}$

  k $\dfrac{3x}{x+2}$ l $1-x$

3 Equations and quadratic functions

Exercise 3A (p. 46)

1 a $x=-\frac{5}{7}$ b $x=\frac{2}{3}$ c $x=3$
  d $x=\frac{1}{2}$ e $x=\frac{1}{3}$ f $x=1$
  g $x=2$ h $y=-\frac{1}{2}$ i $a=5$
  j $b=0$ k $c=7\frac{2}{3}$ l $d=5\frac{1}{2}$
  m $e=0$ n $m=3$ o $v=-2$
  p $x=2\frac{3}{5}$ q $k=\frac{1}{4}$

2 a $a=48$ b $x=1\frac{11}{16}$
  c $a=24$ d $x=2$
  e $x=-1\frac{4}{5}$ f $x=5$
  g $y=-16\frac{2}{3}$ h $x=3$
  i $y=\frac{1}{2}$ j $x=1\frac{2}{3}$
  k $x=1\frac{1}{2}$ l $x=2$
  m $x=15$ n $x=\frac{1}{7}$
  o $x=5$ p $x=4$
  q $x=0.5$ r $x=2$
  s $x=17$ t $a=5$
  u $x=-12$ v $x=4$
  w $x=\frac{2}{3}$ x $x=7$
  y $x=-4$ z $x=\frac{7}{11}$

3 $8.5\,\text{m}$

4 23 balls

5 $x=32$

6 $-2\,\text{m s}^{-2}$

Exercise 3B (p. 55)

1 a $x=2, x=-\frac{1}{2}$ b $x=3, x=4$
  c $y=2, y=3$ d $y=\pm4$
  e $x=0, x=9$ f $x=0, x=1\frac{2}{3}$
  g $x=-3, x=7$ h $x=-4, x=3$
  i $x=3$ j $x=-3, x=-2$
  k $x=1\frac{1}{2}, x=2$ l $e=-2, e=\frac{2}{3}$
  m $d=-1, d=2\frac{2}{3}$ n $e=2, e=-\frac{2}{3}$
  o $f=-\frac{2}{3}, f=\frac{3}{4}$ p $g=-6, g=\frac{1}{5}$
  q $y=\pm7$ r $x=\pm\frac{3}{5}$
  s $x=0, x=6$ t $x=\pm\frac{2}{3}$
  u $y=\pm5$

2 a $a=\frac{2}{3}, a=1$ b $b=1, b=1\frac{1}{2}$
  c $c=\frac{1}{2}, c=3$ d $d=1, d=2\frac{1}{2}$
  e $e=-2, e=-\frac{1}{3}$ f $f=-\frac{1}{2}, f=2$
  g $g=-2, g=1\frac{1}{2}$ h $h=0, h=1$
  i $k=-2\frac{1}{3}, k=2\frac{1}{2}$
  j $l=-2, l=0$ k $m=0, m=4$
  l $t=0, t=\pm10$ m $u=0, u=7$
  n $w=\pm1$ o $x=1, x=2$
  p $y=1$ q $k=-\frac{3}{2}, k=\frac{1}{9}$
  r $x=-4, x=\frac{2}{5}$

3 a $x=-1\pm\sqrt{6}$ b $x=2\pm\sqrt{11}$
  c $t=5\pm3\sqrt{3}$ d $s=-3\pm2\sqrt{3}$
  e $r=-\dfrac{5}{2}\pm\dfrac{\sqrt{29}}{2}$ f $t=\dfrac{3}{2}\pm\dfrac{\sqrt{37}}{2}$

g $a = \frac{1}{2} \pm \frac{\sqrt{5}}{2}$ **h** $b = -\frac{7}{2} \pm \frac{\sqrt{37}}{2}$

i $x = -\frac{1}{2}, x = 1$ **j** $x = 1 \pm \frac{\sqrt{21}}{3}$

4 a $(4x - 1)^2 + 10$ **b** $(3x + \frac{1}{2})^2 + \frac{3}{4}$

5 a $a = 3, b = 1, c = -7$

  b $a = -1, b = 3, c = 14$

  c $a = -2, b = -1, c = -3$

  d $a = 2, b = -\frac{1}{2}, c = \frac{1}{2}$

6 a $x = \frac{1}{2} \pm \frac{\sqrt{5}}{2}$; $x = -0.62, x = 1.62$

  b $x = -3, x = -\frac{1}{2}$

  c $x = \frac{1}{2}, x = 3$

  d $x = -\frac{7}{4} \pm \frac{\sqrt{73}}{4}$; $x = -3.89$, $x = 0.39$

  e $x = \frac{7}{4} \pm \frac{\sqrt{73}}{4}$; $x = -0.39$, $x = 3.89$

  f $x = \frac{7}{4} \pm \frac{\sqrt{73}}{4}$; $x = -0.39$, $x = 3.89$

  g $x = \pm\sqrt{5}$; $x = \pm 2.24$

  h $x = \frac{1}{12} \pm \frac{\sqrt{97}}{12}$; $x = -0.74$, $x = 0.90$

  i $x = \frac{2}{3} \pm \frac{\sqrt{10}}{3}$; $x = -0.39$, $x = 1.72$

  j $x = -\frac{19}{12} \pm \frac{\sqrt{553}}{12}$; $x = -3.54$, $x = 0.38$

7 a $x = 0, x = 1$

  b $x = 0, x = 1$

  c $x = \frac{1}{3}, x = \frac{1}{2}, x = 1$

  d $x = 0, x = \pm 1$

  e $x = \pm 2$

  f $x = -k, x = \pm l$

8 a $x = \pm\sqrt{3}$ **b** $y = -1, y = 27$

  c $z = \frac{1}{4}$ **d** $x = \sqrt[3]{3}, x = \sqrt[3]{4}$

  e $x = -5, x = 4$ **f** $a = \pm 2\sqrt{3}$

9 10 years old

10 $l = 2.11$

11 $r = 5$

12 a $t = \frac{3}{5}$ or $t = 1$

Exercise 3C (p. 59)

1 a $(2, 0), (4, 0); (0, 8); x = 3; (3, -1)$

  b $(-3, 0), (1, 0); (0, 3); x = -1; (-1, 4)$

  c $(0, 0), (2, 0); (0, 0); x = 1; (1, -1)$

  d $(0, 0), (1, 0); (0, 0); x = \frac{1}{2}; (\frac{1}{2}, \frac{3}{4})$

  e $(-6, 0), (4, 0); (0, -48); x = -1; (-1, -50)$

---

f $(-\frac{1}{2}, 0), (1, 0); (0, 1); x = \frac{1}{4}; (\frac{1}{4}, \frac{9}{8})$

g $(-1, 0), (\frac{1}{3}, 0); (0, -5); x = -\frac{1}{3};$ $(-\frac{1}{3}, -\frac{20}{3})$

h $(-\frac{2}{3}, 0), (\frac{2}{3}, 0); (0, 4); x = 0; (0, 4)$

2 a $y = x(x + 3); (-3, 0), (0, 0); (0, 0)$

  b $y = (x - 3)(x - 1); (1, 0), (3, 0); (0, 3)$

  c $y = (2 - x)(2 + x); (-2, 0) (2, 0); (0, 4)$

  d $y = (x - 6)(x + 2); (-2, 0), (6, 0);$ $(0, -12)$

  e $y = (2x - 1)(x + 6); (-6, 0), (\frac{1}{2}, 0);$ $(0, -6)$

  f $y = (2x - 1)(2x + 1); (-\frac{1}{2}, 0), (\frac{1}{2}, 0);$ $(0, -1)$

  g $y = (3 + x)(1 - 4x); (-3, 0), (\frac{1}{4}, 0);$ $(0, 3)$

  h $y = (4x - 3)(2x - 5); (\frac{3}{4}, 0), (\frac{5}{2}, 0);$ $(0, 15)$

3 a $y = (x + 1)^2 - 12; (-1, -12); x = -1$

  b $y = (x - 2)^2 + 3; (2, 3); x = 2$

  c $y = 13 - (3 + x)^2; (-3, -13); x = -3$

  d $y = (x - 4)^2 - 15; (4, -15); x = 4$

  e $y = 2(x + 3)^2 - 21; (-3, -21); x = -3$

  f $y = 5 - 4(1 - x)^2; (1, 5); x = 1$

  g $y = (x - \frac{3}{2})^2 + \frac{11}{4}; (\frac{3}{2}, \frac{11}{4}); x = \frac{3}{2}$

  h $y = 3(x + \frac{1}{3})^2 - \frac{4}{3}; (-\frac{1}{3}, -\frac{4}{3}); x = -\frac{1}{3}$

4 a Min 2, at $x = 1$

  b Min -16, at $x = -3$

  c Max 17, at $x = 3$

  d Max 5, at $x = -1$

  e Min $-3\frac{1}{4}$, at $x = -1\frac{1}{2}$

  f Min $-13\frac{1}{4}$, at $x = 2\frac{1}{2}$

  g Max $3\frac{1}{4}$, at $x = -\frac{1}{2}$

  h Max $11\frac{1}{4}$, at $x = 1\frac{1}{2}$

  i Min -7, at $x = -2$

  j Min $3\frac{11}{12}$, at $x = \frac{1}{6}$

  k Max $6\frac{1}{8}$, at $x = -\frac{1}{4}$

  l Max 7, at $x = 1$

5 a $p = 6, q = -1, r = -5$

  b $p = -2, q = -1, r = 7$

  c $p = 3, q = \frac{1}{3}, r = \frac{11}{3}$

  d $p = 3, q = 2, r = 4$

6 a Min -2; vertex $(1, -2)$, y-intercept $(0, -1)$, $x = 1$

  b Min 5; vertex $(-2, 5)$, y-intercept $(0, 9)$, $x = -2$

  c Max 12; vertex $(3, 12)$, y-intercept $(0, 3)$, $x = 3$

Exercise 3D (p. 62)

1 a 36, 2   **b** $-4, 0$ **c** 41, 2

  d 0, 1   **e** 32, 2   **f** $-8, 0$

  g 400, 2   **h** 1, 2

2 a $k = 1$ **b** $k = \pm 12$ **c** none

  d $k = \frac{4}{5}$ **e** $k = -\frac{25}{8}$ **f** $k = -2$

3 $a = \frac{1}{4}b^2$

6 a $(x - 2)(x - 3) = 0$

  b $(x - 3)(x + 5) = 0$

---

c $x(x + 6) = 0$

d $(x - p)(x - q) = 0$

e $(3x + 2)(4x - 1) = 0$

7 $k = -2, k = -\frac{2}{3}$

Exercise 3E Review (p. 63)

1 a $x = \frac{1}{5}$

  b Any value

  c $x = \pm 2$

  d $x = 0, x = 4$

  e $x = 0, x = \pm 2$

  f $x = 0, x = 7$

  g $x = -2, x = 3$

  h $x = -1, x = 6$

  i $x = \pm 1, x = \pm 4$

  j $x = \frac{1}{2}, x = 3$

  k $x = 0, x = \pm 1$

  l $x = -\frac{3}{2}, x = \frac{2}{3}$

2 a $(-6, 0), (0, 0); (0, 0); x = -3;$ $(-3, -9)$

  b $(-4, 0), (2, 0); (0, 24); x = -1;$ $(-1, 27)$

  c $(-2, 0), (5, 0); (0, -10); x = \frac{3}{2};$ $(\frac{3}{2}, -\frac{49}{4})$

  d $(-10, 0), (1, 0); (0, 10); x = -\frac{9}{2};$ $(-\frac{9}{2}, \frac{121}{4})$

3 a 20, 2 **b** $-7, 0$ **c** 0, 1 **d** 1, 2

6 $p = 5, q = \frac{1}{5}, r = \frac{4}{5}$, min $\frac{4}{5}$ when $x = \frac{1}{5}$

7 $k = \frac{1}{6}$

8 a $x = \frac{b}{a - 2}$ **b** $x = \pm\sqrt{r^2 - y^2}$

  c $x = \frac{dk - b}{a - ck}$ **d** $x = \frac{1}{2l^2}$

9 a $x = 0, x = -k$ **b** $x = \pm a$

  c $x = 0, x = -k$ **d** $x = a \pm b$

10 $x - 3x^{\frac{1}{3}} - 2$

4 Inequalities

Exercise 4A (p. 65)

1 True for a, b, c, e, g
   False for d, f

2 True for a, b, c, f, g
   False for d, e

Exercise 4B (p. 67)

1 a $x < -\frac{2}{3}$ **b** $x \geqslant -\frac{1}{2}$ **c** $x \leqslant 1$

  d $x < -3$ **e** $a > \frac{11}{4}$ **f** $b < -\frac{1}{3}$

  g $c \geqslant -\frac{2}{3}$ **h** $d \leqslant \frac{11}{3}$ **i** $e \geqslant -2$

  j $f > 1$ **k** $h < 0$ **l** $x \leqslant 2$

  m $x > -2$ **n** $x < \frac{17}{12}$

2 a $g > 5$ **b** $x > -1$

  c $x > \frac{5}{3}$ **d** $x \leqslant \frac{18}{13}$

3 a $-1 < x < 5$ **b** $1 \leqslant x \leqslant 2$

  c $-3 < x < -2$ **d** $-\frac{4}{5} < x \leqslant \frac{4}{5}$

  e $-\frac{1}{2} < x \leqslant \frac{8}{3}$ **f** $-1 < x < \frac{3}{2}$

Exercise 4C (p. 70)

1 a $2 < x < 3$
  b $x < -4$ or $x > 5$
  c $x \leqslant -\frac{1}{2}$ or $x \geqslant 1$
  d $x \leqslant \frac{3}{2}$ or $x \geqslant \frac{5}{2}$
  e $x < 0$ or $x > 1$
  f $x < -2$ or $x > 3$
  g $1 < x < 2$
  h $x < -2$ or $x > 9$
  i $x \leqslant -2$ or $x \geqslant 3$
  j $x < 0$ or $x > \frac{4}{3}$
  k $y < 3$ or $y > 4$
  l $y \leqslant -1$ or $y \geqslant 5$
  m $\frac{1}{2} \leqslant y \leqslant \frac{2}{3}$
  n $-1 < y < \frac{3}{8}$
  o $-3 \leqslant u \leqslant -\frac{1}{2}$
  p $-4 < v < \frac{1}{2}$
  q $x = -\frac{1}{2}$
  r No solution

2 a $-\frac{1}{3} < x < 2$
  b $x < -\frac{5}{9}$ or $x > 1$
  c $1 < x < 4$
  d $x = 2$
  e $x \leqslant 0$ or $x \geqslant \frac{1}{5}$
  f $-8 < x < 2$

3 a $k < 1$
  b $k < -2\sqrt{14}$ or $k > 2\sqrt{14}$
  c $-1 < k < 1$

4 a $k > \frac{1}{12}$
  b $-2\sqrt{21} < k < 2\sqrt{21}$
  c $\frac{4}{3} < k < 4$

5 a $x \leqslant -3, 1 \leqslant x \leqslant 2$
  b $x < \frac{1}{2}, 1 < x < 4$
  c $x < \frac{1}{5}$
  d $-2 \leqslant x \leqslant -1, 1 \leqslant x \leqslant 2$

Exercise 4D Review (p. 71)

1 a $x \geqslant \frac{7}{4}$
  b $x > -1$
  c $x < 4$
  d $x \geqslant 8$
  e $x < \frac{8}{9}$
  f $y > 9$

2 a $x < -2$ or $x > 2$
  b $-10 \leqslant x \leqslant 10$
  c $x < -2$ or $x > 1$
  d $-2 < x < 5$
  e $x < 5 - \sqrt{3}$ or $x > 5 + \sqrt{3}$
  f $x \leqslant -4$ or $x \geqslant \frac{1}{2}$

3 a $k < \frac{9}{4}$
  b $k < -2\sqrt{6}$ or $k > 2\sqrt{6}$
  c $-\frac{1}{4}\sqrt{2} < k < \frac{1}{4}\sqrt{2}$

4 a $k > 9$
  b $-2\sqrt{2} < k < 2\sqrt{2}$
  c $-\frac{3}{4} < k < 3$

5 Simultaneous equations

Exercise 5A (p. 75)

1 a $x = 4, y = 2$ **b** $c = 8, d = 7$
  c $x = 7, y = 4$ **d** $g = 3, h = 3$
  e $x = 5, y = 1$ **f** $r = 4, s = 3$
  g $t = 3, u = 2$ **h** $x = -4, y = 7$
  i $x = 4, y = -1$ **j** $x = 1, y = 2$
  k $n = 4, p = \frac{1}{2}$ **l** $x = 2, y = \frac{1}{2}$

2 a $s = 2, t = 3$ **b** $a = \frac{7}{4}, y = \frac{1}{4}$
  c $b = -1, x = 2$ **d** $y = -1, z = 1$
  e $x = 2, y = -1$ **f** $q = -1, r = 2$
  g $p = 3, q = 2$ **h** $r = -1, s = -1$
  i $x = 3, y = 0$ **j** $x = 3, y = 1$

Exercise 5B (p. 78)

1 a $x = \frac{1}{4}, y = \frac{3}{2}$
  b $x = 7, y = 4$
  c $x = -1, y = 3$
  d $x = 10, y = -1$
  e $x = 9, y = 1$
  f $s = 8, x = 5$
  g $x = \frac{11}{19}, y = -\frac{18}{19}$
  h $x = \frac{10}{29}, y = \frac{7}{29}$

2 a $x = -1, y = -2$
  b $x = \frac{23}{13}, y = -\frac{11}{13}$
  c $x = 0, y = \frac{7}{2}$
  d $x = 3, y = -2,$
  e $x = \dfrac{dm - bn}{ad - bc}, y = \dfrac{an - cm}{ad - bc}$

3 a $\left(\frac{1}{2}, \frac{5}{2}\right)$ **b** $(1, 9)$
  c $(-1, 1)$ **d** $\left(\frac{11}{2}, \frac{9}{2}\right)$

Exercise 5C (p. 83)

1 a $x = 3, y = 9; x = -2, y = 4$
  b $x = 1, y = 1; x = -3, y = 9$
  c $x = 0, y = 0; x = 6, y = 12$
  d $x = 6, y = 5; x = 1, y = 0$
  e $x = 3, y = 3; x = -3, y = -3$
  f $x = 4, y = 2; x = -4, y = -2$

2 a $(-8, -1), (1, 8)$
  b $(-2, -2), (1, 4)$
  c $(-4, -3), (3, 4)$
  d $(0, 1), (2, 0)$
  e $(12, -4), \left(-\frac{7}{4}, \frac{3}{2}\right)$
  f $\left(0, -\frac{1}{2}\right), (-1, -1)$

3 a $x = 2, y = -3; x = -1, y = 3$
  b $x = -2, y = 10$
  c $x = 5, y = -9; x = -2, y = 5$
  d $x = 0, y = -10; x = 5, y = -5$

Exercise 5D (p. 87)

1 a 2 **b** 0 **c** 2
  d 1 **e** 0 **f** 1
2 $k = \frac{1}{10}$
3 $k < 13.5$
5 $k = 2$

6 $30\,\text{cm}$ by $20\,\text{cm}$

7 $8\,\text{cm}$ and $15\,\text{cm}$

8 $a = 3, b = 4$

Exercise 5E Review (p. 87)

1 a $x = 1, y = -1$
  b $x = 2, y = 3$
  c $x = 3, y = 4; x = 4, y = 3$
  d $x = \frac{52}{11}, y = \frac{101}{11}; x = -2; y = -11$
  e $x = \frac{21}{11}, y = \frac{13}{11}$
  f $x = 7, y = 1; x = -\frac{5}{8}, y = -\frac{21}{5}$

2 $\left(\frac{1}{4}, 0\right)$

3 $(0, -1), (1, 0)$

5 $\left(2\frac{1}{3}, \frac{1}{3}\right), (3, 1)$

6 a $k > -11$ **b** $k = -11$

7 $k < -\frac{2}{5}\sqrt{5}$ or $k > \frac{2}{5}\sqrt{5}$

8 $k = -2$ or $k = -14$

9 $k < -2\sqrt{5}$ or $k > 2\sqrt{5}$

Exam practice 1 (p. 89)

1 a $a = 1, b = 2$ **b** $c = \frac{1}{9}, d = \frac{2}{9}$

2 a $x = -\frac{1}{2}, y = 2\frac{1}{2}$ **b** 8

3 a 3 **b** 1.2

4 a $y = \frac{9}{2}x + \frac{3}{2}$ **b** $x = -\frac{1}{9}$

5 a $7 + 3\sqrt{5}$ **b** $26 - 2\sqrt{5}$

6 b $k = \pm 9$

7 $x = \frac{1}{16}, x = 64$

8 a $1 \pm \sqrt{13}$
  b $x < 1 - \sqrt{13}, x > 1 + \sqrt{13}$

9 a $x = -5, x = 4$ **b** $x < -5, x > 4$

10 a $x = -\frac{3}{2}$ **b** $y = -\dfrac{ax + c}{b}$

11 a $-k \pm \sqrt{7 + k^2}$
  c $-\sqrt{2} - 3$ or $-\sqrt{2} + 3$

12 b $k \leqslant 0, k \geqslant \frac{8}{25}$ **c** $0, \frac{8}{25}$

13 b $x = -\frac{8}{3}, y = -\frac{1}{3}; x = 4, y = 3$

14 a $(2x - 3)^2 + 6$ **b** $\left(\frac{3}{2}, 6\right)$

15 $x < -\frac{3}{2}$ or $x > 4$

16 $-\sqrt{6} + 2\sqrt{3} + \sqrt{2}$
  (or $p = -1, q = 2, r = 1$)

17 a $x > \frac{21}{2}$ **b** $x < -2$ or $x > 7$

18 $(3, 1), \left(-\frac{11}{5}, -\frac{8}{5}\right)$

19 a $1 + x^2 + x^3 + x^5$
  b $1.000\,001\,001\,000\,001$

20 a $4x - 10 > 32$ **b** $x(x - 5) < 104$
  c $10.5 < x < 13$

21 c $10y^2 - 10001y + 1000 = 0$
d $x = -1$ or $x = 3$

22 a $\dfrac{6}{y}$ **b** $t = 1$ or $t = -216$

23 a $x < 2\frac{1}{2}$ **b** $\frac{1}{2} < x < 5$
c $\frac{1}{2} < x < 2\frac{1}{2}$

6 Coordinate geometry and the straight line

Exercise 6A (p. 96)

1 a 13   **b** $\sqrt{41}$
c $2\sqrt{37}$ **d** $\sqrt{(p-r)^2 + (q-s)^2}$
e $3\sqrt{10}$ **f** 4
g 4   **h** $\sqrt{74}$
i 13

2 a $\left(5\frac{1}{2}, 8\right)$ **b** $\left(1\frac{1}{2}, 5\right)$
c $(5, -4)$ **d** $\left(\dfrac{p+r}{2}, \dfrac{q+s}{2}\right)$
e $\left(-1\frac{1}{2}, -2\frac{1}{2}\right)$ **f** $(6, 11)$
g $(3, 2)$ **k** $\left(-\frac{1}{2}, -4\frac{1}{2}\right)$
l $\left(0, 3\frac{1}{2}\right)$

3 17

4 $PQ = PR$; $\left(-\frac{5}{2}, \frac{9}{2}\right)$

5 $LM = LN$; $\left(-\frac{3}{2}, -\frac{3}{2}\right)$

6 a AC; 25; $\frac{1}{2}$
b AB; $25\frac{1}{2}$; 3

7 $h = 9\frac{1}{4}$

8 A, B and D; $5\sqrt{2}$

9 $a = 2, b = -3$

10 $k = -3$ or $k = 5$

11 $3\left(\sqrt{2} + \sqrt{10}\right)$

13 a BCED
b ACFD
c ACED, BCFD

14 P, R and S

15 13; $6\frac{1}{2}$

Exercise 6B (p. 105)

1 a $\frac{9}{4}$ **b** $\frac{3}{2}$ **c** $-\frac{4}{5}$
d $-\frac{10}{11}$ **e** 0   **f** $\dfrac{s-q}{r-p}$
g -1 **h** $\dfrac{b}{a}$ **i** $-\frac{1}{14}$
j $-\frac{1}{6}$

2 a $-\frac{1}{3}$ **b** -4 **c** $\frac{1}{6}$
d $\frac{3}{2}$ **e** $-\dfrac{1}{2m}$ **f** $\dfrac{a}{b}$
g $\dfrac{2}{m}$

---

3 Parallel **a**, **f**
Perpendicular **b**, **d**, **g**
Neither **c**, **e**

4 Yes **a**, **e**, **f**
No **b**, **c**, **d**

5 a $(2, 5)$ **b** $(-3, 20)$ **c** $(0, -1)$

6 a $x = 2$ **b** $x = 0$ **c** $x = -2\frac{1}{2}$

7 a $(2, 0)$; $(0, -6)$ **b** $\left(1\frac{3}{5}, 0\right)$; $(0, 4)$
c $(3, 0)$; $(0, -4)$ **d** $(0, 0)$; $(0, 0)$

9 a $(-3, 0), (0, 2)$ **b** $\left(-2, 0\right), \left(0, \frac{1}{2}\right)$
c $(-2, 0), (0, -6)$ **d** $\left(\frac{5}{7}, 0\right), \left(0, -\frac{5}{3}\right)$

10 a $y = \frac{1}{3}x$ **b** $y = -2x$ **c** $y = mx$

11 a $y = \frac{1}{4}x$; gradient $\frac{1}{4}$
b $y = -\frac{5}{4}x$; gradient $-\frac{5}{4}$
c $y = \frac{3}{2}x$; gradient $\frac{3}{2}$
d $y = \frac{7}{4}x$; gradient $\frac{7}{4}$
e $y = \frac{q}{p}x$; gradient $\frac{q}{p}$

12 a $y = 3x + 2$ **b** $y = 3x - 1$
c $y = \frac{1}{5}x + 2$ **d** $y = \frac{1}{5}x + 4$

13 a $y = \frac{2}{3}x + 2$; gradient $\frac{2}{3}$; intercept 2
b $y = \frac{1}{4}x + \frac{1}{2}$; gradient $\frac{1}{4}$; intercept $\frac{1}{2}$
c $y = -3x - 6$; gradient -3; intercept -6
d $y = \frac{7}{3}x - \frac{5}{3}$; gradient $\frac{7}{3}$; intercept $-\frac{5}{3}$
e $y = -4$; gradient 0; intercept -4
f $y = -\dfrac{l}{m}x - \dfrac{n}{m}$; gradient $-\dfrac{l}{m}$; intercept $-\dfrac{n}{m}$

14 a $y = 4x - 1$
b $y = 3x + 11$
c $x - 3y - 17 = 0$
d $3x + 4y - 41 = 0$
e $3x - 6y - 4 = 0$
f $y = ax - a$

15 a $3x - 4y + 21 = 0$
b $5x + 4y - 23 = 0$
c $3x + 11y - 35 = 0$
d $x - 5y - 19 = 0$
e $2x + 3y - 7 = 0$
f $hx + ky - 4hk = 0$

16 a $3x - 4y + 1 = 0$
b $5x - 2y + 16 = 0$
c $7x - y - 28 = 0$
d $3x - 4y - 6 = 0$

17 $\frac{8}{21}$

18 $a = 8$

19 $a:b = 3:2$

21 $5\sqrt{2}$; $\left(3\frac{1}{2}, 4\frac{1}{2}\right)$

22 $\frac{1}{2}\sqrt{34}$; $\left(\frac{3}{4}, -1\frac{3}{4}\right)$

Exercise 6C (p. 109)

1 a $(7, -7)$ **b** $\left(-1\frac{1}{2}, -5\frac{1}{2}\right)$
c $\left(1\frac{4}{7}, -1\frac{6}{7}\right)$ **d** $(4, -7)$

---

2 $y = \frac{1}{8}x$

3 a $(5, 6)$; $\frac{1}{2}$; -2; $y = -2x + 16$
b $(-3, 6)$; -2; $\frac{1}{2}$; $y = \frac{1}{2}x + 7\frac{1}{2}$
c $\left(-1\frac{1}{2}, -1\right)$; 8; $-\frac{1}{8}$; $y = -\frac{1}{8}x - 1\frac{3}{16}$
d $(4p, 2q)$; $\dfrac{q}{3p}$; $-\dfrac{3p}{q}$;
$3px + qy - 12p^2 - 2q^2 = 0$

4 $2x - 5y + 19 = 0$

5 $26x + 4y - 21 = 0$

6 a $4x - 3y - 13 = 0$ **b** $(4, 1)$
c 5

7 $2\sqrt{5}$

8 a $2x + 7y - 14 = 0$
b $2x - 7y - 14 = 0$

9 $\sqrt{85}$, $6x + 7y - 85 = 0$

10 $(1, 8)$; 52

12 b 12; 78

13 $2x + y - 17 = 0$; $72\frac{1}{4}$

14 $x - y = 0$, $2x + 2y - 9 = 0$;
$\left(\frac{9}{4}, 3\right)$, $\left(3, \frac{3}{2}\right)$; $\frac{3}{2}\sqrt{5}$

16 $a = -2, b = 9$ or $a = \frac{5}{2}, b = -9$

17 $a = -2, b = 4$

Exercise 6D Review (p. 111)

1 a $3x - 5y + 14 = 0$
b $3x + 5y = 14$
c $2x + 5y + 14 = 0$

2 a $2x - 3y + 12 = 0$
b $(3, 6)$

3 a $x + 2y = 0$ **b** $(2, -1)$
c $\sqrt{5}$

4 b $PQR = 26.6°$

6 a $y = x$ **b** $\left(3\frac{1}{2}, 3\frac{1}{2}\right)$

7 a $(2h, -2k)$ **b** $-\dfrac{3k}{h}$
c $hx - 3ky - 2h^2 - 6k^2 = 0$

8 $x + 3y - 17s = 0$

9 a $(0, 3), (4, 0)$
b $-4y + 3x = 12$
$4y - 3x = 12$
$3y + 4x = 12$
c 6

10 a $(0, -12)$; $(4, 0), (-3, 0)$
b $(0, 2)$; $\left(\frac{2}{3}, 0\right), \left(\frac{1}{2}, 0\right)$
c $(0, 9)$; touches x-axis at $(3, 0)$
d $(0, 0)$; $(9, 0)$; touches x-axis at $(0, 0)$
e $(0, 25)$; $(-1, 0)$; touches x-axis at $(5, 0)$
f $(0, 9)$; $(1, 0), (-1, 0), (3, 0), (-3, 0)$

11 a $(6, 40), (-1, 5)$ **b** $7\sqrt{26}$

12 $PQ = 11\sqrt{10}$

13 a $3\sqrt{10} + 5\sqrt{2}$

7 Graphs of functions

Exercise 7A (p. 118)

6 a Translation $\begin{pmatrix} 0 \\ 6 \end{pmatrix}$

b Translation $\begin{pmatrix} 0 \\ -5 \end{pmatrix}$

c Stretch in y direction, s.f. $\frac{1}{2}$

d Stretch in x direction, s.f. $\frac{1}{4}$

7 a Translation $\begin{pmatrix} -3 \\ 0 \end{pmatrix}$

b Reflection in x-axis

c Stretch in y direction, s.f. 5, reflection in x-axis

d Reflection in x-axis

Exercise 7B (p. 121)

1 a $x = 0, y = 2$ **b** $x = 1, y = 0$

2 a $x = 0, y = 0$ **b** $x = 0, y = 0$

4 A: $(2, -3), (-2, 3)$
B: $(4, 0), (-4, 0)$

5 O: $(0, 0), (3, 0), (0, 0), (0, 0)$
A: $(3, -2), (6, 2), (-3, 2), (\frac{3}{2}, 2)$
B: $(5, 0), (8, 0), (-5, 0), (\frac{5}{2}, 0)$

6 b A: $(-4, 0), (-4, 0)$
B: $(-3, -4), (-3, 2)$
C: $(2, 4), (2, -2)$
D: $(4, 2), (4, -1)$
O: $(0, 0), (0, 0)$

c Stretch in y direction, sf 2; reflection in x-axis

7 b A: $(-3, 0), (-4, 3)$
B: $(-2, -2), (-3, 1)$
C: $(3, 2), (2, 5)$
D: $(5, 1), (4, 4)$
O: $(1, 0), (0, 3)$

c Translation $\begin{pmatrix} 1 \\ 0 \end{pmatrix}$; translation $\begin{pmatrix} 0 \\ 3 \end{pmatrix}$

8 $(x + 1)^2 + 7$;
vertex $(-1, 7)$, intercept $(0, 8)$

9 a $(x - 1)^2 + 3$;
vertex $(1, 3)$, intercept $(0, 4)$

b $(x + 2)^2 - 9$;
vertex $(-2, -9)$, intercept $(0, -5)$

c $(x - 3)^2 - 6$;
vertex $(3, -6)$, intercept $(0, 3)$

d $\left(x + \frac{1}{2}\right)^2 + \frac{3}{4}$;
vertex $\left(-\frac{1}{2}, \frac{3}{4}\right)$, intercept $(0, 1)$

e $8 - (x + 1)^2$;
vertex $(-1, 8)$, intercept $(0, 7)$

10 a $y = x^2 - 1$
b $y = (x + 1)^2$
c $y = 4x^2$
d $y = -3x^2$

11 a $y = 4x^3$
b $y = -x^3$
c $y = 27x^3$
d $y = (x - 2)^3$

Exercise 7C (p. 127)

1 a $(-4, -4), (2, -4)$
b $(2, -5)$
c $(-1, 5), (\frac{1}{3}, 5)$
d $(-6, -72), (-4, -52)$
e $(-3, 1), (2, -14)$
f None

2 a $(0, 100)$ **b** $(1, 4)$
c $(1, 2), (7, 8)$ **d** $(0, 4), (2, 4)$
e $\left(-\frac{3}{2}, \frac{11}{4}\right)$ **f** $(-1, 6)$

4 a $(-3, 5), (1, 5)$
b $(-5, 17), (1, 5)$
c $(-1, 1), (0, 2), (2, 10)$

5 $7\sqrt{2}$

6 $-10, 2$

7 $-11, 13$

8 $-\frac{1}{3}, 8$

9 $\left(-\frac{2}{3}, \frac{19}{9}\right), (\frac{3}{2}, 5)$

10 b 1.3

11 b $0.2, 1.6$
c $y = 2x + 1$; $0.5, 1.5$

12 b $-2.5, 0.2, 2.4$
c $y = 7$; 2.8
d $y = 2x - 1$

13 $x = 0$ or 1; $(0, 0), (1, 1)$

14 $(-1, 1), (0, 0), (1, 1)$

15 $(-1, -1), (0, 0), (1, 1)$

Review Exercise 7D (p. 128)

1 a $(-5, 3), (2, 3)$
b $(-1, -4), (\frac{2}{3}, -\frac{2}{3})$
c $(-1, 8), (\frac{5}{6}, \frac{101}{36})$
d $(-4, -64), (0, 0), (5, 125)$

4 $\frac{16}{5}$

5 b -2.3 **c** 2.3 **e** 1.7

8 Sequences and series; arithmetic series

Exercise 8A (p. 133)

1 a $3, 6, 9, 12$; 101st
b $3, 9, 27, 81$; 6th
c $1, 5, 9, 13$; 62nd
d $6, 4, 2, 0$; 20th
e $1, 16, 81, 256$; 11th
f $0, 3, 8, 15$; 29th

2 a $31, 36, 41$; $u_n = 5n + 6$

b $\frac{1}{256}, \frac{1}{1024}, \frac{1}{4096}$; $u_n = \frac{1}{4^{n-1}}$

c $8, 5, 2$; $u_n = 23 - 3n$
d $10, 12, 14$; $u_n = 2n$
e $32, 64, 128$; $u_n = 2^n$

f $\frac{6}{5}, \frac{7}{6}, \frac{8}{7}$; $u_n = \frac{n + 1}{n}$

g $25, 36, 49$; $u_n = n^2$
h $125, 216, 343$; $u_n = n^3$
i $250, 432, 686$; $u_n = 2n^3$

3 a $3, 7, 11, 15, 19, 23$
b $64, 32, 16, 8, 4, 2$
c $1, 4, 9, 16, 25, 36$
d $6, 3.5, 2.25, 1.625, 1.3125, 1.15625$

4 8; 3.52

5 $\frac{3}{4}$; $30, 34.5$

6 a $1010, 1020.4, 1031.2$
b $640, 265.6, -123.8$
c 40

Exercise 8B (p. 135)

1 a $1^3 + 2^3 + 3^3 + 4^3$
b $2^2 + 3^2 + \cdots + n^2$
c $1^2 + 1 + 2^2 + 2 + \cdots + n^2 + n$
d $\dfrac{1}{1 \times 2} + \dfrac{1}{2 \times 3} + \dfrac{1}{3 \times 4}$
e $2^2 + 2^3 + 2^4 + 2^5$
f $-1^2 + 2^2 - 3^2 + 4^2$
g $1^1 + 2^2 + 3^3 + \cdots + n^n$
h $-\frac{1}{3} + \frac{1}{4} - \frac{1}{5} + \frac{1}{6}$
i $n(n - 1) + (n + 1)n + (n + 2)(n + 1)$
j $\dfrac{n - 2}{n - 1} + \dfrac{n - 1}{n} + \dfrac{n}{n + 1}$

2 a $\displaystyle\sum_{1}^{n} r$ **b** $\displaystyle\sum_{1}^{n+1} r^4$

c $\displaystyle\sum_{1}^{5} \frac{1}{r}$ **d** $\displaystyle\sum_{2}^{5} 3^r$

e $\displaystyle\sum_{1}^{5} \frac{r}{3^{r-1}}$ **f** $\displaystyle\sum_{1}^{5} \frac{r(2r + 1)}{2(r + 1)}$

g $\displaystyle\sum_{1}^{6} (-1)^r r$ **h** $\displaystyle\sum_{2}^{6} r(r + 5)$

i $\displaystyle\sum_{1}^{6} (-1)^{r+1} 2^{r-1}$

j $\displaystyle\sum_{1}^{5} (-1)^{r+1} r(2r + 1)$

3 a 21 **b** 45 **c** 44
d $3\frac{11}{20}$ **e** $-4\frac{9}{10}$ **f** 150
g -80 **h** $\frac{37}{180}$ **i** 3
j 0 **k** 24 **l** 40

4 a $15k^2$ **b** $55n$ **c** $56a$
d $-k$ **e** $\dfrac{25a}{12}$ **f** $\dfrac{4a}{m}$

Exercise 8C (p. 138)

1 a $1\frac{1}{2}$ **b** -3 **c** 0.1
e $\frac{1}{3}$ **g** n **i** $1\frac{1}{8}$
j -7 **l** -0.2

2 a 75; 147 **b** -34; -82

c $7\frac{1}{8}$; $\dfrac{5n - 3}{8}$ **d** -148; $52 - 2n$

e $-13\frac{1}{2}$; $\dfrac{15 - n}{2}$ **f** 799; $3 + 4n$

3 a 23   b 13   c 31
   d 21   e 91   f 13

4 a 2601   b 420
   c 250.5   d $121x$
   e $\frac{1}{2}n(2a + (n-1)d)$

5 a 444   b -80
   c 20 100   d -520
   e $2n(n+2)$ f $\frac{1}{8}n(11-n)$

Exercise 8D (p. 143)

1 a 36   b 17   c 22   d 10
   e 18   f $2n$ g n h n

2 a 90   b 123   c 163   d -187
   e 53   f 82   g $41x$ h $a+40d$

3 a 632   b 288
   c $60\frac{1}{2}$ d -55
   e $\frac{n}{2}(2a+n-1)$ f $\frac{n}{2}(2a-(n-1)d)$

4 a 165   b 0   c 49
   d $2n^2$ e 9   f pr^2

5 a 75   b 75
   c 16   d 22
   e $n(n+2)$ f $\frac{1}{4}(n-5)(10-n)$

6   1; 2

7   9; 268

8   14; 4

10   60

11 a 5, 7, 9   b 60

12   1; 64

13   -13; 42

14   $n = 6$

15 a 6, 8, 10

16   7500

17   7650

18   $3\frac{1}{2}$; $\frac{1}{10}$; $148\frac{1}{2}$

19 a 12; 0.2   b 22

20   $x = 4$

21   14, 18

22   $z = -2$; 15

23   13 rows; 9 tins

24 a 22 rows
   b i 52 trees   ii 31 trees
   c 9.8%

25 a 228 seats   b 34.2%

26 a 5 pieces   b 48 cm

Exercise 8E Review (p. 146)

1   2550

2   4234

3   18 terms

4 a 4, -8, 16, -32, 64
   b $2\frac{1}{2}$, $2\frac{3}{4}$, $2\frac{7}{8}$, $2\frac{15}{16}$, $2\frac{31}{32}$
   c 2, 5, 14, 41, 122
   d 5, 2, $\frac{7}{2}$, $\frac{17}{7}$, $\frac{52}{17}$

5 a 30   b -15 c 10   d 0

6   -9; 5

7   2, 4, 6, 8, 10

8   0.95, 2

9   $n = 35$

10   $\dfrac{ar+b}{r+1}$; $\dfrac{br+a}{r+1}$

11 a 74   b -990

9   Differentiation

Exercise 9A (p. 154)

1 a 1   b 5   c $4x$ d $2x$
   e 0   f $-\dfrac{2}{x^3}$ g $4x^3$

Exercise 9B (p. 158)

1 a $4x^3$ b $12x^3$ c $20x^3$
   d $12x^{11}$ e $21x^6$ f 5
   g 5   h $10x-3$ i 0
   j $-5x^{-6}$ k $-12x^{-4}$ l $-\dfrac{2}{x^3}$
   m $-\dfrac{2}{x^2}$ n $\dfrac{6}{x^3}$ o $-\dfrac{1}{x^4}$
   p $\dfrac{4}{x^5}$ q $-\dfrac{15}{4x^6}$ r $\frac{1}{3}x^{-\frac{2}{3}}$
   s $-\frac{3}{2}x^{-\frac{3}{2}}$ t $\frac{1}{2}x^{-\frac{1}{2}}$ u $\frac{1}{4}x^{-\frac{1}{4}}$
   v $-\frac{1}{2}x^{-\frac{3}{2}}$ w $x^{-\frac{4}{3}}$ x $6\sqrt{x}$
   y $-\frac{3}{4}x^{-\frac{3}{2}}$

2 a $12x^3 - 6x^2 + 2x - 1$
   b $8x^3 + x^2 - \frac{1}{2}x$
   c $6x^5 - x^{-2} + \frac{3}{2}x^{-\frac{3}{2}}$
   d $3ax^2 + 2bx + c$
   e $18x^2 - 8$
   f $15x^2 + 3x$
   g $-6x^{-2} - 6x^{-3} - \frac{3}{2}x^{-\frac{5}{2}}$
   h $1 - x^{-2}$

3 a -1
   b 0
   c $12x^2 - 3$
   d $ax - 2b$
   e $4x + 2$
   f $-6x^{-3} - 9x^{-4}$
   g $\frac{1}{2}x^{-\frac{1}{2}} + \frac{1}{3}x^{-\frac{2}{3}} + \frac{1}{4}x^{-\frac{3}{4}}$
   h $\frac{3}{2}x^{-\frac{1}{2}} - 2x^{-\frac{3}{2}} - 5x^{-2}$
   i $x^2 - 1$
   j $2x + 3$
   k $\frac{1}{2}x^{\frac{1}{2}} + \frac{5}{6}x^{-\frac{1}{2}} + \frac{1}{2}x^{-\frac{3}{2}}$
   l $\frac{1}{5}x^{-\frac{1}{2}} + \frac{1}{10}x^{-\frac{3}{2}} - \frac{3}{10}x^{-\frac{5}{2}}$

4 a $36x + 6$ b $-\frac{10}{9}x^{-\frac{5}{3}}$
   c 0   d $-x^{-\frac{3}{2}} - \frac{9}{4}x^{-\frac{5}{2}}$

5 a $5 - 20t$, -20
   b $9t^2 - 8t + 7$, $18t - 8$
   c $3 + 10t^{-3}$, $-30t^{-4}$
   d $u + at$, a

6 a 1, 2   b 1, 1   c 3, -4
   d -5, 4   e 4, 4   f 9, -24
   g 2, 0   h 18, $8\frac{1}{4}$

7 a $(4, 16)$ b $(-2, -8)$, $(2, 8)$
   c $(0, 0)$ d $\left(\frac{3}{2}, -\frac{5}{4}\right)$
   e $(-1, 8)$, $(1, 6)$
   f $(8, 2)$
   g $\left(-\frac{1}{3}, \frac{4}{27}\right)$, $(1, 0)$
   h $(1, 4)$, $(3, 0)$
   i $\left(\frac{1}{2}, -\frac{3}{2}\right)$, $\left(-\frac{1}{2}, \frac{3}{2}\right)$
   j $\left(4, \frac{2}{3}\right)$

8   $a = 3$, $b = -4$

9   $c = 4$, $d = -\frac{1}{2}$

10 a 4   b 14   c $\frac{1}{2}$

12   $y' = 68$; $y'' = 31\frac{1}{2}$

13   9.2 cm/yr; 2 cm/yr

14 a -2.13°C cm^{-1} b 6.04°C cm^{-2}

15 a 5.76 g l^{-1} s^{-1}

Exercise 9C (p. 161)

1 a $y = 4x - 4$
   b $y = 24x - 46$
   c $y = -\frac{1}{2}x + \frac{3}{2}$
   d $y = 1 - x$
   e $x - 16y + 52 = 0$
   f $y = 4 - 2x$

2 a $x + 4y = 18$
   b $x + 24y = 1204$
   c $y = 2x - \frac{7}{2}$
   d $y = x + 1$
   e $32x + 2y = 135$
   f $2x - 4y + 11 = 0$

3   $a = -1$; $b = 7$

4 a $x = 1$, $x = -2$ b $y = -4$, $y = 23$

5 a 6   b $y = 6x + 2$ c $\left(-\frac{2}{5}, -\frac{2}{5}\right)$

6 a $(0, 0)$, $(1, 0)$, $(2, 0)$
   b $y = 2x$, $y = 1 - x$, $y = 2x - 4$

7 a $(1, 3)$, $(-4, -12)$
   b $y = -2x + 5$, $y = 8x + 20$
   c $\left(-\frac{3}{2}, 8\right)$

8 a $\left(-\frac{1}{3}, -\frac{11}{9}\right)$ b $y = -\frac{11}{9}$; $x = -\frac{1}{3}$

9   $y = 2x - 10$

10   $h = \frac{8}{3}$, $k = \frac{16}{5}$

11 a $2\frac{1}{2}$ b $2\frac{1}{2}$

12   $\left(7\frac{1}{2}, -12\right)$

13   $a = -1$; $b = 3$

14   $a = -4$; $b = 1$

15 a $(-2, 0)$, $(2, 0)$
   b $y = \frac{1}{2}x + 1$, $y = -\frac{1}{2}x + 1$

16 a $y = x + 3$, $y = \frac{1}{2}x + 6$

Exercise Review 9D (p. 163)

1 a $50x^4$ **b** $3x^3$

  c $-\dfrac{5}{x^2}$ **d** $-\dfrac{1}{2x^2}$

  e $-\dfrac{2}{5x^2}$ **f** $\frac{1}{2}x^{-\frac{1}{2}}$

  g $-6x^{\frac{1}{2}}$ **h** $-\dfrac{6}{x^3}$

  i $-\dfrac{1}{4x^{\frac{3}{2}}}$ **j** $\dfrac{1}{x^{\frac{5}{3}}}$

2 a $12x^3 + 36x^2 - 4x$

  b $-\dfrac{3}{x^2} + \dfrac{2}{3x^3} + \dfrac{1}{2\sqrt{x}}$

3 a $10x^3 - 2x^2 + 1 + \dfrac{1}{x^2}, \ 30x^2 - 4x - \dfrac{2}{x^3}$

  b $2x^{-\frac{1}{2}} + \frac{3}{2}x^{\frac{1}{2}}(= \frac{2}{\sqrt{x}} + \frac{3}{2}\sqrt{x})$,

   $-x^{-\frac{3}{2}} + \frac{3}{4}x^{-\frac{1}{2}}(= -\dfrac{1}{x\sqrt{x}} + \dfrac{3}{4\sqrt{x}})$

4 a $9x^2 + \dfrac{4}{x^2}$ **b** 13

5 a $(2, 8), (-2, -8)$
  b $(1, -10), (-1, 10)$
  c $(4, 96)$
  d $(2, 8), (-1, -1)$

6 $y - 5x + 3 = 0, \ 5y + x - 11 = 0$

7 a $y = 4 - x; \ y = x$ **c** $(-2, -2)$

8 $a = 3, b = 8$ or $a = -7, b = -32$

9 $-1\,\text{cm s}^{-1}; \ 2\,\text{cm s}^{-1}$

10 $y = 3x \pm 8$

10 Integration

Exercise 10A (p. 169)

1 a $x^3 + c$ **b** $\frac{3}{2}x^2 + c$

  c $\frac{3}{5}x^5 + c$ **d** $3x + x^2 + c$

  e $\frac{1}{2}x^2 - \frac{1}{3}x^3 + c$ **f** $2x + c$

  g $\frac{1}{2}ax^2 + bx + c$ **h** $mx + c$

  i $\frac{1}{2}x^2 + c$

2 a $-\frac{1}{2}x^{-2} + c$ **b** $-x^{-1} + c$

  c $-\frac{2}{3}x^{-3} + c$ **d** $-3x^{-1} + c$

  e $\frac{1}{8}x^{-2} + c$ **f** $-\frac{1}{2}x^{-3} + c$

  g $\frac{2}{3}x^{\frac{3}{2}} + c$ **h** $\frac{3}{4}x^{\frac{4}{3}} + c$

  i $\frac{3}{5}x^{\frac{5}{3}} + c$ **j** $-2x^{\frac{1}{2}} + c$

  k $2x^{\frac{1}{3}} + c$ **l** $-x^{\frac{1}{4}} + c$

3 a $x^4 + c$
  b $6x + \frac{1}{2}x^2 + c$
  c $\frac{4}{3}x^3 + 6x^2 + 9x + c$
  d $\frac{1}{3}x^3 + 2x - x^{-1} + c$
  e $\frac{2}{5}x^{\frac{5}{2}} + 2x^{\frac{3}{2}} + c$
  f $\frac{1}{4}x^4 + \frac{2}{3}x^3 + \frac{1}{2}x^2 + c$

4 a $3x + \frac{5}{2}x^2 + c$
  b $2x^3 + c$
  c $-x^{-1} + c$

  d $\frac{1}{2}x^2 + \frac{4}{3}x^{\frac{3}{2}} + x + c$
  e $\frac{1}{2}x^2 + \frac{1}{3}x^3 + c$
  f $\frac{20}{3}x^{\frac{3}{4}} + \frac{4}{7}x^{\frac{7}{4}} + c$
  g $\frac{3}{2}x^{\frac{1}{2}} + c$
  h $\frac{1}{5}x^5 + \frac{4}{3}x^3 + 4x + c$

5 a $\frac{2}{3}x^6 + c$ **b** $-\frac{1}{16}x^{-4} + c$
  c $4x + c$ **d** $12x^{\frac{1}{2}} + x + c$

6 a $\frac{1}{2}at^2 + c$
  b $\frac{1}{12}t^4 + c$
  c $\frac{1}{3}t^3 + 2t^2 + 3t + c$
  d $-\dfrac{t^{-n}}{n} + c$
  e $-t^{-1} - t^{-3} - \frac{1}{2}t^{-2} + c$
  f $\frac{2}{5}t^{\frac{5}{2}} + \frac{2}{3}t^{\frac{3}{2}} + c$

7 a $\frac{1}{2}ax^2 + c$
  b $\frac{1}{3}ay^3 + c$
  c $-kx^{-1} + c$
  d $\frac{1}{3}y^3 - y + 6y^{-1} + c$
  e $\frac{2}{7}x^{\frac{7}{2}} + \frac{1}{3}x^3 + c$
  f $\frac{5}{6}y^{\frac{6}{5}} + c$
  g $4x^{\frac{3}{4}} - 3x^{\frac{2}{3}} + c$
  h $\frac{6}{11}x^{\frac{11}{6}} + \frac{6}{5}x^{\frac{5}{6}} + c$
  i $x^3 + x - 4\sqrt{x} + c$

Exercise 10B (p. 171)

1 $y = x^2 + 5x - 25$

2 $y = x^3 + \dfrac{1}{x} - \dfrac{17}{2}$

3 $f(x) = x^2 + \dfrac{1}{x} - 1$

4 $f(x) = x^3 - \frac{2}{3}x^{\frac{3}{2}} + 4$

5 a $y = x^3 - 4x^2 + 3x$
  b $(1, 0), (3, 0)$

6 a $y = 4x^2 - x^3$ **b** $(4, 0)$
  c $y = \frac{256}{27}$

7 a $y = 2x^3 + \dfrac{1}{2x^2} + \dfrac{1}{2}$

  b $s = \dfrac{3t^2}{2} + \dfrac{8}{t} - 8$

8 a $A = \frac{1}{2}x^4 + \frac{1}{3}x^3 - x^2 - x + \frac{7}{6}$
  b $\frac{4}{3}$

9 a $y = \dfrac{3x^2}{2} + c; \ y = \dfrac{3x^2}{2} + 3$

  b $y = -\dfrac{4}{x} + 5x + c$

   $y = -\dfrac{4}{x} + 5x + 9$

  c $v = 3t^2 + t^3 + c; \ v = 3t^2 + t^3 - 8$

  d $y = x + \dfrac{1}{x} + c; \ y = x + \dfrac{1}{x} + 2$

10 $y = x^3 - \frac{1}{2}x^2 - 6x + 6$

11 $s = ut + \frac{1}{2}at^2$

Exercise 10C Review (p. 173)

1 a $\dfrac{x^5}{5} + x^3 - 5x + c$

  b $2x^3 + \frac{5}{2}x^2 + x + c$

  c $\dfrac{x^4}{4} + \dfrac{x^3}{3} + \dfrac{x^2}{2} + x - x^{-1} - \dfrac{x^{-2}}{2} + c$

  d $\dfrac{x^4}{2} + \dfrac{x^2}{6} - \dfrac{x^{3.5}}{7} + c$

  e $2x^{\frac{1}{2}} + 2x^{-\frac{1}{2}} + c$

  f $\frac{1}{3}x^3 + \frac{2}{5}x^{\frac{5}{2}} - \frac{2}{3}x^{\frac{3}{2}} - x + c$

  g $\frac{2}{7}x^{\frac{7}{2}} + \frac{2}{3}x^{\frac{3}{2}} + c$

  h $3x^{\frac{1}{3}} + \frac{3}{5}x^{\frac{5}{3}} + \frac{8}{3}x^{\frac{3}{2}} + c$

  i $-x^{-2} + \frac{1}{2}x^{-1} - 8x^{-\frac{1}{2}} + c$

2 a $y = \frac{1}{3}x^3 + 2x - 9$
  b $y = -6\frac{2}{3}$

3 $f(x) = x^3 - \frac{5}{2}x^2 - 2x + 2$

4 $y = 2x^2 - 3x - 8$

5 $f(x) = -\dfrac{2}{x} - \dfrac{x^2}{2} + 6$

6 a $a = 6; b = -1$ **b** $y = 3x^2 - x$

11 Proof

Exercise 11A (p. 175)

1 a $\Leftarrow$ **b** $\Leftarrow$ **c** $\Leftrightarrow$ **d** $\Rightarrow$
  e $\Rightarrow$ **f** $\Leftrightarrow$ **g** $\Leftrightarrow$ **h** $\Leftarrow$
  i $\Leftrightarrow$ **j** $\Rightarrow$ **k** $\Rightarrow$ **l** $\Rightarrow$

Exam practice 2 (p. 181)

1 a $2\sqrt{221}$ **b** $11x - 10y + 19 = 0$

2 a $3x + 4y - 10 = 0$

3 a $2y - 3x = -3$ **b** $(3, 3)$

4 a $\left(\frac{5}{2}, -\frac{5}{2}\right)$ **b** $2x + 3y = 1$
  c $\left(\frac{14}{13}, -\frac{5}{13}\right)$

5 a $y = 5x - x^{-1} + c$
  b $y = 12\frac{1}{2}$

6 a $26\,733$ **b** $53\,467$

7 a $\frac{1}{6}x^4 + \frac{15}{7}x^{\frac{7}{3}} + c$
  b $\frac{8}{5}x^{\frac{5}{2}} + \frac{8}{3}x^{\frac{3}{2}} + 2x^{\frac{1}{2}} + c$

8 a $71\,071$ **b** $71\,355$

9 $6x^2 + \frac{1}{2}x^{-\frac{1}{2}} - 2x^{-2}$

10 a $7x + 5y - 18 = 0$

b $\dfrac{162}{35}$

11 a $15x^{\frac{1}{2}}$ **b** $7x + 4x^{\frac{5}{4}} + c$

12 a $2x - \frac{1}{2}x^{-\frac{3}{2}}$ **b** $2 + \frac{3}{4}x^{-\frac{5}{4}}$

c $\frac{1}{3}x^3 + 2x^{\frac{1}{2}} + c$

13 b $k = 4$ **c** 1751.5

14 a $x + 2y = 16$ **b** $y = -4x$

c $\left(\frac{6}{7}, \frac{53}{7}\right)$

15 a 9 **c** $y = 3x - 1$

d $\left(\frac{1}{3}, 0\right)$ **e** $k = 1$

16 a i 455 **ii** $13\,230$

b 11

17 a i $(0, 0), (2, 0)$ **ii** $\left(1, \frac{5}{2}\right)$

b i $(-3, 0), (-1, 0), (0, 1)$ **ii** $\left(-2, \frac{5}{2}\right)$

18 a $3x^2 - 10x + 5$ **b** $\frac{1}{3}$

c $y = 2x - 7$ **d** $\frac{7}{2}\sqrt{5}$

19 a $y = x^3 - 10x^2 + 29x - 20$

c $y = x + 4$ **d** $x = \frac{14}{3}$

20 a $f(x) = \frac{1}{2}x^2 - \dfrac{1}{x} + c$

b $y = \frac{1}{2}x^2 - \dfrac{1}{x} + \dfrac{7}{2}$

21 a £2450 **b** £59 000 **c** 30

22 a l_1: $(0, 0)$, l_2: $(1.5, 0)$, $(0, -3)$

b $x = \frac{4}{3}, y = -\frac{1}{3}$

c $3y - 12x + 17 = 0$

23 a $\frac{1}{2}$ **b** 6 **c** $4\sqrt{5}$

d 10 **e** $2x + y - 16 = 0$

f $(4, 8)$

24 a $-\frac{4}{3}, \frac{3}{4}$ **b** $\sqrt{(k^2 - 8k + 80)}$

c 10 **d** $(11, 6)$

25 a $7x - 2y = -25$ **b** $19\frac{1}{2}$

26 a $(0, 0), (5, 0)$

b $(-1, 0), (4, 0); \left(0, -\frac{3}{2}\right)$

27 a $-\frac{5}{16}$ **b** $8x - 16y - 7 = 0$

c $\left(0, -\frac{7}{16}\right)$ **d** $32x + 16y - 3 = 0$

e $\left(0, \frac{3}{16}\right)$ **f** $\frac{5}{64}$

28 a $(-1, 0), (-4, 0), (-2, 3)$

b $\left(\frac{1}{2}, 0\right), (2, 0), (1, 3)$

c $(1, 0), (4, 0), (2, 9)$

d $(2, 0), (8, 0), (4, 3)$

29 $3 < x < 4$

30 a $u_3 = -4k + 15$

b $u_4 = -8k^2 + 30k - 30$

c $-\frac{1}{4}, 4$

31 b 4 **c** $y = \frac{1}{4}x - 1$

d $\left(-\frac{1}{4}, -\frac{17}{16}\right)$

C1 sample exam paper (p. 184)

1 $\dfrac{5}{7} + \dfrac{4}{7}\sqrt{2}$

2 $2x^2 + \frac{3}{4}x^{\frac{4}{3}} + c$

3 a $8, 13, 18$

b 5

c 1110

4 a 2^{4x}

b $x = 0$ or 4

5 a $3x - 5y + 2 = 0$

6 b $x = 2, y = \frac{3}{2}$

7 a i $x_2 = a - 3$

ii $x_3 = \dfrac{9 - 2a}{a - 3}$

b $3 < a < 3\frac{3}{4}$

8 a $x = -k \pm \sqrt{k^2 + 5}$

c $x = -\sqrt{3} \pm 2\sqrt{2}$

9 b $y = 20x - 28$

10 a $a = 4, b = -\frac{1}{2}, c = -4$

b $\left(\dfrac{1}{2}, -4\right)$

11 a $f'(x) = x^3 + \dfrac{4}{x^3} - 2$

b $f(x) = \dfrac{x^4}{4} - 2x - \dfrac{2}{x^2} + 10.75$

d $x + 3y - 22 = 0$

12 More algebra

Exercise 12A (p. 190)

1 a $A = 2, B = 3$

b $A = 1, B = 1$

c $A = 2, B = 3$

d $A = 6, B = 1$

e $A = 1, B = -1$

f $A = 1, B = 2$

g $A = 2, B = 1$

2 a $A = 1, B = 2, C = 3$

b $A = 2, B = -1, C = 3$

c $A = 2, B = -1, C = 3$

d $A = 3, B = -\frac{1}{2}, C = -\frac{1}{2}$

e $A = -2, B = 1, C = 1$

f $A = 2, B = 1, C = -1$

4 a $A = 1, B = -1, C = 2$

b $A = 1, B = -2, C = 3$

c $A = 1, B = -3, C = 4$

d $A = 3, B = -7$

e $A = 5, B = -3, C = 7$

f $A = 1, B = 4, C = 5$

5 a $A = 2, B = 3, C = 5$

b $A = 7, B = -2, C = 1$

c $A = 4, B = 5, C = -50$

d $A = -4, B = \frac{3}{8}, C = \frac{25}{16}$

6 a $3(x - 1)^2 - 7; (1, -7)$

b $5(x + 2)^2 - 22; (-2, -22)$

c $3\left(x + \frac{8}{3}\right)^2 - \frac{55}{3}; \left(-\frac{8}{3}, -\frac{55}{3}\right)$

d $-2\left(x + \frac{1}{4}\right)^2 + \frac{25}{8}; \left(-\frac{1}{4}, \frac{25}{8}\right)$

7 a $R = -2$ **b** $R = 12$

8 a $A = -1, B = 1, C = 2$

b $A = 4, B = 3, C = -2$

c $A = 2, B = 3, C = -1$

Exercise 12B (p. 195)

1 a $x^2 + 5x + 6$ **b** $x^2 + x - 12$

c $x^2 + x + 1$ **d** $x^2 + x - 6$

e $2x^2 + 5x - 3$ **f** $x^2 + 2x + 1$

g $x^2 + x + 2$ **h** $2x^2 + 2x + 1$

i $x^3 - 2x^2 + 4x - 3$

j $2x^3 - 3x^2 + 4x - 3$

k $x^3 + 4x^2 - 2x - 2$

l $x^3 - 1$

2 a $x^2 + 2x + 4, -3$

b $x^2 + 4x - 10, 8$

c $x^2 - 5x + 6, 0$

d $x^2 - 5x + 9, -19$

e $2x^2 - 6x + 17, 0$

f $2x^2 - 5x + 6, 0$

g $x^2 + 3x - 1, 11$

h $2x^2 - 3x + 5, -16$

3 a $4x^2 + 6x - 3$

b $x^2 + 4x - 1$

c $x^3 - x^2 + 5x - 1, 7$

d $x^2 + 3x - 1, 11$

e $x^2 - 3x + 4$

f $6x^3 - 3x^2 - 2x + 1$

g $x^2 + 2x + 4, -3$

h $x^2 + 4x - 10, 8$

Exercise 12C (p. 199)

1 a $-12, -12, -6, 0, 0; x + 2$ and $x - 2$

b $-1, 0, -2, 19, -21; x - 1$

c $0, 6, -2, 88, -24; x$

d $3, 0, 0, 3, 3; x - 1$ and $x + 1$

3 a $a = -10$ **b** $a = 4$ **c** $a = 5$

4 $x - 2, x + 3; x = -3, x = 1$ or $x = 2$

5 a $(x - 1)(x - 2)(x + 2)$

b $(x + 1)(x - 2)(x + 3)$

c $(x - 1)(x + 2)(x - 3)$

d $(x + 2)(x - 2)(2x + 1)$

6 a $x = -2, x = 1$ or $x = 2$

b $x = -3, x = -1$ or $x = 2$

c $x = -2, x = 1$ or $x = 3$

d $x = -2, x = -\frac{1}{2}$ or $x = 2$

7 $x + 3, 2x - 1; x = 2, x = -3$ or $x = \frac{1}{2}$

8 $a = 4, b = 1; (x - 1)(x + 2)(x + 3)$

9 $a = 0, b = 0; x(x + 1)^2(2x - 1)$

10 a $(x-4)(x+1)$
   c $x = 4$, $x = -1$ or $x = \pm 2\sqrt{2}$

Exercise 12D (p. 200)

1 a 2   **b** 18   **c** -11
  d -1 **e** 2   **f** $-\frac{5}{2}$

2 a $a = -3$ **b** $a = 2$
  c $a = 4$ **d** $a = 2$

3 $a = 4$, $b = -5$

4 $a = -3$

5 $a = 2$

6 $a = 3$, $b = -1$, $c = -2$

7 $a = 2$, $b = -1$, $c = -2$

Review Exercise 12E (p. 201)

1 $2x^3 + 2x^2 - 3x + 1$

2 $x^2 + 8x + 18$

3 $x^3 - x + 3$

4 $x^3 - 3x^2 + 4x + 1$

5 $A = 12$, $B = -7$

6 $A = 2$, $B = 1$, $C = -3$

7 $x^2 + 2x - 5$
  a $x = 2$
  b $x = 2$
  c $x = 2$ or $x = -1 \pm \sqrt{6}$

8 $a = 1$, $b = 4$

9 $a = -5$

10 $a = -9$, $b = 9$

11 $x(x+1)(x-2)(2x+1)$; $x = 0$,
  $x = -1$, $x = 2$ or $x = -\frac{1}{2}$

12 b $x = 2$, $x = -1 \pm \sqrt{7}$

13 $a = -1$

14 a $a = 3$

15 $a = -3$, $b = 6$, $c = 5$

---

13 Coordinate geometry of the circle

Exercise 13A (p. 209)

1 a $(5, -1)$, 4   **b** $(-6, -2)$, 10
  c $(4, 7)$, 1   **d** $(-m, n)$, $\sqrt{c}$

2 a $(-3, 1)$, 5   **b** $(2, 4)$, 8
  c $(-1, -1)$, 2   **d** $(6, -5)$, 7

3 a $(x-3)^2 + (y-5)^2 = 4$
  b $(x-4)^2 + (y-7)^2 = 9$
  c $(x-5)^2 + (y+2)^2 = 25$
  d $(x+4)^2 + (y-3)^2 = \frac{1}{4}$
  e $x^2 + (y-6)^2 = 3$

f $(x-4)^2 + (y+3)^2 = \frac{5}{4}$
g $(x+1)^2 + (y+7)^2 = 100$
h $\left(x - \frac{1}{4}\right)^2 + \left(y + \frac{1}{2}\right)^2 = \frac{9}{4}$

4 a $(2, 3)$, 4   **b** $(2, 1)$, 3
  c $(-5, 7)$, 8   **d** $(2, -1)$, $\sqrt{6}$
  e $(-4, 2)$, $\sqrt{20}$ **f** $\left(-\frac{1}{2}, \frac{3}{2}\right)$, $\sqrt{7}$
  g $\left(\frac{3}{5}, -\frac{1}{2}\right)$, 1   **h** (a, a), $\sqrt{2}a$

5 a, d, g (if $a > 0$), **h** (if $c < 0$)

6 $(x-2)^2 + (y-5)^2 = 25$

7 $(x-3)^2 + (y+5)^2 = 13$

8 $(x-6)^2 + (y+1)^2 = 25$

9 a $x^2 + y^2 = 25$
  b $(x-1)^2 + (y+3)^2 = 61$
  c $(x+2)^2 + (y-5)^2 = 41$
  d $(x-4)^2 + (y+1)^2 = 13$

10 $(x-7)^2 + (y-7)^2 = 49$

11 $(x-4)^2 + y^2 = 25$; $(x+4)^2 + y^2 = 25$

12 $x^2 + y^2 = c^2$; $\dfrac{k}{h+c}$, $\dfrac{k}{h-c}$

13 $x^2 + y^2 = 169$; $\left(\frac{17}{2}, \frac{17}{2}\right)$

Exercise 13B (p. 214)

1 $(2, 3)$, $\frac{1}{4}$; $4x + y - 28 = 0$,
  $x - 4y + 10 = 0$

2 a $2x + 3y - 9 = 0$
  b $y = \frac{3}{5}x$ **c** $y = \frac{1}{2}x - 7$
  d $4x + 9y + 5 = 0$

3 a $x - 4y - 7 = 0$ **b** $x - 3y + 4 = 0$

4 5, $(-4, 0)$, 12

5 a $\sqrt{10}$ **b** $\sqrt{15}$
  c $\sqrt{29}$ **d** $2\sqrt{7}$

6 5

7 $x - y - 1 = 0$, $x + y - 5 = 0$

8 $(1, 2)$

9 $(-2, 5)$

10 $(4, -2)$; $(-1, 3)$; -1

11 a 0, $\frac{12}{5}$ **b** $0 < m < \frac{12}{5}$
  c $m < 0$ or $m > \frac{12}{5}$

12 $\pm 5\sqrt{2}$

14 $(1, 1)$, $(5, 5)$

Exercise 13C Review (p. 216)

1 a $(-2, -3)$, $\sqrt{30}$ **b** $(3, 4)$, 5

2 $(5, 2)$, $2\sqrt{34}$; $(x-5)^2 + (y-2)^2 = 34$

3 a $(x-4)^2 + (y+3)^2 = 18$
  b $(x+2)^2 + (y-5)^2 = 65$
  c $(x-5)^2 + (y-2)^2 = 29$

4 $(-2, -1)$

5 7

6 $8x + y - 26 = 0$

7 $(-1, 4)$, $(3, 8)$, $x + y = 7$

---

14 Binomial expansion

Exercise 14A (p. 219)

1 a $1 + 4x + 6x^2 + 4x^3 + x^4$
  b $1 - 5x + 10x^2 - 10x^3 + 5x^4 - x^5$
  c $1 + 6x + 15x^2 + 20x^3 + 15x^4$
    $+ 6x^5 + x^6$
  d $1 - 6x + 15x^2 - 20x^3 + 15x^4$
    $- 6x^5 + x^6$

2 a $1 + 8x + 24x^2 + 32x^3 + 16x^4$
  b $1 - 15y + 90y^2 - 270y^3 + 405y^4$
    $- 243y^5$
  c $1 + 12z + 48z^2 + 64z^3$
  d $1 - \dfrac{3x}{2} + \dfrac{3x^2}{4} - \dfrac{x^3}{8}$
  e $a^7 + 7a^6b + 21a^5b^2 + 35a^4b^3$
    $+ 35a^3b^4 + 21a^2b^5 + 7ab^6 + b^7$
  f $a^{10} - 5a^8b^2 + 10a^6b^4 - 10a^4b^6$
    $+ 5a^2b^8 - b^{10}$
  g $a^6 - 3a^4b^2 + 3a^2b^4 - b^6$

3 a $x^4 + 4x^3y + 6x^2y^2 + 4xy^3 + y^4$
  b $x^5 + 5x^4y + 10x^3y^2 + 10x^2y^3$
    $+ 5xy^4 + y^5$
  c $81 + 108x + 54x^2 + 12x^3 + x^4$
  d $x^3 - 6x^2 + 12x - 8$
  e $8 - 6x + \dfrac{3x^2}{2} - \dfrac{x^3}{8}$
  f $\dfrac{1}{x^3} + 3 + 3x^3 + x^6$
  g $16 - 32x^3 + 24x^6 - 8x^9 + x^{12}$

4 a i $a^3 + 3a^2b + 3ab^2 + b^3$
    ii $a^3 - 3a^2b + 3ab^2 - b^3$
  b i 14
    ii $10\sqrt{2}$
    iii $40\sqrt{2}$

5 a 194   **b** $160\sqrt{6}$ **c** 98

Exercise 14B (p. 222)

1 a 6   **b** 24   **c** 120   **d** 90
  e 210   **f** 1320   **g** 330   **h** $\frac{1}{28}$
  i 20   **j** 5   **k** 28   **l** 210

2 a 479 001 600   **b** 2.09×10^{13}
  c 1820   **d** 28
  e 2 629 575   **f** 1
  g 111 930   **h** 111 930

Exercise 14C (p. 226)

1 a $1 + 3y + 3y^2 + y^3$;
    $1 - 9y + 27y^2 - 27y^3$
  b $z^4 + 4z^3 + 6z^2 + 4z + 1$;
    $16z^4 + 32z^3 + 24z^2 + 8z + 1$

2 a $1 + 4x + 6x^2 + 4x^3 + x^4$
  b $1 - 5x + 10x^2 - 10x^3 + 5x^4 - x^5$
  c $1 + 6x + 15x^2 + 20x^3 + 15x^4 + 6x^5$
    $+ x^6$

d $1 - 7x + 21x^2 - 35x^3 + 35x^4 - 21x^5 + 7x^6 - x^7$

3 a $1 + 8x + 28x^2$ **b** $1 - 9x + 36x^2$
c $1 + 10x + 45x^2$ **d** $1 - 11x + 55x^2$

4 a $1 + 9x + 27x^2 + 27x^3$
b $1 - 10y + 40y^2 - 80y^3 + 80y^4 - 32y^5$
c $1 + 16z + 96z^2 + 256z^3 + 256z^4$
d $1 - x + \dfrac{x^2}{3} - \dfrac{x^3}{27}$
e $1 + 6x + \dfrac{27}{2}x^2 + \dfrac{27}{2}x^3 + \dfrac{81}{16}x^4$
f $x^5 - 5x^4 + 10x^3 - 10x^2 + 5x - 1$
g $8x^3 + 12x^2 + 6x + 1$
h $\dfrac{x^4}{16} - \dfrac{x^3}{2} + \dfrac{3}{2}x^2 - 2x + 1$

5 a $x^4 + 4x^3y + 6x^2y^2 + 4xy^3 + y^4$
b $a^6 + 6a^5b + 15a^4b^2 + 20a^3b^3 + 15a^2b^4 + 6ab^5 + b^6$
c $x^8 - 4x^6y^2 + 6x^4y^4 - 4x^2y^6 + y^8$
d $16 - 32x + 24x^2 - 8x^3 + x^4$
e $27 - \dfrac{27}{2}x + \dfrac{9x^2}{4} - \dfrac{x^3}{8}$
f $81 - 108x^3 + 54x^6 - 12x^9 + x^{12}$
g $81a^4 + 432a^3b + 864a^2b^2 + 768ab^3 + 256b^4$
h $x^3 - 3x + \dfrac{3}{x} - \dfrac{1}{x^3}$
i $1 - 3x^2 + 3x^4 - x^6$
j $1 + \dfrac{8}{x} + \dfrac{24}{x^2} + \dfrac{32}{x^3} + \dfrac{16}{x^4}$

6 a $1 + 21x + 189x^2$
b $256 - 1024x + 1792x^2$
c $1 + \dfrac{10}{3}x + 5x^2$
d $1 - 6x + \dfrac{33}{2}x^2$

7 a $45 + 29\sqrt{2}$
b $17 - 12\sqrt{2}$
c $1801 - 1527\sqrt{2}$

8 $(a + b)^3 = a^3 + 3a^2b + 3ab^2 + b^3$
$(a - b)^3 = a^3 - 3a^2b + 3ab^2 - b^3$
a 20 **b** $12\sqrt{3}$ **c** $30\sqrt{3}$ **d** $30\sqrt{6}$

9 a 56 **b** $144\sqrt{5}$
c 194 **d** $216\sqrt{2}$

10 a $x^2 = 0.0001; x^3 = 0.000001;$
$x^4 = 0.00000001$
b $x^2 = 0.000001; x^3 = 0.000000001$
$x^4 = 0.000000000001$

11 a $1 - 7x + 21x^2 - 35x^3 + 35x^4$
b 0.93207

12 a $1 + x + \dfrac{3}{8}x^2$ **b** 1.104

13 a $32 + 80x + 80x^2$
b 32.08008

14 $c = 2, n = 12$

15 $a = -5, n = 4$

16 $k = -3, c = 2160$

17 $n = 8$

Exercise 14D Review (p. 227)

1 a $1 + 4x + 6x^2$ **b** $1 - 6x + 12x^2$

c $1 + \dfrac{5x}{3} + \dfrac{10x^2}{9}$ **d** $16 - 32x + 24x^2$
e $\dfrac{1}{x^6} + \dfrac{6}{x^4} + \dfrac{15}{x^2}$

2 a $1 + 4x + 6x^2 + 4x^3 + x^4$
c $x = 0$

3 $a^{11} + 11a^{10}b + 55a^9b^2 + 165a^8b^3$

4 $a^5 - 5a^4b + 10a^3b^2 - 10a^2b^3 + 5ab^4 - b^5$, 77 400

5 $40\sqrt{6}$

6 $a = -7, n = 4$

7 a $1 - 18x + 144x^2 - 672x^3$
b 0.8337

8 a $n = 16$ **b** $1 + 8x + 30x^2 + 70x^3$

9 a $1024 + 1280x + 720x^2 + 240x^3$
b 1159

10 a $x^3 + 3x + \dfrac{3}{x} + \dfrac{1}{x^3}$
b $x^3 + 6x + \dfrac{12}{x} + \dfrac{8}{x^3}$

15 Applications of differentation

Exercise 15A (p. 231)

1 a $f'(x) = 2x + 3, x > -\frac{3}{2}$
b $f'(x) = -4 - 2x, x < -2$
c $f'(x) = 6x^2 - 18x + 12, x < 1$
or $x > 2$
d $f'(x) = 10x, x > 0$
e $f'(x) = 3x^2 - 48, x < -4$ or $x > 4$
f $f'(x) = 18 + 3x^2$, for all values of x.

2 a $x < \frac{1}{2}$ **b** $x > \frac{3}{14}$
c never decreasing
d $0 < x < \frac{2}{3}$
e $-1 < x < 1$
f $x < -3$ or $x > \frac{1}{2}$

Exercise 15B (p. 236)

1 a $(2, -6)$
b $(-3, -2)$
c $\left(\frac{1}{3}, -\frac{1}{3}\right)$
d $(2, 7), (-2, -1)$
e $\left(-2\frac{1}{2}, -30\frac{1}{4}\right)$
f $(0, 5), (1, 4), (-1, 4)$
g $(2, 41), (-2, -23)$
h $(1, 2)$

2 c $\left(\frac{3}{2}, \frac{3}{4}\right)$

3 a $(0, 0)$ min
b $(2, 8)$ max, $(3, 7)$ min
c $(-2, 0)$ min
d $(1, 1)$ min, $(3, 5)$ max
e $(0, 0)$ min
f $\left(\frac{2}{5}, 20\right)$ min, $\left(-\frac{2}{5}, -20\right)$ max
g $(2, 4\sqrt{2})$ min

4 a $x = 3$ or $x = -7$
b $x = \pm 1$
c $x = \frac{1}{2}$

d $x = -1$
e $x = \sqrt[3]{\dfrac{10}{\pi}}$

5 a $(0, 0)$ inflexion
b $(0, 0)$ inflexion, $(3, -162)$ min, $(-3, 162)$ max
c $(0, 0)$ inflexion, $\left(-2\frac{1}{4}, -8\frac{139}{256}\right)$ min
d $(3, 47)$ max, $\left(-5, -38\frac{1}{3}\right)$ min
e $\left(\frac{1}{3}, -54\right)$ min, $\left(-\frac{1}{3}, 54\right)$ max
f $(-2, -12)$ inflexion

6 a $f'(x) = 6x(1 - x)$
b $f(0) = 0, f(1) = 1$
d $0 < x < 1$
e $x < 0, x > 1$
g 2 roots

7 a $f'(x) = 2x - 1$
b $x = \frac{1}{2}, f(x) = -2\frac{1}{4}$, min
c increasing: $x > \frac{1}{2}$; decreasing: $x < \frac{1}{2}$
e 2 roots

8 i $f'(x) = 3x^2 - 12;$
$x = -2, f(x) = 16$, max;
$x = 2, f(x) = -16$, min;
incr: $x < -2, x > 2$;
decr: $-2 < x < 2$; 3 roots
ii $f'(x) = -2 - 2x;$
$x = -1, f(x) = 4$, max;
incr: $x < -1$;
decr: $x > -1$; 2 roots
iii $f'(x) = 3x(2 - x);$
$x = 0, f(x) = 0$, min;
$x = 2, f(x) = 4$, max;
incr: $0 < x < 2$;
decr: $x < 0, x > 2$; 2 roots
iv $f'(x) = -3x^2;$
$x = 0, f(x) = 3$, inflexion;
incr: none;
decr: $x < 0, x > 0$; 1 root
v $f'(x) = 3x^2 - 10x + 3;$
$x = \frac{1}{3}, f(x) = 2\frac{13}{27}$, max;
$x = 3, f(x) = -7$, min;
incr: $x < \frac{1}{3}, x > 3$;
decr: $\frac{1}{3} < x < 3$; 3 roots

9 a $(1, 2)$ min, $(-1, -2)$ max
b $\left(\frac{1}{4}, \frac{1}{4}\right)$ max
c $\left(\frac{1}{2}, 3\right)$ min
d $(2, 12)$ min

10 $a = -22, b = \frac{11}{5}$

11 a $a = 3, b = -5$

12 a 2 s **b** 0.9 s, 6.05 m

Exercise 15C (p. 241)

1 a $100 - x$
b $A = 100x - x^2$
c $\dfrac{\mathrm{d}A}{\mathrm{d}x} = 100 - 2x; x = 50$
d 50 m by 50 m, 2500 m²

2 125 000 m², 250 m by 500 m

3 a $50\,\text{m}$ **b** $0 < t < 5$

4 a Height $= \dfrac{500}{x^2}$

External surface area $= x^2 + \dfrac{2000}{x}$

   b $x = 10$ **c** $300\,\text{m}^2$

5 a $S = \pi r^2 + \dfrac{54\pi}{r}$ **b** $r = 3$

   c $27\pi\,\text{cm}^2$

6 a $5x$ **d** $V = \frac{15}{16}x(1620 - 15x^2)$

   e $x = 6$ **f** $6075\,\text{cm}^3$

7 $2\,\text{cm}, 3\,\text{cm}$

8 $18\,\text{cm}^3, x = 1$

9 $7\frac{11}{27}\,\text{cm}^3, x = \frac{2}{3}$

10 a $A = 2\pi r^2 + 2\pi rh, V = \pi r^2 h$

   b $h = \dfrac{12 - r^2}{r}, V = \pi r(12 - r^2)$

   c $r = 2$

11 a £6

   b $£np = £1000p(84 + 12p - p^2)$

Exercise 15D (p. 248)

1 a $(-\frac{\sqrt{3}}{2}, 0), (-\frac{1}{2}, 1), (\frac{1}{2}, -1), (\frac{\sqrt{3}}{2}, 0)$

   c $2, 1$ **d** $-1 < k < 1$

2 a $(3, 0)$; touches x-axis at $(0, 0)$;
$y \to \mp\infty$; min $(0, 0)$, max $(2, 4)$

   b $(6, 0)$; touches x-axis at $(0, 0)$;
$y \to \pm\infty$; max $(0, 0)$, min $(4, -32)$

   c $(0, 9), (\pm 1, 0), (\pm 3, 0)$;
about y-axis; $y \to +\infty$;
max $(0, 9)$, min $(\pm\sqrt{5}, -16)$

   d $(0, 0), \left(\pm\sqrt{\frac{5}{3}}, 0\right)$; $180°$ rotational
symmetry about $(0, 0)$; $y \to \pm\infty$;
inflexion $(0, 0)$, max $(-1, 2)$,
min $(1, -2)$

3 a $(0, -18), (3, 0)$; max $(-1, -16)$,
min $(1, -20)$; $y \to \pm\infty$

   b $(0, 0), (\pm 2\sqrt{3}, 0)$; min $(-2, -16)$,
max $(2, 16)$; $y \to \mp\infty$

   c $(0, 8), (2, 0)$; inflexion $(2, 0)$;
$y \to \mp\infty$

   d $(1, 0), (0, -4)$; none; $y \to \pm\infty$

4 b $-\frac{1}{16} < k < 0; k = 0$;
$k = -\frac{1}{16}$ or $k > 0$; none;
$k < -\frac{1}{16}$

5 b None; $c < 432; c = 432; c > 432$

8 Even: **a** and **c**
Odd: **b** and **d**
Neither: **e**

Review Exercise 15E (p. 249)

1 a $(-1, 2)$ max, $(1, -2)$ min

   b $(-2, -16)$ min, $(0, 0)$ max,
$(2, -16)$ min

   c $(1, 3)$ max

   d $(1, 4)$ min

2 a $\geqslant 0$

3 a Sometimes increasing,
sometimes decreasing

   b Always increasing

   c Always decreasing

4 a 6, min   **b** 7, max

   c $\frac{7}{8}$, min   **d** $\frac{17}{2}$, max

5 b $\frac{7}{2}\pi; 3\pi$

6 $-3\,\text{cm s}^{-1}$, dec; $7\,\text{cm s}^{-1}$, inc

7 a $x < 2, x > 3$ **b** $2 < x < 3$

8 $y = 1, y = 1, y = 0$

9 a cuts at $(0, 0)$, touches at $(1, 0)$;
$y \to \pm\infty$; min $(1, 0)$, max $(\frac{1}{3}, \frac{4}{27})$

   b touches at $(0, 0), (2, 0)$; $y \to +\infty$;
min $(0, 0)$ and $(2, 0)$, max $(1, 1)$

   c touches at $(-1, 0)$, cuts at
$(2, 0), (0, 2)$; $y \to \mp\infty$;
min $(-1, 0)$, max $(1, 4)$

   d cuts at $(0, 0), \left(-\sqrt[3]{4}, 0\right)$; $y \to +\infty$;
min $(-1, -3)$

10 $320\,\text{m}$; $8\,\text{sec}$

11 a $46\,\text{cm}$; $8\,\text{sec}$ **b** $254\,\text{cm}$

12 $256\,\text{cm}^3$

13 $32\,\text{m}$

14 a $a = 12, 288\,\text{cm}^3$ **b** $10\,\text{cm}^3/\text{day}$

15 a $13\,\text{min}$ **b** $11\,\text{am}$

16 a $n = 10; 118\,\text{mg}$

   b $15\,\text{mg/day}; 4\frac{1}{2}\,\text{days}$

16 Trigonometric functions

Exercise 16A (p. 255)

1 $60°, 90°$; $\dfrac{\pi}{6}, \dfrac{\pi}{4}, \dfrac{2\pi}{3}, 2\pi$

2 a $90°$ **b** $45°$ **c** $60°$ **d** $120°$
   e $30°$ **f** $270°$ **g** $450°$ **h** $720°$
   i $900°$ **j** $240°$ **k** $630°$ **l** $135°$

3 a 2π **b** $\dfrac{\pi}{2}$ **c** $\dfrac{\pi}{4}$

   d $\dfrac{\pi}{12}$ **e** $\dfrac{\pi}{3}$ **f** $\dfrac{2\pi}{3}$

   g $\dfrac{5\pi}{3}$ **h** $\dfrac{3\pi}{2}$ **i** 3π

   j $\dfrac{\pi}{6}$ **k** $\dfrac{5\pi}{6}$ **l** $\dfrac{5\pi}{2}$

4 a 2 **b** 1 **c** 1 **d** 3
   e 3 **f** 3 **g** 3 **h** 2
   i 4 **j** 4 **k** 4 **l** 3

Exercise 16B (p. 266)

1 a $\sin 10°$ **b** $-\tan 60°$
   c $-\cos 20°$ **d** $-\sin 50°$
   e $\cos 20°$ **f** $-\sin 35°$

   g $\tan 40°$ **h** $-\cos 16°$
   i $-\tan 37°$ **j** $-\cos 50°$
   k $-\sin 70°$ **l** $-\tan 50°$
   m $\cos 67°$ **n** $\sin 50°$
   o $-\sin 80°$ **p** $\tan 4°$

2 a $\dfrac{1}{2}$ **b** $\dfrac{\sqrt{3}}{2}$ **c** $\dfrac{\sqrt{2}}{2}$ **d** $\dfrac{\sqrt{3}}{3}$

   e 1 **f** $-\dfrac{\sqrt{2}}{2}$ **g** $-\dfrac{1}{2}$ **h** $-\sqrt{3}$

   i $-\dfrac{\sqrt{2}}{2}$ **j** $\dfrac{1}{2}$ **k** $\dfrac{\sqrt{3}}{3}$ **l** $\dfrac{\sqrt{3}}{2}$

   m $\dfrac{\sqrt{2}}{2}$ **n** $-\dfrac{\sqrt{2}}{2}$ **o** $\dfrac{\sqrt{3}}{3}$ **p** $-\dfrac{\sqrt{3}}{2}$

3 a -1 **b** 1 **c** 0 **d** 0
   e 0 **f** 0 **g** 0 **h** -1

4 a $\dfrac{\sqrt{2}}{2}$ **b** $\dfrac{1}{2}$ **c** $\dfrac{\sqrt{3}}{2}$

   d 0 **e** $\dfrac{\sqrt{2}}{2}$ **f** $\sqrt{3}$

   g $-\dfrac{\sqrt{3}}{2}$ **h** $\dfrac{1}{2}$ **i** 1

   j $\dfrac{\sqrt{3}}{2}$ **k** $-\dfrac{\sqrt{3}}{3}$ **l** $-\dfrac{\sqrt{3}}{2}$

5 a $30°, 150°, 390°, 510°$

   b $140°, 220°, 500°, 580°$

   c $-200°, -20°, 160°, 340°$

   d $\dfrac{5\pi}{4}, \dfrac{7\pi}{4}, \dfrac{13\pi}{4}, \dfrac{15\pi}{4}$

   e $-\pi, \pi$

   f $\dfrac{\pi}{3}, \dfrac{4\pi}{3}, \dfrac{7\pi}{3}, \dfrac{10\pi}{3}$

7 a Translation $\binom{0}{1}$, $360°$

   b Translation $\binom{-60°}{0}$, $360°$

   c Stretch s.f. $\frac{1}{2}$ parallel to x-axis, $180°$

   d Stretch s.f. 2 parallel to x-axis, $720°$

   e Reflection in x-axis, $360°$

   f Reflection in y-axis, $360°$

8 a Stretch s.f. 3 parallel to y-axis, $360°$

   b Translation $\binom{0}{1}$, $180°$

   c Stretch s.f. $\frac{1}{2}$ parallel to x-axis, $180°$

   d Translation $\binom{20°}{0}$, $360°$

   e Reflection in either axis, $180°$

   f Translation $\binom{-60°}{0}$, $360°$

9 a Stretch s.f. 2 parallel to x-axis, 4π

   b Stretch s.f. $\frac{3}{2}$ parallel to x-axis, 3π

   c Reflection of $\sin\frac{x}{2}$ in either
axis, 4π

   d Translation $\binom{0}{-1}$, 2π

   e Translation $\binom{\frac{\pi}{4}}{0}$, 2π

   f Stretch s.f. $\frac{1}{2}$ parallel to x-axis, $\frac{\pi}{2}$

11 a 2π **b** π **c** π

   d $\dfrac{\pi}{3}$ **e** 2π **f** 4π

   g $\dfrac{2\pi}{k}$ **h** $\dfrac{\pi}{k}$

12 No roots, $k < -4, k > 4$; 1 root, none;
2 roots, $k = \pm 4$; 3 roots, none;
4 roots, $-4 < k < 4, k \neq 0$;
5 roots, $k = 0$

13 3

14 a A; s.f. 6; $x = 6\sqrt{2}$
  b B; s.f. 4; $x = 4\sqrt{3}$, $y = 8$
  c B; s.f. 2; $x = 2$, $y = 2\sqrt{3}$
  d B; s.f. 3; $x = 3\sqrt{3}$
  e B; s.f. $\sqrt{3}$; $x = \sqrt{3}$, $y = 3$
  f A; s.f. $2\sqrt{2}$; $x = 2\sqrt{2}$, $y = 4$
  g B; s.f. $\frac{3}{2}$; $x = \frac{3}{2}$, $y = \frac{3\sqrt{3}}{2}$

Exercise 16C (p. 275)

1 a $\cos\theta$ **b** $\sin\theta$ **c** $\tan\theta$ **d** $\sin\theta$
  e $\tan\theta$ **f** 1 **g** 1

2 a $\frac{\pi}{6}$ **b** $\frac{\pi}{3}$ **c** $\frac{\pi}{4}$
  d 0.46 **e** 0.62 **f** 0.96
  g $\frac{\pi}{6}$ **h** 0.62

3 a $\sin\theta = \frac{3}{5}$, $\tan\theta = \frac{3}{4}$
  b $\cos\theta = -\frac{12}{13}$, $\tan\theta = -\frac{5}{12}$
  c $\sin\theta = -\frac{7}{25}$, $\cos\theta = \frac{24}{25}$

4 a 60°, 300° **b** 45°, 225°
  c 90° **d** 150°, 210°
  e 120°, 300° **f** 36.9°, 143.1°
  g 0°, 240°, 360° **h** 4.3°, 184.3°

5 a $\frac{\pi}{6}, \frac{7\pi}{6}$ **b** 0.78, 2.37
  c $0, \frac{4\pi}{3}, 2\pi$ **d** $\frac{2\pi}{3}$
  e 2.19, 4.10 **f** 0.20, 3.34

6 a $\frac{\pi}{6}, \frac{5\pi}{6}, \frac{7\pi}{6}, \frac{11\pi}{6}$
  b $\frac{\pi}{6}, \frac{5\pi}{6}, \frac{7\pi}{6}, \frac{11\pi}{6}$
  c $\frac{\pi}{12}, \frac{5\pi}{12}, \frac{13\pi}{12}, \frac{17\pi}{12}$
  d $\frac{3\pi}{8}, \frac{7\pi}{8}, \frac{11\pi}{8}, \frac{15\pi}{8}$
  e $\frac{\pi}{18}, \frac{11\pi}{18}, \frac{13\pi}{18}, \frac{23\pi}{18}, \frac{25\pi}{18}, \frac{35\pi}{18}$
  f $\frac{\pi}{2}, \frac{7\pi}{6}, \frac{11\pi}{6}$

7 a $\pm 45°, \pm 135°$
  b $-108.4°, 71.6°$
  c $\pm 15°, \pm 45°, \pm 75°, \pm 105°, \pm 135°, \pm 165°$
  d $\pm 37.8°, \pm 142.2°$
  e $-168.4°, -131.6°, 11.6°, 48.4°$
  f $-156.1°, -96.1°, -36.1°, 23.9°, 83.9°, 143.9°$

8 a 0°, 30°, 150°, 180°, 360°
  b 60°, 180°, 300°
  c 45°, 225°
  d 0°, 135°, 180°, 315°, 360°
  e 60°, 90°, 270°, 300°
  f 0°, 180°, 199.5°, 340.5°, 360°
  g 90°, 210°, 330°
  h 120°, 180°, 240°

  i 60°, 90°, 120°, 240°, 270°, 300°
  j 0°, 180°, 360°
  k 40.9°, 139.1°, 220.9°, 319.1°
  l 63.4°, 116.6°, 243.4°, 296.6°

9 a $0, \pm\frac{\pi}{3}$
  b $-\frac{\pi}{2}$
  c $-\frac{\pi}{4}, \frac{3\pi}{4}, -2.21, 0.93$
  d $-\frac{3\pi}{4}, \frac{\pi}{4}, 0.73, 2.41$
  e $\frac{\pi}{6}, \frac{5\pi}{6}, \pm 1.91$
  f $-\frac{7\pi}{8}, -\frac{3\pi}{8}, \frac{\pi}{8}, \frac{5\pi}{8}$

10 a 19.5°, 30°, 150°, 160.5°
   b −116.6°, −45°, 63.4°, 135°
   c 0°, ±120°
   d −135°, 0°, 45°, ±180°
   e −30°, −150°, ±180°
   f −104.0°, 0°, 76.0°, ±180°
   g −14.0°, 0°, 166.0°, ±180°
   h 0°, ±35.3°, ±144.7°, ±180°
   i 0°, ±180°
   j 0°, ±30°, ±150°, ±180°

11 a $\frac{\pi}{3}, \frac{5\pi}{3}$
   b $\frac{\pi}{6}, \frac{5\pi}{6}, \frac{7\pi}{6}, \frac{11\pi}{6}$
   c $\frac{\pi}{3}, \frac{5\pi}{3}$
   d $\frac{3\pi}{2}$
   e $\frac{\pi}{2}, 0.73, 2.41$
   f $\frac{\pi}{6}, \frac{5\pi}{6}, \frac{7\pi}{6}, \frac{11\pi}{6}$
   g 0.84, 5.44
   h 0.68, 1.73, 2.77, 3.82, 4.87, 5.91, $\frac{\pi}{12}, \frac{5\pi}{12}, \frac{3\pi}{4}, \frac{13\pi}{12}, \frac{17\pi}{12}, \frac{7\pi}{4}$

12 a $\frac{x^2}{a^2} + \frac{y^2}{b^2} = 1$
   b $(x-1)^2 + (y-1)^2 = 1$
   c $(x-4)^2 + (y-2)^2 = 1$

14 No roots, $k < -2$, $k > 2$;
   1 root, $k = -2$;
   2 roots, $-2 < k \leqslant 2$

15 $x = 0.8$

16 $x = 123°$

Exercise 16D Review (p. 278)

1 a 3 **b** 2 **c** 0

2 a Reflect in x-axis; 360°
  b Stretch s.f. $\frac{1}{2}$ parallel to x-axis; 180°
  c Stretch s.f. 3 parallel to y-axis; 360°
  d Translation $\binom{-40°}{0}$; 360°
  e Translation $\binom{0}{2}$; 360°

  f Reflect in y-axis and translate $\binom{20°}{0}$ or reflect in $x = 10°$; 360°
  g Stretch s.f. 2 parallel to x-axis, then translation $\binom{1}{0}$; 720°

3 a 30°, 210°
  b 84.3°, 275.7°
  c 135°, 315°
  d 26.6°, 153.4°, 206.6°, 333.4°
  e 83.1°, 336.9°
  f 225°, 315°
  g 0°, 60°, 120°, 180°, 240°, 300°
  h 29.0°, 331.0°

4 a $\pm\frac{\pi}{3}, \pm\frac{2\pi}{3}$
  b −0.59, 2.55
  c $-\frac{\pi}{2}, 0$
  d $-\frac{3\pi}{4}, -2.03, \frac{\pi}{4}, 1.11$
  e $-\frac{5\pi}{6}, -\frac{\pi}{6}, 0.73, 2.41$
  f 0
  g $\pm 1.23, \pm\frac{2\pi}{3}$
  h 0.67, 2.48

5 a 2.42, 3.86 **b** −90°, 30°
  c 112.5°, 202.5° **d** 30°, 150°
  e $-\frac{11\pi}{12}, -\frac{5\pi}{12}, \frac{\pi}{12}, \frac{7\pi}{12}$

6 $x = \pm 120°$

7 E.g. $\left(\frac{2\pi}{15}, \sqrt{3}\right)$

8 a $(x-1)(x-2)(2x-1)$
  b $x = \frac{1}{2}$, $x = 1$ or $x = 2$
  c i $x = \frac{\pi}{4}, \frac{5\pi}{4}$; 1.1, 4.2; 0.5, 3.6
    ii $x = \frac{\pi}{2}$; $\frac{\pi}{6}, \frac{5\pi}{6}$
    iii $x = 0, 2\pi$; $\frac{\pi}{3}, \frac{5\pi}{3}$

9 a $\frac{3 + 2\sqrt{3}}{6}$ **b** $\frac{4\sqrt{3}}{3}$
  c $\frac{\sqrt{2}+1}{2}$ **d** $6 + 3\sqrt{3}$
  e $\sqrt{3}$ **f** 0

17 Applications of trigonometry

Exercise 17A (p. 284)

1 $\frac{10\pi}{9}$ radians; 200°

2 a 7 cm, 35 cm² **b** 1.04 cm, 2.704 cm²
  c 10π cm, 30π cm²

3 8 cm

4 6 cm

5 $\frac{4}{5}$ rad

6 $3\,\text{cm}^2$

7 $4\,\text{rad}$

8 $\left(20 + \dfrac{10\pi}{3}\right)\text{cm}$; $\dfrac{50\pi}{3}\,\text{cm}^2$

9 $15\,\text{s}$

11 $\dfrac{\pi}{8}\,\text{m}^2$

12 $12\,\text{cm}$

13 $3:\pi$

14 $6.44\,\text{cm}$

15 $1.6:1$

16 a $\dfrac{3l^2}{8}\,\text{m}^2$ **b** $\dfrac{14l}{15}\,\text{m}$

Exercise 17B (p. 289)

1 a $6.50\,\text{cm}^2$ **b** $72.4\,\text{cm}^2$ **c** $32.2\,\text{cm}^2$
  d $70.9\,\text{m}^2$ **e** $9\,\text{cm}^2$

2 a $45.5\,\text{cm}^2$ **b** 55.5
  c $23.7\,\text{cm}^2$ **d** $24.6\,\text{cm}^2$

3 a $4\,\text{cm}^2$ **b** $1.82\,\text{cm}^2$ **c** $2.18\,\text{cm}^2$

4 $\dfrac{32\pi}{3}\,\text{cm}^2$; $16\sqrt{3}\,\text{cm}^2$; $\dfrac{16}{3}(2\pi - 3\sqrt{3})\,\text{cm}^2$
  b $92.7\,\text{cm}^2$; $48\,\text{cm}^2$; $44.7\,\text{cm}^2$

5 a $151\,\text{cm}^2$ **b** $62.4\,\text{cm}^2$ **c** $364\,\text{cm}^2$

6 b $\frac{1}{2}r^2(2\pi - \theta + \sin\theta)$

7 a $\sqrt{3}r\,\text{cm}$ **b** $\sqrt{3}r\,\text{cm}$

8 a $\dfrac{2\pi}{3}\,\text{rad}$ **b** area $= \dfrac{\pi r^3}{3}$

9 $\frac{25}{2}(2\sqrt{3} - \pi)\,\text{cm}^2$

10 a $6r, 4\sqrt{3}r$

Exercise 17C (p. 291)

No triangle possible: 1b, 2d, 2f, 2g

Just one triangle possible: 1a, 2b, 2c, 2e, 2i, 2j

More than one triangle possible: 2a, 2h

Exercise 17D (p. 295)

1 a $A = 48°$, $b = 13.8\,\text{cm}$, $c = 15.4\,\text{cm}$
  b $E = 95°$, $d = 1.40\,\text{cm}$, $f = 1.80\,\text{cm}$
  c $a = 19.7\,\text{m}$, $A = 73.3°$, $B = 58.7°$ or
    $a = 3.84\,\text{m}$, $A = 10.7°$, $B = 121.3°$
  d $B = 56.1°$, $a = 6.53\,\text{cm}$, $c = 5.04\,\text{cm}$
  e $X = 19.7°$, $x = 4.63\,\text{cm}$, $z = 8.29\,\text{cm}$
  f $A = 24.3°$, $C = 26.7°$, $a = 4.18\,\text{cm}$

2 a $v = 6\sqrt{2}$ **b** $x = y = 1$

Exercise 17E (p. 299)

1 a $a = 13\,\text{cm}$, $B = 32.2°$, $C = 87.8°$
  b $A = 38.2°$, $B = 81.8°$, $C = 60°$
  c $D = 38.0°$, $E = 112.4°$, $F = 29.5°$
  d $q = 68.0\,\text{cm}$, $P = 14.2°$, $R = 65.8°$
  e $m = 2.65\,\text{cm}$, $L = 40.9°$, $N = 19.1°$

2 a $120°$ **b** $5\sqrt{2}$

Exercise 17F (p. 301)

1 a $C = 45.1°$, $a = 231\,\text{cm}$, $b = 213\,\text{cm}$
  b $A = 54.6°$, $B = 78.1°$, $C = 47.2°$
  c $A = 35°$, $b = 232$, $c = 162$
  d $A = 31.3°$, $B = 44.7°$, $c = 57.9\,\text{m}$
  e $B = 54.8°$, $C = 97.2°$, $c = 18.0\,\text{m}$ or
    $B = 125.2°$, $C = 26.8°$, $c = 8.17\,\text{m}$
  f $A = 17.9°$, $B = 120°$, $C = 42.1°$

2 $1.43\,\text{km}$

3 $25.8\,\text{m}$

4 $1.04°$

5 $347.3°$, 3.64 nautical miles

6 $200\,\text{m}$

7 a $22.0°$ **b** $0.21\,\text{m}$

8 $63\,\text{m}$

9 $312\,\text{m}$

10 a $12.7\,\text{m}$ **b** $39.2\,\text{m}$ **c** $62.9\,\text{m}$

11 $313\,\text{m}$

Exercise 17G Review (p. 303)

1 a $D = 104.5°$, $E = 46.6°$, $F = 29.0°$;
    $2.90\,\text{cm}^2$
  b $p = 5.24\,\text{cm}$, $Q = 34.7°$, $R = 49.3°$;
    $5.97\,\text{cm}^2$
  c $B = 38.2°$, $C = 110.8°$, $c = 9.07\,\text{cm}$;
    $14.0\,\text{cm}^2$ or $B = 141.8°$, $C = 7.2°$,
    $c = 1.21\,\text{cm}$; $1.87\,\text{cm}^2$

2 a $\frac{15}{2}\,\text{m}^2$ **b** $3\,\text{cm}^2$

3 $10\pi\,\text{cm}$, $75\pi\,\text{cm}^2$

4 $\dfrac{35\pi}{2}\,\text{cm}$, $\dfrac{25}{2}\left(\pi - 2\sqrt{2}\right)\,\text{cm}^2$

5 $x = 16$

6 a $210\,\text{cm}^2$ **b** $14.5\,\text{cm}$

7 $10.5\,\text{cm}$

8 $45.1\,\text{m}^2$

Exam practice 3 (p. 305)

1 b $5, 25$

2 a $(5, -3)$ **b** 7

3 $-19.5, -160.5, 90$

4 b $(2x + 1)(x - 2)(x + 2)$
  c $-2, -\frac{1}{2}, 2$

5 $x = -72.6°, -17.4°, 107.4°, 162.6°$

6 a $p = -20$ **b** 1

7 b $81.0\,\text{m}$ **c** $26.7\,\text{m}$ **d** $847\,\text{m}^2$

8 a $210°$, $330°$ (exact); $48.6°$, $131.4°$
    (1 d.p.)

9 a $(x - 3)^2 + (y - 4)^2 = 18$
  b $(2 + 2\sqrt{2}, 5 + 2\sqrt{2})$,
    $(2 - 2\sqrt{2}, 5 - 2\sqrt{2})$
  c 8

10 a $1 - 20x + 180x^2 - 960x^3$
  b $0.817\,04$

11 b $6.77\,\text{cm}^2$ **c** $15.7\,\text{cm}^2$ **d** $22.5\,\text{cm}$

12 a $(\frac{3}{2}, 0)$, $(-5, 0)$, $(0, 15)$
  c $(-\frac{7}{4}, 21\frac{1}{8})$

13 a $n = 8$ **b** $\frac{35}{8}$

14 a $1 + 3nx + \frac{9}{2}n(n-1)x^2 + \frac{9}{2}n(n-1)$
    $(n-2)x^3$
  b 12 **c** $40\,095$

15 a $a + b = 2$ **b** $a = 3$, $b = -1$

16 $A = 8192$, $B = -53\,248$, $C = 159\,744$

17 a $81 + 216x + 216x^2 + 96x^3 + 16x^4$
  b $81 - 216x + 216x^2 - 96x^3 + 16x^4$
  c 1154

18 $76.3°$, $283.7°$

19 a $7\frac{31}{32}$ **b** $\text{f}(x) = \dfrac{1}{3}x^3 - 2x - \dfrac{1}{x} - \dfrac{8}{3}$

20 $x^5 - 5x^3 + 10x - \dfrac{10}{x} + \dfrac{5}{x^3} - \dfrac{1}{x^5}$

21 a $v = 20$ **b** $C = 12$; cost $= £30$

22 i $46.6°$, $173.4°$
  ii b $\cos A = \sqrt{\frac{8}{9}} = \frac{2}{3}\sqrt{2}$

23 a $2x + 4y + \pi x$ **b** $4xy + \frac{1}{2}\pi x^2$
  d $14, 7.0$ **e** 700

24 b $(x + 3)(2x^2 - x - 5)$
  c $-1.35, 1.85$

25 a $\dfrac{\text{d}y}{\text{d}x} = 4x^3 - 16x$
  b c $(0, 3)$ max, $(2, -13)$ min,
    $(-2, -13)$ min
  d $x - 12y - 49 = 0$

26 b $\text{f}(n) = (n + 2)(n + 1)(n + 3) + 3$

27 $0°$, $131.8°$, $228.2°$

28 $n = 9$, $p = -2$

29 a $a = -2$, $b = 5$

30 b $(0, 0.5)$, $(150, 0)$, $(330, 0)$
  c $180°$, $300°$

31 b $(x + 3)(x - 2)(x - 4)$

32 a $225, 345$
  b $22.2, 67.8, 202.2, 247.8$

33 a $(3, -4), 10$ **b** $(9, 4)$

34 a $(4, 8)$; 17 **b** $\dfrac{4 - x}{y - 8}$ **c** $x = 21$

35 c 1200

36 a $2\sqrt{3}$ **b** 2π

37 a $k = 4$
  b $(1, -9)$ min, $(-\frac{4}{3}, \frac{100}{27})$ max
  c $(-2, 0)$, $(-\frac{1}{2}, 0)$, $(2, 0)$; $(0, -4)$

38 $p = \frac{3}{4}$, $A = -144$

18 Exponentials and logarithms

Exercise 18A (p. 313)

1 Bases:
  a 10   **b** 10   **c** 3
  d 4   **e** 2   **f** $\frac{1}{2}$
  g a
  Logarithms:
  a 2   **b** 1.6021   **c** 2
  d 3   **e** 0   **f** -3
  g b

2 a $\log_2 16 = 4$ **b** $\log_3 27 = 3$
  c $\log_5 125 = 3$
  d $\log_{10} 1\,000\,000 = 6$
  e $\log_{12} 1728 = 3$ **f** $\log_{16} 64 = \frac{3}{2}$
  g $\log_{10} 10\,000 = 4$ **h** $\log_4 1 = 0$
  i $\log_{10} 0.01 = -2$ **j** $\log_2\left(\frac{1}{2}\right) = -1$
  k $\log_9 27 = \frac{3}{2}$ **l** $\log_8\left(\frac{1}{4}\right) = -\frac{2}{3}$
  m $\log_{\frac{1}{3}} 81 = -4$ **n** $\log_e 1 = 0$
  o $\log_{16}\left(\frac{1}{2}\right) = -\frac{1}{4}$ **p** $\log_{\frac{1}{8}} 1 = 0$
  q $\log_{81} 27 = \frac{3}{4}$ **r** $\log_{\frac{1}{16}} 4 = -\frac{1}{2}$
  s $\log_a c = 5$ **t** $\log_a b = 3$
  u $\log_p r = q$

3 a $2^5 = 32$ **b** $3^2 = 9$
  c $5^2 = 25$ **d** $10^5 = 100\,000$
  e $2^7 = 128$ **f** $9^0 = 1$
  g $3^{-2} = \frac{1}{9}$ **h** $4^{\frac{1}{2}} = 2$
  i $e^0 = 1$ **j** $27^{\frac{1}{3}} = 3$
  k $a^2 = x$ **l** $3^b = a$
  m $a^c = 8$ **n** $x^y = z$
  o $q^p = r$

4 a 6   **b** 2   **c** 7
  d 2   **e** $\frac{1}{3}$ **f** 0

5 a $\frac{1}{3}$ **b** 3   **c** 3
  d -1 **e** $\frac{1}{2}$ **f** 2

Exercise 18B (p. 316)

1 a $\log a + \log b$ **b** $\log a - \log c$
  c $-\log b$ **d** $2\log a + \frac{3}{2}\log b$
  e $-4\log b$
  f $\frac{1}{3}\log a + 4\log b - 3\log c$
  g $\frac{1}{2}\log a$ **h** $\frac{1}{3}\log b$
  i $\frac{1}{2}\log a + \frac{1}{2}\log b$ **j** $1 + \lg a$
  k $-2 - 2\lg b$ **l** $\frac{1}{2}\log a - \frac{1}{2}\log b$

2 a $\log 6$ **b** $\log 2$
  c $\log 6$ **d** $\log 2$
  e $\log(ac)$ **f** $\log\left(\frac{xy}{z}\right)$
  g $\log\left(\frac{a^2}{b}\right)$ **h** $\log\left(\frac{a^2 b^3}{c}\right)$
  i $\log\sqrt{\frac{x}{y}}$ **j** $\log\left(\frac{p}{\sqrt[3]{q}}\right)$
  k $\lg(100a^3)$ **l** $\lg\left(\frac{10a}{\sqrt{b}}\right)$

m $\lg\left(\frac{a^2}{2000c}\right)$ **n** $\lg\left(\frac{10x^3}{\sqrt{y}}\right)$

o $\log\left(\frac{b^a}{a^b}\right)$

3 a 3   **b** 2   **c** 2
  d 1   **e** $\log 2$ **f** $\log 7$
  g $-\log 2$ **h** 0   **i** 0
  j 3   **k** 2   **l** $\frac{2}{3}$

Exercise 18C (p. 319)

1 a 2   **b** y **c** 5   **d** n
  e 7   **f** n **g** 5   **h** 5
  i x **j** 0   **k** $2x+1$ **l** e
  m 1   **n** 1

2 a 1.086   **b** -0.631
  c -3.170 **d** 1.585
  e -1.292 **f** 2.059
  g 0.861   **h** 6.456

Exercise 18D (p. 322)

1 a $x = 6$ **b** $x = -5$
  c $x = 3.79$ **d** $x = -3$
  e $x = 0.158$ **f** $x = 5$
  g $x = 5.03$ **h** $x = 0.403$
  i $x = 0$ **j** 16
  k 2   **l** $\frac{1}{2}$

2 a $x = 0.756$ **b** $x = -7.06$
  c $x = 1.16$ **d** $x = -2.58$
  e $x = 6.84$ **f** $x = 0$

3 a $x = 1.58$
  b $x = 1$ or $x = 2$
  c $x = 0$ or $x = 2$
  d $x = 2$
  e $x = -1$ or $x = 1$
  f $x = -1$ or $x = 0$
  g $x = 2$
  h $x = 3$
  i $x = 0.631$ or $x = 1.26$
  j $x = 0$ or $x = 2.32$

4 a $x > 3$ **b** $x < 3.32$
  c $x > 6.64$ **d** $x > 5.64$
  e $x > -3.53$ **f** $x < 11.9$

5 13

6 6

7 4, 6

8 a $2k$ **b** $k + 2$
  c $\frac{1}{2}k - 1$ **d** 9

10 a Reflection in y-axis; (0, 1)
  b Stretch in x direction s.f. $\frac{1}{2}$; (0, 1)
  c Stretch in y direction s.f. 3; (0, 3)

11 a (1, 9)

Exercise 18E Review (p. 323)

1 a $\log_2 32 = 5$ **b** $\log_{10} 100 = 2$
  c $\log_a c = b$ **d** $\log_p q = 3$
  e $\log_{27} 3 = \frac{1}{3}$ **f** $\log_3 \frac{1}{3} = -1$

2 a $2^3 = 8$ **b** $6^2 = 36$
  c $a^c = b$ **d** $d^f = c^4$
  e $p^4 = 8$ **f** $c^q = 3$

3 a 7   **b** 3   **c** 4
  d 4   **e** $\frac{1}{4}$ **f** -2

4 a $\log a + 2\log b - \log c$
  b $\frac{1}{2}\log a + \frac{1}{2}\log b$
  c $4\log a - 2\log b - \log c$
  d $2\log a + 3\log b + 4\log c$

5 a $\log\left(\frac{a}{b}\right)$ **b** $\log(\sqrt{x}y^3)$
  c $\log 62.5$ **d** $\log x^{3a}$

6 a 2   **b** 6   **c** $\log 8$
  d $\log 3$ **e** 4   **f** $\frac{1}{3}$

7 a $n = -10$ or $n = 100$
  b $x = \frac{9}{8}$
  c $x = 1.26$
  d $x = \frac{1}{2}$
  e $x = 0$ or $x = 1$
  f $y = -8$ or $y = 1$

8 a $x < 5.64$ **b** $x > 10$

9 a $\frac{1}{2}$, 3   **b** 0.792

10 -0.396

11 a 4.6%

19 Further integration and the trapezium rule

Exercise 19A (p. 330)

1 a $3\frac{3}{4}$ **b** 2   **c** -2 **d** $36\frac{137}{144}$

2 a 9   **b** 81   **c** 0   **d** $\frac{1}{2}$

3 a 0   **b** $52\frac{1}{2}$ **c** $29\frac{1}{2}$
  d $8\frac{2}{3}$ **e** -14 **f** $\frac{264}{5}$
  g $\frac{4}{5}$ **h** $-\frac{3}{2}$ **i** -4

4 a $\frac{45}{32}$ **b** $\frac{14}{3}$ **c** $\frac{45}{4}$ **d** 21
  e $\frac{23}{36}$ **f** $\frac{15}{4}$ **g** $48\frac{24}{35}$ **h** $4\frac{8}{81}$
  i 23

5 a $\frac{42}{5}$ **b** $\frac{4}{3}$ **c** $\frac{32}{3}$ **d** $\frac{28}{3}$

6 a 26   **b** $\frac{175}{3}$ **c** $\frac{53}{6}$ **d** $\frac{5}{2}$

7 $\frac{1}{6}$

8 50

9 a $\frac{7}{3}$ **b** 4

10 a $\frac{16}{3}$ **b** $\frac{64}{3}$ **c** 12   **d** $\frac{4}{3}$

12 a 9   **b** $\frac{45}{4}$
  c 12   **d** $2(\sqrt{3} - \sqrt{2})$

13 a 18   **b** $\frac{16}{3}$ **c** 4   **d** $\frac{9}{2}$

Exercise 19B (p. 337)

1 a $\frac{5}{4}$ **b** $\frac{28}{3}$

2 b $x + 2y = 11$
  c (11, 0)
  d $29\frac{1}{3}$

3 a $x + 4y = 2$
 c $\frac{11}{64}$

4 b $(3, 0), (0, 3)$
 c $\frac{5}{6}$
 d $\frac{1}{3}$

5 a $(1, 3), (3, 3); \frac{4}{3}$
 b $(1, 5), (-2, 5); \frac{9}{2}$
 c $(0, 0), (4, 8); \frac{16}{3}$
 d $(-2, 12), (1, 3); \frac{27}{2}$
 e $(-1, 0), (3, 4); \frac{32}{3}$
 f $(0, 3), (4, 5); \frac{4}{3}$

6 a $1\frac{1}{2}$ **b** $\frac{11}{6}$

7 $6\frac{3}{4}$

8 0

9 $\frac{7}{4}$

10 $121\frac{1}{2}$

11 a $A(1, 3), B(-1, 3)$ **b** $\frac{8}{3}$

12 a $\frac{32}{3}$ **b** $\frac{2}{15}$ **c** $\frac{32}{3}$ **d** 10

Exercise 19C (p. 344)

1 a $9, 16, 25; 39.5$
 b 39

2 $7.992, 7.937, 7.786, 7.483, 6.955, 6.083;$ 22.6

3 5.63; overestimate

4 242 litres

5 $86\,\text{m}$

6 a $1, 1.005, 1.020, 1.044, 1.077, 1.118,$ $1.166, 1.221, 1.281$
 b 0.879

7 a 2.02 **b** 0.223
 c 1.82 **d** 1.09

8 a $1 + 10x^3 + 45x^6 + 120x^9$
 b i 0.204 **ii** 0.204

Exercise 19D Review (p. 346)

1 a $21\frac{1}{3}$ **b** 68 **c** $-\frac{1}{6}$

2 a $16\frac{2}{3}$ **b** $2\frac{4}{15}$ **c** $13\frac{1}{3}$ **d** $149\frac{1}{3}$

3 $85\frac{1}{3}$

4 $\frac{3}{4}$

5 a $25\frac{2}{3}$ **b** $20\frac{5}{6}$ **c** $10\frac{2}{3}$ **d** $20\frac{5}{6}$

6 a 0 **b i** $\frac{1}{4}$ **ii** $\frac{1}{4}$

7 a $21\frac{1}{3}$ **b i** 22 **ii** 21.5

8 0.38

9 a 6.45 **b** 6.35

10 a $A(1, \frac{3}{2}), B(4, 3)$ **b** $AB = \dfrac{3\sqrt{5}}{2}$

11 Area A:Area B $= 5:1$

12 $\frac{64}{3}$

13 a $74\frac{2}{3}$ **b** $6\frac{2}{3}$

20 Geometric series

Exercise 20A (p. 356)

1 a 3 **b** $\frac{1}{4}$ **c** -2 **d** -1
 f a **g** 1.1 **j** 6

2 a $5 \times 2^{10}; 5 \times 2^{19}$
 b $10 \times \left(\frac{5}{2}\right)^6; 10 \times \left(\frac{5}{2}\right)^{18}$
 c $\frac{2}{3} \times \left(\frac{9}{8}\right)^{11}; \frac{2}{3} \times \left(\frac{9}{8}\right)^{n-1}$
 d $3 \times \left(-\frac{2}{3}\right)^7; 3 \times \left(-\frac{2}{3}\right)^{n-1}$
 e $\frac{2}{7} \times \left(-\frac{3}{2}\right)^8; \frac{2}{7} \times \left(-\frac{3}{2}\right)^{n-1}$
 f $3 \times \left(\frac{1}{2}\right)^{18}; 3 \times \left(\frac{1}{2}\right)^{2n-1}$

3 a $\frac{242}{81}$ **b** $\frac{43}{16}$ **c** $-\frac{266}{243}$ **d** 683

4 a 9 **b** 8 **c** 7
 d 8 **e** $n+1$ **f** n

5 a $2^{10} - 2$ **b** $\frac{1}{2}\left(3^5 - \frac{1}{27}\right)$
 c $0.03(2^7 - 1)$ **d** $-\frac{16}{405}\left[\left(\frac{3}{2}\right)^8 - 1\right]$
 e $5(2^{n+1} - 1)$ **f** $a\left(\dfrac{1 - r^n}{1 - r}\right)$

6 a 1092 **b** 5193.98 **c** $\frac{31}{32}$

7 $2; 2\frac{1}{2}; 157\frac{1}{2}$

8 $3, \frac{2}{3}; -3, -\frac{2}{3}$

9 $n = 6; 13\frac{1}{2}$

10 £$10\,700\,000$

11 $6\frac{3}{4}$

12 12

13 8 terms

14 a $k^8; k^{n-5}$ **b** $2; 2p^{5-n}$

15 $\frac{5}{2}, -\frac{1}{3}$

16 $\frac{5}{2}(3^n - 1); 16$

17 a 1.4% **b** $134\,000, 144\,000$
 c $1\,241\,000$

18 a 1.8% **b** 2.2%
 c £8428 million

19 a $1500 \times 1.08 + 1500 \times 1.08^2$ $+ \cdots + 1500 \times 1.08^{10}$
 b £$23\,468.23$

Exercise 20B (p. 361)

1 a i $\frac{3}{2}\left(1 - \left(\frac{1}{3}\right)^n\right)$ **ii** $\frac{3}{2}$
 b i $24\left(1 - \left(\frac{1}{2}\right)^n\right)$ **ii** 24
 c i $\frac{1}{3}\left(1 - \left(\frac{1}{10}\right)^n\right)$ **ii** $\frac{1}{3}$
 d i $\frac{13}{99}\left(1 - \left(\frac{1}{100}\right)^n\right)$ **ii** $\frac{13}{99}$
 e i $\frac{5}{9}(1 - (0.1)^n)$ **ii** $\frac{5}{9}$
 f i $\frac{6}{11}(1 - (0.01)^n)$ **ii** $\frac{6}{11}$
 g i $\frac{2}{3}\left(1 - \left(-\frac{1}{2}\right)^n\right)$ **ii** $\frac{2}{3}$
 h i $\frac{81}{2}\left(1 - \left(-\frac{1}{3}\right)^n\right)$ **ii** $\frac{81}{2}$

2 a 4 **b** $\frac{15}{2}$ **c** $\frac{81}{64}$ **d** $\dfrac{p}{1024}$

3 $\frac{2}{3}$

4 $2; \frac{1}{2}; \frac{1}{4}$

5 $\frac{3}{5}, 40; \frac{2}{5}, 60$

6 43.957

7 125

8 a 18.0
 b $\frac{2}{3}, 8$

Exercise 20C Review (p. 362)

1 a $\frac{1}{4}, \frac{1}{8}, 8\left(1 - \dfrac{1}{2^n}\right)$; converges; $S_\infty = 8$
 b $4.3923, 4.83\,153; 30(1.1^n - 1)$; does not converge
 c $256, -1024; \frac{1}{5}(1 - (-4)^n)$; does not converge
 d $0.096, -0.0192; 50(1 - (-0.2)^n)$; converges; $S_\infty = 50$

2 a 102 **b** 457
 c 10.7 **d** 1590 or 75.9
 e $1\,110\,000$ **f** 362

3 $\frac{3}{4}, -\frac{3}{2}, 3$

4 $\frac{3}{4}$

5 a $\frac{3}{5}k$ **b** -73

6 a 6.72 **b** 0.123

7 10

8 7

11 432

12 $x = 5; y = -1$

13 a $\frac{8}{9}$ **b** $\frac{4}{33}$ **c** $3\frac{2}{9}$
 d $2\frac{23}{33}$ **e** $1\frac{1}{225}$ **f** $2\frac{317}{330}$

15 $4, 15\frac{7}{8}; -12, 57\frac{7}{8}$

16 a £5674 **b** 12 years
 c £254; £340

17 1.8×10^{19}

Exam practice 4 (p. 365)

1 b $x = 2.32$

2 $4\frac{1}{2}$

3 a $-1, 3$ **b** $10\frac{2}{3}$

4 a $\dfrac{p}{4}$ **b** $\frac{3}{4}p + 1$

5 b 0.023, (or -0.023)
 c $S_n = \dfrac{1200(1 - (-0.25)^n)}{1 - (-0.25)} = 960(1 - (-0.25)^n)$

6 a $-\dfrac{x^4}{4} + \dfrac{27x^2}{2} - 34x + c$
 b 34 **c** 12

7 a $1\,\text{hr}\,21\,\text{min}$ **c** $3\,\text{hr}\,51\,\text{min}$

8 a $\frac{3}{2}x^{-\frac{1}{2}} + 2x^{-\frac{3}{2}}$ **b** $2x^{\frac{3}{2}} - 8x^{\frac{1}{2}} + c$
 c $A = 6, B = -2$

9 a $u_1 = 76$, $u_2 = 60.8$
  b $u_{21} = 0.876$
  c 367
  d 380

10 a $\frac{2}{5}$; 200 **b** $\frac{1000}{3}$ **c** 8.9×10^{-4}

11 a (1, 2), (4, 5) **b** 4.5

12 a $\frac{x+3}{x} = 1 + \frac{3}{x}$ **b** $\frac{1}{5}$

13 134

14 b $r = 0.6$, $a = 5$ **c** 12.5

15 a 1.14
  b $A = 4$, $B = 9$, $C = -7$
  c $\frac{91}{8}$

16 a $p = 1.357$, $q = 1.382$
  b 2.59

17 a 0.8 **b** 10 **c** 50 **d** 0.189

18 a $u_1 = 1$, $u_2 = \frac{1}{3}$, $u_3 = -\frac{1}{9}$
  b 5.986

19 b $\frac{243}{4}$ **c** $\frac{729}{4}$ **d** 3.16

20 $1\frac{1}{2}$

21 a $p = -3$ **b** -1 **c** $-3 - \frac{1}{2}t$

22 b $d = 3700$ **c** £91 000 **d** £96 700

23 a $x = 4$, $y = 7.80$; $x = 6$,
      $y = 7.80$; $x = 8$, $y = 6.13$
  b $55.7\,\text{m}^2$ **c** $84.3\,\text{m}^2$
  d over-estimate

24 a $x + y - 3 = 0$ **c** $\frac{4}{3}$

25 b $\alpha = \frac{1}{4}$, $\beta = \frac{3}{2}$
  d 0.585

C2 Sample exam paper (p. 372)

1 a $\dfrac{dy}{dx} = 2 - \dfrac{10}{x^2}$
  b $a = 5$, $b = 4$
  c Minimum

2 a $v = 0$, 11.27, 21.06, 29.39, 36.24
  b 400 m

3. b Radius $= 2.5\,\text{m}$
      Centre $= (6.5, 3)$
  c $(x - 6.5)^2 + (y - 3)^2 = 2.5^2$

4 $n = 7$, $b = 4$

5 b 2
  c 14348906

6 b $657\,\text{m}^2$
  c $210\,\text{m}^2$

7 b $a = 2$, $b = 3$
  c $x = 0.825$, $y = 4.40$

8 a $\theta = 171.8°$, $351.8°$
  b $\theta = 211.6°$, $328.4°$

9 b $(x - 3)(x - 2)(x + 1)$
  c $7\frac{1}{3}$

'London Qualifications Ltd' accepts no responsibility whatsoever for the accuracy or method of working in the answers given.

Glossary

abscissa
The x-coordinate

algorithm
A systematic procedure for solving a problem

altitude of a triangle
The perpendicular distance from a vertex to the base

angle of elevation or depression
The angle between the line of sight and the horizontal

arithmetic series
A series whose consecutive terms have a common difference

ascending powers of x
In order, smallest power of x first, e.g.
$a_0 + a_1x + a_2x^2 + \cdots$

asymptote
A line to which a curve approaches

base of a log
b as in $\log_b x$

base of a power
b as in b^x; *see also* exponent, index

bearing
A direction measured from the North clockwise

binomial
An expression consisting of two terms, e.g. $x + y$

bisect
Cut into two equal parts

Cartesian equation
Relationship connecting x and y coordinates

chord
A straight line joining two points on a curve

circumcentre of a triangle
The centre of the circle which passes through all three vertices of the triangle

coefficient
The numerical factor in a term containing variables, e.g. -5 in $-5x^2y$

collinear
Lying on the same straight line

complex number
A number of the form $a + ib$ where $i = \sqrt{-1}$

composite number
A positive integer with factors other than 1 and itself

congruent
Identical in shape and size

constant
A quantity whose value is fixed

constant term
The term in an expression which has no variable component , e.g. -3 in $x^2 - 5x - 3$

continuous curve or function
One whose graph has no break in it

convergent
Approaching closer and closer to a limit

converse
The converse of $a \Rightarrow b$ is $b \Rightarrow a$

coordinate
A magnitude used to specify a position

coprime numbers
Two positive integers whose HCF is 1

corollary
A result which follows directly from one proved

cubic equation
An equation of the form $ax^3 + bx^2 + cx + d = 0$, the highest power of x being 3

definite integral
An integral with limits

degree of a polynomial
The highest power of the variable, e.g. 2 in
$x^2 - 5x - 3$

denominator
'Bottom' of a fraction – the divisor – remember D for Down

descending powers of x
In order, largest power of x first, e.g.
$a_4x^4 + a_3x^3 + a_2x^2 + \cdots$

difference of squares
$x^2 - y^2 = (x + y)(x - y)$

discontinuous curve or function
One whose graph has a break in it

discriminant of a quadratic equation
The value of $b^2 - 4ac$ for the equation
$ax^2 + bx + c = 0$

displacement
Change in position from a given point

divergent
Not convergent

dividend
A number (or expression) which is divided by a divisor to produce a quotient and possibly a remainder

divisibility tests
A number is divisible by
  2 if the last digit is even
  3 if the digit sum is divisible by 3
  4 if the number formed by the last two digits is divisible by 4
  5 if the last digit is 0 or 5
  8 if the number formed by the last three digits is divisible by 8

9 if the digit sum is divisible by 9

11 if the sum of the digits in the odd positions differs from the sum of the digits in the even positions by 0 or any multiple of 11

For a composite number, such as 6 ($= 2 \times 3$) use tests for 2 and 3

divisor

A number (or expression) by which another is divided to produce a quotient and possibly a remainder

even function

A function where $f(x) = f(-x)$; the function is symmetrical about the y-axis

exponent

In the power 3^4, 4 is the exponent; also called index

exponential function

A function of the form $a^{f(x)}$ where a is constant, e.g. 2^x; and e^x, *the* exponential function

foot of a perpendicular

The point where the perpendicular meets a specified line

frustum of a cone or pyramid

The part remaining when the top is cut off by a plane parallel to the base

general solution

A solution, given in terms of a variable, which generates all required solutions

geometric series

A series whose consecutive terms have a common ratio

HCF

Highest common factor

heptagon

A seven-sided 2D figure

hypotenuse

The side of a right-angled triangle opposite the right angle

identity

An equation which is true for all values of the variable(s)

improper fraction (algebraic)

Fraction where the degree of the numerator is greater than or equal to the degree of the denominator

improper fraction (numerical)

$\frac{p}{q}$ where $p > q$; p, q are positive integers; *see also* proper fractions

incentre of a triangle

The centre of the circle which touches all three sides of the triangle

included angle

The angle between two given sides

increment

A small change in the value of a quantity

indefinite integral

An integral without limits

index (pl. indices)

In 3^4, 4 is the index; also called exponent

infinity (∞)

The concept of 'without end'

integer

A whole number, +ve or −ve or zero

irrational number

A real number which is not rational, e.g $\sqrt{2}$, π, e

isosceles trapezium

A trapezium with an axis of symmetry through the mid-points of the parallel sides

kite

A quadrilateral with one diagonal an axis of symmetry

LCM

Lowest (or least) common multiple

LHS

Left-hand side, for example, of an equation

limit

The value to which a sequence converges

line segment

A finite part of an infinite line

ln

Napierian or natural log, to base e

logarithm (log) of a number

The power to which a base must be raised to obtain the number

lowest terms

In its lowest terms, a fraction which cannot be cancelled, the numerator and denominator having no common factor

major

The larger arc, sector or segment

median of a triangle

A line joining a vertex to the mid-point of the opposite side

minor

The smaller arc, sector or segment

monomial

An expression consisting of one term

Napierian or natural log

log to the base e, ln

normal at a point

A line which passes through the point and is perpendicular to the curve at that point

numerator

'Top' of a fraction; the dividend

odd function

A function where $f(x) = -f(-x)$; the function has $180°$ rotational symmetry about the origin

ordinate

The y-coordinate

oscillating sequence

A sequence which neither converges to a limit, nor diverges to $+\infty$ or $-\infty$

parallelogram

A quadrilateral with both pairs of opposite sides parallel

period

The smallest interval (or number of terms) after which a function (or sequence) regularly repeats

periodic function or sequence

One which repeats at regular intervals

perpendicular

At right angles

perpendicular bisector of AB

The line which bisects AB at right angles; the set of points equidistant from A and B

point of contact

The point at which a tangent touches a curve

polygon

A plane figure with many sides

polynomial (of degree n)

A sum of terms of the form
$a_0 + a_1 x + a_2 x^2 + \cdots + a_n x^n$

power

For example, $81 = 3^4$ is the fourth power of 3; *see also* base, exponent, index

prime number

A positive integer which is divisible only by itself and 1
NB: 1 is not included in the set of prime numbers

prism

A solid with uniform cross-section

produce

Extend, as of a line

proper fraction (algebraic)

Fraction where the degree of the numerator is less than the degree of the denominator

proper fraction (numerical)

$\frac{p}{q}$ where , $p < q$; p, q are positive integers; *see also* improper fraction

quadrant

One of the four parts into which the plane is divided by the coordinate axes

quadratic equation

An equation of the form $ax^2 + bx + c = 0$, the highest power of x being 2

quartic equation

An equation of the form
$ax^4 + bx^3 + cx^2 + dx + e = 0$,
the highest power of x being 4

quotient

The result of dividing one number or expression (dividend) by another (divisor) – there may be a remainder

radian

Measure of an angle; 1 radian = angle subtended at the centre of a circle radius r by an arc of length r; 1 radian $\approx 57°$

rational number

A number which can be expressed as $\frac{p}{q}$ where p and q are integers, $q \neq 0$

real number

A number corresponding to some point on the number line

reciprocal of $\frac{a}{b}$

$\frac{b}{a}$; and vice versa, reciprocal of $\frac{b}{a}$ is $\frac{a}{b}$

reductio ad absurdum

Proof by assuming the result is not true and arriving at a contradiction

regular polygon

A polygon with all sides and all angles equal

respectively

In the order mentioned

rhombus

A parallelogram with four equal sides; the diagonals bisect each other at $90°$

RHS

Right-hand side, for example, of an equation

right cone or pyramid

The vertex being vertically above the centre of the base

root of a number

$\sqrt[n]{a}$ (nth root of a); if nth root of a is b, then $b^n = a$

root of an equation

A solution of the equation

scale factor

The number by which corresponding lengths are multiplied in similar figures or in a transformation

scalene

A triangle with three unequal length sides

sector of a circle

Part of a circle bounded by an arc of the circle and two radii

segment of a circle

Part of a circle bounded by an arc of the circle and a chord

sequence

An ordered list of numbers or terms, e.g. 1, 2, 4, 8, ...

series

The sum of a sequence, e.g. $1 + 2 + 4 + 8 + \cdots$

sigma (Σ)

Symbol indicating summation,

e.g. $\displaystyle\sum_{1}^{n} r = 1 + 2 + \cdots + n$

similar

Having the same shape (all corresponding lengths being multiplied by the same scale factor)

slant height of a cone

The distance from the vertex to a point on the circumference of the base

solution

A value (or values) which satisfies the given problem

standard form

A number in the form $a \times 10^n$ where $1 \leqslant a < 10$ and n is an integer

subtended angle

Angle subtended by the line segment AB at C is the angle ACB

surd

An irrational root, e.g. $\sqrt{2}$, $\sqrt{7}$, $\sqrt[3]{11}$

tangent at a point

A line which passes through the point and touches the curve at that point

term (of a sequence)

One of a sequence, e.g. 4 in 1, 2, 4, 8, . . .

term (of an expression)

Part of an expression. e.g. x^2, $-5x$ or -3 in $x^2 - 5x - 3$

trapezium

A quadrilateral with one pair of sides parallel

trinomial

An expression consisting of three terms

unknown

A letter which represents a specific value or values

variable

A letter which represents various values

vertex (pl. vertices) of a parabola

The turning point of a parabola

vertex (pl. vertices) of a polygon

The point where two sides meet

vertex (pl. vertices) of a solid

The point of a cone or the point where faces of the solid meet

Index

SINGLE USER LICENCE AGREEMENT FOR AS CORE FOR EDEXCEL CD-ROM
IMPORTANT: READ CAREFULLY

WARNING: BY OPENING THE PACKAGE YOU AGREE TO BE BOUND BY THE TERMS OF THE LICENCE AGREEMENT BELOW.

This is a legally binding agreement between You (the user or purchaser) and Pearson Education Limited. By retaining this licence, any software media or accompanying written materials or carrying out any of the permitted activities You agree to be bound by the terms of the licence agreement below.

If You do not agree to these terms then promptly return the entire publication (this licence and all software, written materials, packaging and any other components received with it) with Your sales receipt to Your supplier for a full refund.

YOU ARE PERMITTED TO:

- Use (load into temporary memory or permanent storage) a single copy of the *Live Player* software on only one computer at a time. If this computer is linked to a network then the *Live Player* software may only be installed in a manner such that it is not accessible to other machines on the network.

- Transfer the *Live Player* software from one computer to another provided that you only use it on one computer at a time.

- Print a single copy of any PDF file from the CD-ROM for the sole use of the user.

YOU MAY NOT:

- Rent or lease the software or any part of the publication.

- Copy any part of the documentation, except where specifically indicated otherwise.

- Make copies of the software, other than for backup purposes.

- Reverse engineer, decompile or disassemble the software.

- Use the software on more than one computer at a time.

- Install the software on any networked computer in a way that could allow access to it from more than one machine on the network.

- Use the software in any way not specified above without the prior written consent of Pearson Education Limited.

- Print off multiple copies of any PDF file.

ONE COPY ONLY

This licence is for a single user copy of the software

PEARSON EDUCATION LIMITED RESERVES THE RIGHT TO TERMINATE THIS LICENCE BY WRITTEN NOTICE AND TO TAKE ACTION TO RECOVER ANY DAMAGES SUFFERED BY PEARSON EDUCATION LIMITED IF YOU BREACH ANY PROVISION OF THIS AGREEMENT.

You only own the disk on which the software is supplied.

Pearson Education Limited warrants that the diskette or CD-ROM on which the software is supplied are free from defects in materials and workmanship under normal use for ninety (90) days from the date You receive them. This warranty is limited to You and is not transferable. Pearson Education Limited does not warrant that the functions of the software meet Your requirements or that the media is compatible with any computer system on which it is used or that the operation of the software will be unlimited or error free.

You assume responsibility for selecting the software to achieve Your intended results and for the installation of, the use of and the results obtained from the software. The entire liability of Pearson Education Limited and its suppliers and your only remedy shall be replacement of the components that do not meet this warranty free of charge.

This limited warranty is void if any damage has resulted from accident, abuse, misapplication, service or modification by someone other than Pearson Education Limited. In no event shall Pearson Education Limited or its suppliers be liable for any damages whatsoever arising out of installation of the software, even if advised of the possibility of such damages. Pearson Education Limited will not be liable for any loss or damage of any nature suffered by any party as a result of reliance upon or reproduction of or any errors in the content of the publication.

Pearson Education Limited does not limit its liability for death or personal injury caused by its negligence.

This licence agreement shall be governed by and interpreted and construed in accordance with English law.